Engineering Statics Labs
with SOLIDWORKS Motion 2015

Huei-Huang Lee

Department of Engineering Science
National Cheng Kung University, Taiwan

Publications

SDC Publications
P.O. Box 1334
Mission, KS 66222
913-262-2664
www.SDCpublications.com
Publisher: Stephen Schroff

ISBN-13: 978-1-58503-941-8
ISBN-10: 1-58503-941-1

Printed and bound in the United States of America.

Contents

Preface

A New Way of Thinking in Engineering Mechanics Curricula

Figure 1. illustrates how engineering mechanics curricula are implemented nowadays. Engineering students learn physics and mathematics in their high school years and their first college year. Based on this foundation, the students go further into studying engineering mechanics courses such as Statics, Dynamics, Mechanics of Materials, Heat Transfer, Fluid Mechanics, etc. This paradigm has been practiced for as long as any university professor can remember. I've grown up with this paradigm too. More than 30 years has passed since I graduated from college, and even the contents of the textbooks remain essentially identical. The only difference is that we have CAD and CAE courses now (as shown in the figure). So, what are the problems of this conventional paradigm of engineering mechanics curricula?

First, conventional curricula relies too much on mathematics to teach the concepts of engineering mechanics. Many students are good at engineering thinking but not good at mathematical thinking. For most of students, especially in their junior years, mathematics is an inefficient tool (a nightmare, some would say). As a matter of fact, very few students enjoy mathematics as a tool of learning engineering ideas and concepts. Nowadays, CAE software has matured to a point that it can be used as a tool to learn engineering ideas, concepts, and even formulas. We'll show this through each section of this book. Often, mathematics is not the only way to show engineering concepts, or to explain formulas. Using graphics-based CAE tools is often a better way. It is possible to reduce the dependency on mathematics by a substantial extent.

Second, as shown in the figure, the CAD course is usually taught as a standalone subject that doesn't serve as part of foundation for engineering mechanics courses. The 3D modeling techniques learned in the CAD course can be a powerful tool. For example, modern CAD software usually allows you to build a mechanism and study the motion of parts. However, our engineering mechanics textbooks haven't illuminated these advantages yet.

Third, the CAE course is usually taught in the senior or graduate years, because CAE textbooks require some background knowledge of engineering mechanics. It is my long-term observation that the CAE course should be taught as early as junior years, for the following reasons: (a) If a student begins to learn CAE in his junior years, he will have many years to become proficient at this critical engineering skill. (b) After knowing what CAE is and how it can help him solve problems, a student would be more knowledgeable and confident about what he should concentrate on when learning engineering mechanics courses. (c) As mentioned earlier, CAE can be used as a learning tool, like mathematics, for the ensuing subjects. It'll largely facilitate the learning of engineering mechanics courses.

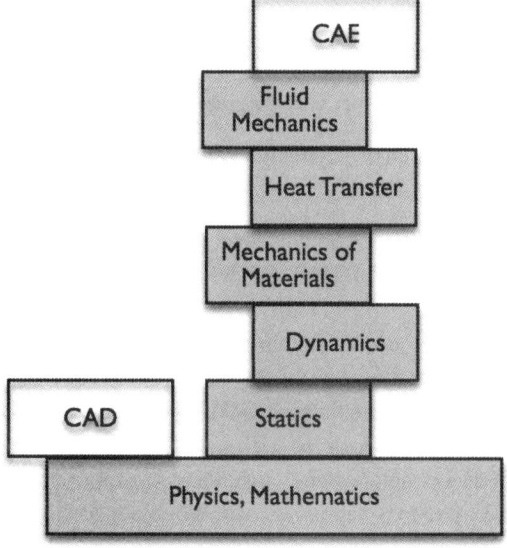

Figure 1. Conventional Paradigm of Engineering Mechanics Curricula

Figure 2. shows an idea that I'd like to propose for engineering mechanics curricula; this book is developed based on this idea. Engineering students usually learn CAD tools in their junior years. For example, among many CAD tools, **SOLIDWORKS** has been popularized in many colleges. Naturally **SOLIDWORKS** might serve as a "virtual laboratory" for the ensuing engineering mechanics courses. The idea is simple, the benefits should be appreciated, but the implementation needs much more elaboration.

First, a series of well-designed lab exercise books are crucial to the success of this idea. These software-based lab books must map their contents to contemporary textbooks.

Second, a CAD/CAE software platform must be chosen to serve as the virtual laboratory. We (Mr. Stephen Schroff of SDC Publications and I) have chosen **SOLIDWORKS** together with its rigid-body mechanics add-in **Motion** as the platform for this book, for the following reasons: (a) As mentioned, the software has been popularized in many colleges. Many students are familiar with this software. (b) The **Motion** is an integrated part of **SOLIDWORKS** and a natural extension of **SOLIDWORKS**. (c) The licensing of the **Motion** is included with a **SOLIDWORKS** license. (d) Compared with other CAD/CAE software I've investigated, it is friendly enough for college juniors. (e) Finally, after a thorough investigation, I've concluded that it has capabilities to implement all of the ideas I want to cover in this book.

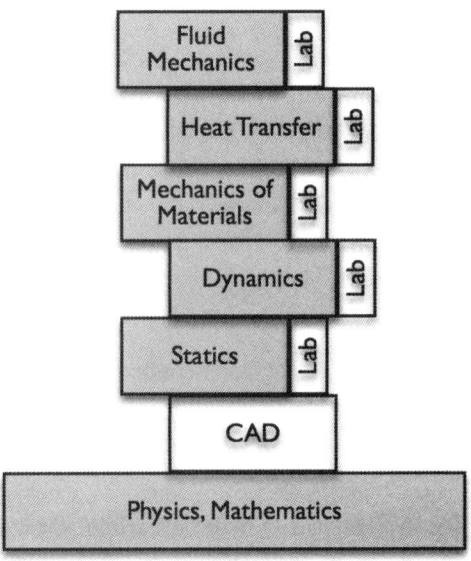

Figure 2. Proposed Paradigm of Engineering Mechanics Curricula

Use of This Book

This book is designed as a software-based lab book to complement a standard textbook in an Engineering Statics course, which is usually taught in junior undergraduate years.

There are 8 chapters in this book. Each chapter contain two sections. Each section is designed for a student to follow the exact steps in that section and learn a concept or topic of Engineering Statics. Typically, each section takes 15-30 minutes to complete the exercises.

Companion Webpage

A webpage is maintained for this book:

http://myweb.ncku.edu.tw/~hhlee/Myweb_at_NCKU/SWS2015.html

The webpage contains links to following resources: (a) videos that demonstrate the steps of each section in the book, (b) finished **SOLIDWORKS** files of each section, (c) a 121-page PDF tutorial, *Part and Assembly Modeling with SOLIDWORKS 2015*.

This book contains the instructions needed to complete all the exercises. But, whenever you have difficulties following the steps in the book, the videos might be used to resolve your questions.

As for the finished **SOLIDWORKS** files, if everything works smoothly, you may not need them at all. Every model can be built from scratch by following the steps described in the book. I provide these files just in case you need them. For example, when you run into trouble and you don't want to redo it from the beginning, you may find these files useful. Or you may happen to have trouble following the steps in the book; you can then look up the details in the files. Another reason I provide these finished files is as follows. It is strongly suggested that, in the beginning of a section when previously saved **SOLIDWORKS** files are needed, you use my files rather than your own files so that you are able to obtain results that have minimum deviations in numerical values from those in the book.

I provide the 121-page PDF tutorial (*Part and Assembly Modeling with SOLIDWORKS 2015*), for those students who have no experience at all in **SOLIDWORKS** and want to acquire some, to feel more comfortable working on the exercises in this book.   Please note that this book (*Engineering Statics Labs with SOLIDWORKS Motion 2015*) is self-contained and requires no pre-existing experience in geometric modeling with **SOLIDWORKS**.

Companion Disc

For each hardcopy of the book, we also provide a disc containing all of the resources in the webpage to save your time downloading the files.

Notations

Chapters and sections are numbered in a traditional way.   Each section is further divided into subsections.   For example, the first subsection of the second section of Chapter 3 is denoted as "3.2-1."   Textboxes in a subsection are ordered with numbers, each of which is enclosed by a pair of square brackets (e.g., [4]).   We refer to that textbox as "3.2-1[4]."   When referring to a textbox from the same subsection, we drop the subsection identifier.   For example, we simply write "[4]." Equations are numbered in a similar way, except that the equation number is enclosed by a pair of round brackets rather than square brackets.   For example, "3.2-1(2)" refers to the 2nd equation in the subsection 3.2-1.   Notations used in this book are summarized as follows (see page 6 for more details):

3.2-1	Numbers after a hyphen are subsection numbers.
[1], [2], ...	Numbers with square brackets are textbox numbers.
(1), (2), ...	Numbers with round brackets are equation numbers.
SOLIDWORKS	**SOLIDWORKS** terms are boldfaced to facilitate the readability of text.
Round-cornered textboxes	A round-cornered textbox indicates that mouse or keyboard actions are needed.
Sharp-cornered textboxes	A sharp-cornered textbox is used for commentary only; i.e., mouse or keyboard actions are not needed in that step.
#	A symbol # is used to indicate the last textbox of a subsection.

Huei-Huang Lee

Associate Professor
Department of Engineering Science
National Cheng Kung University, Tainan, Taiwan
e-mail: hhlee@mail.ncku.edu.tw
webpage: myweb.ncku.edu.tw/~hhlee

Chapter 1
Equilibrium of a Body

Engineering Statics

In Engineering Statics, we study the forces among rigid bodies at rest.

Since the bodies are not moving, the forces must satisfy Newton's equations of static equilibrium. The static equilibrium equations must be satisfied for a single body, a group of bodies in the system, or the entire bodies system. It is fair to say that Engineering Statics is the study of the static equilibrium equations and their application.

In "solving" a structural problem, we mean finding the forces acting on each structural member. In many cases, static equilibrium equations are enough to solve the problems. In these cases, we called these problems **statically determinate problems**. In other cases, we may need to seek other conditions, such as deformation conditions, to solve the problems. In these cases, we called these problems **statically indeterminate problems**. In Engineering Statics, we study statically determinate problems only.

Rigid Body Assumption

In the real world, all solid bodies are more or less deformable. There are no such things as **rigid bodies**. However, if the deformation of a body is not our concern and if the deformation is negligible, we can treat the body as a **rigid body**. In Engineering Statics, we assume all bodies studied in this book are **rigid bodies**. If deformation needs to be considered, we'll use a spring to represent a body. In Engineering Statics, springs are the only elements that are deformable.

Rigid body assumption might introduce errors. However, in many cases, the errors are negligible. In statically determinate problems, rigid body assumption usually introduce negligible errors.

Section 1.1

Supported Block: A 2D Case

1.1-1 Introduction

[1] In this section, we consider a block [2] supported by a hinge [3] and a roller [4] and find the reaction forces at the supports. Before we proceed to solve the problem using **SOLIDWORKS**, let's manually calculate the reaction forces.

From the free-body diagram [5], taking the moment equilibrium about A, we have

$$\sum M_A = 0$$

$$(150 \text{ N})(1 \text{ m}) + (300 \text{ N})(0.5 \text{ m}) - B_Y(1.5 \text{ m}) = 0$$

$$B_Y = 200 \text{ N}$$

Force equilibrium in Y direction,

$$\sum F_Y = 0$$

$$A_Y + B_Y - (300 \text{ N}) = 0 \quad (\text{where } B_Y = 200 \text{ N})$$

$$A_Y = 100 \text{ N}$$

Finally, force equilibrium in X direction,

$$\sum F_X = 0$$

$$A_X + (150 \text{ N}) = 0$$

$$A_X = -150 \text{ N}$$

Note that the negative sign of A_X indicates that it is opposite to the assumed direction shown in [5]. Now, let's solve this problem with **SOLIDWORKS**. If you know how to solve a simple problem like this, you may be able to solve a much more complicated problem.

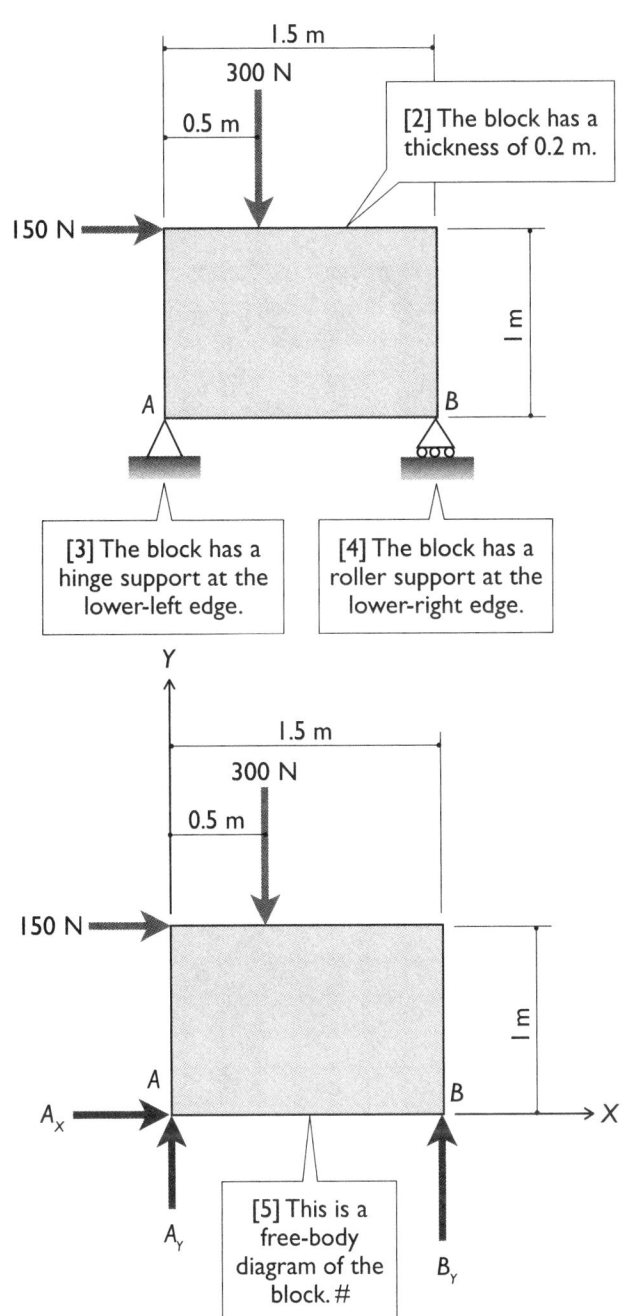

1.5 m

300 N

0.5 m

[2] The block has a thickness of 0.2 m.

150 N

1 m

A

B

[3] The block has a hinge support at the lower-left edge.

[4] The block has a roller support at the lower-right edge.

Y

1.5 m

300 N

0.5 m

150 N

1 m

A

B

A_X

X

A_Y

[5] This is a free-body diagram of the block. #

B_Y

1.1-2 Launch **SOLIDWORKS** and Create a New Part

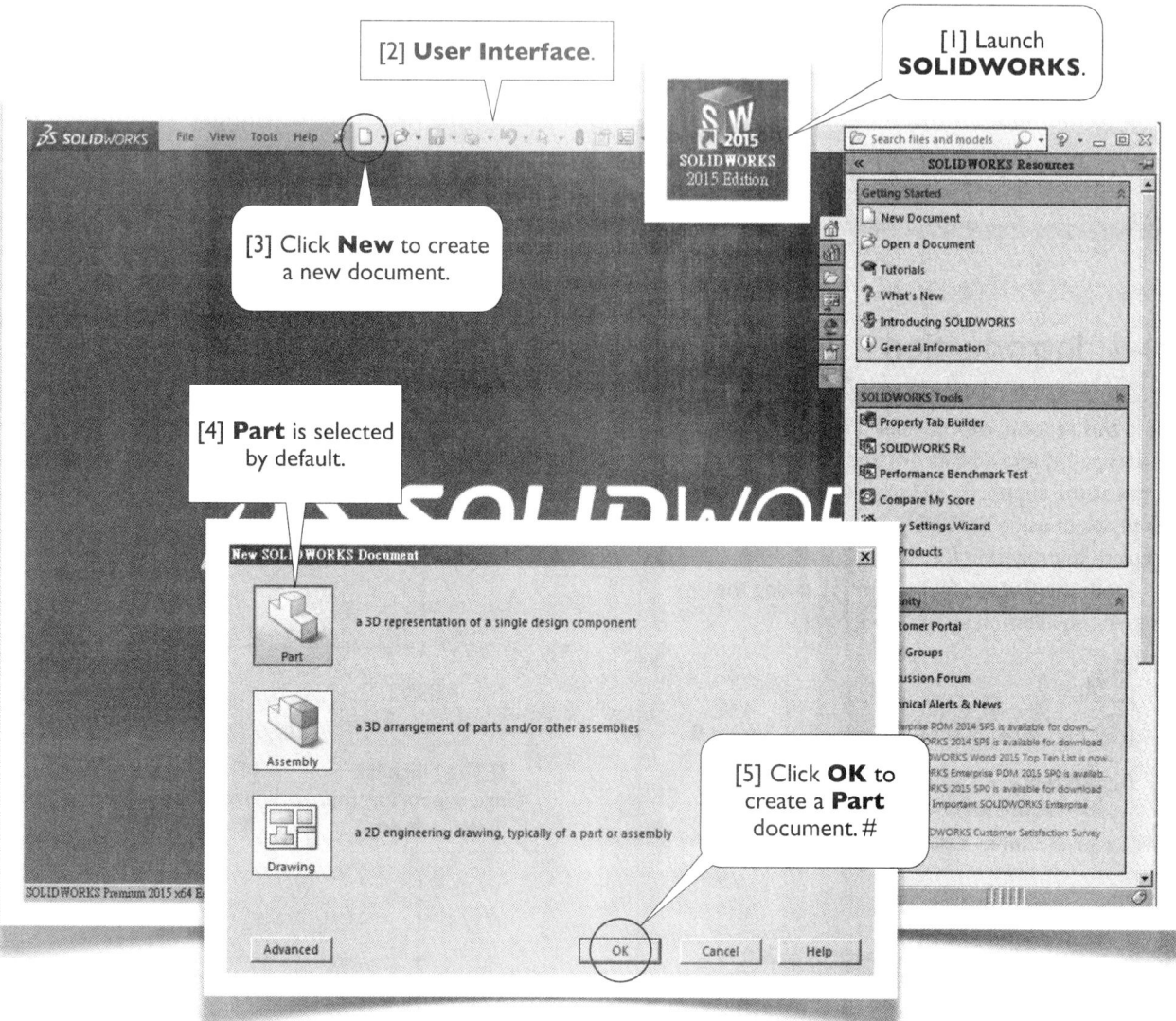

[1] Launch **SOLIDWORKS**.

[2] **User Interface**.

[3] Click **New** to create a new document.

[4] **Part** is selected by default.

[5] Click **OK** to create a **Part** document. #

About the Textboxes

1. Within each subsection (e.g., 1.1-2), textboxes are ordered with numbers, each of which is enclosed by a pair of square brackets (e.g., [1]). When you read the contents of a subsection, please follow the order of the textboxes.

2. The textbox numbers are also used as reference numbers. Inside a subsection, we simply refer to a textbox by its number (e.g., [1]). From other subsections, we refer to a textbox by its subsection identifier and the textbox number (e.g., 1.1-2[1]).

3. A textbox is either round-cornered (e.g., [1, 3, 5]) or sharp-cornered (e.g., [2, 4]). A round-cornered textbox indicates that **mouse or keyboard actions** are needed in that step. A sharp-cornered textbox is used for commentary only; i.e., mouse or keyboard actions are not needed in that step.

4. A symbol # is used to indicate the last textbox of a subsection [5], so that you don't leave out any textboxes.

SOLIDWORKS Terms

In this book, terms used in the **SOLIDWORKS** are boldfaced (e.g., **Part** in [4, 5]) to facilitate the readability.

1.1-3 Set Up Unit System

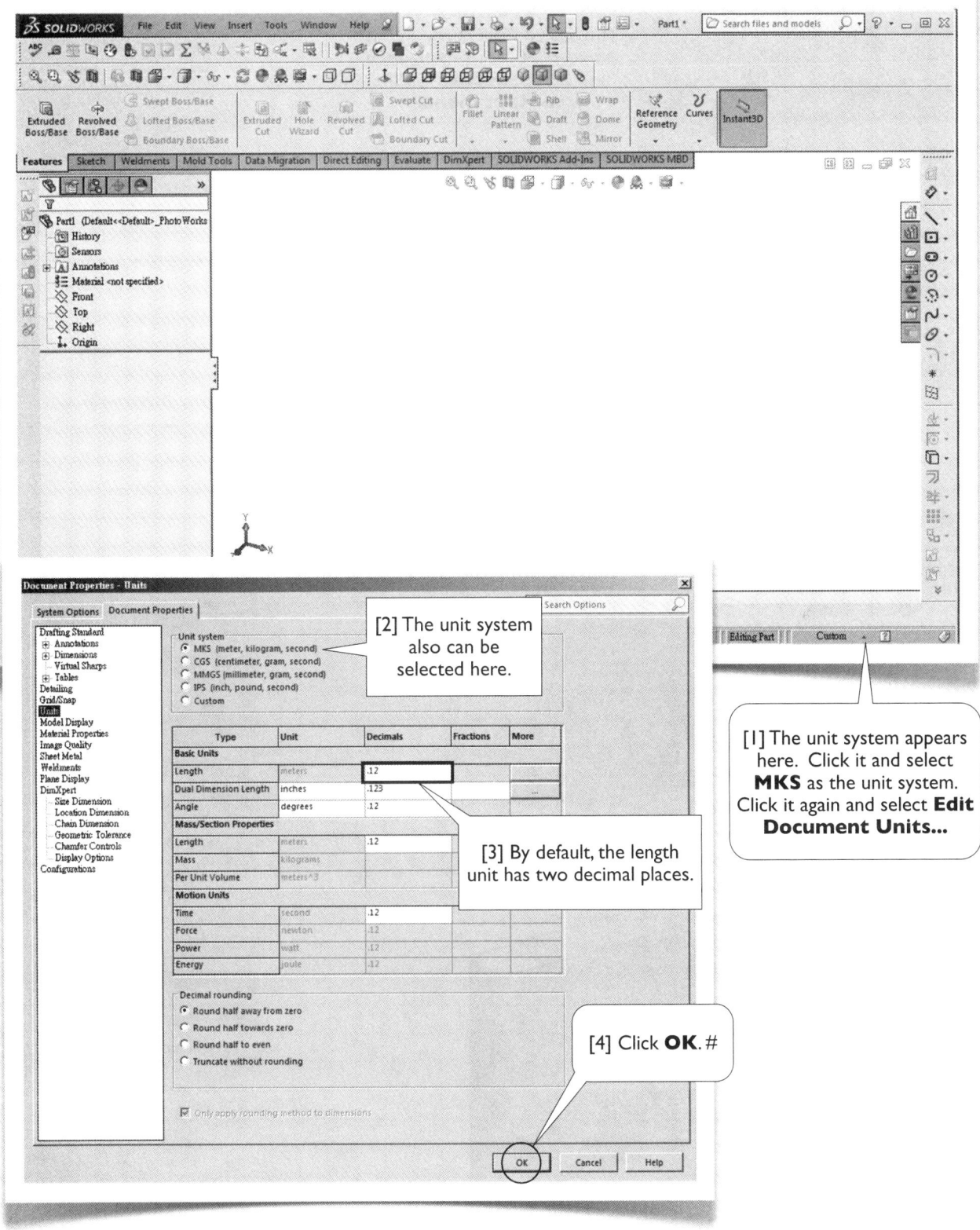

[1] The unit system appears here. Click it and select **MKS** as the unit system. Click it again and select **Edit Document Units...**

[2] The unit system also can be selected here.

[3] By default, the length unit has two decimal places.

[4] Click **OK**. #

1.1-4 Create a Part: **Block**

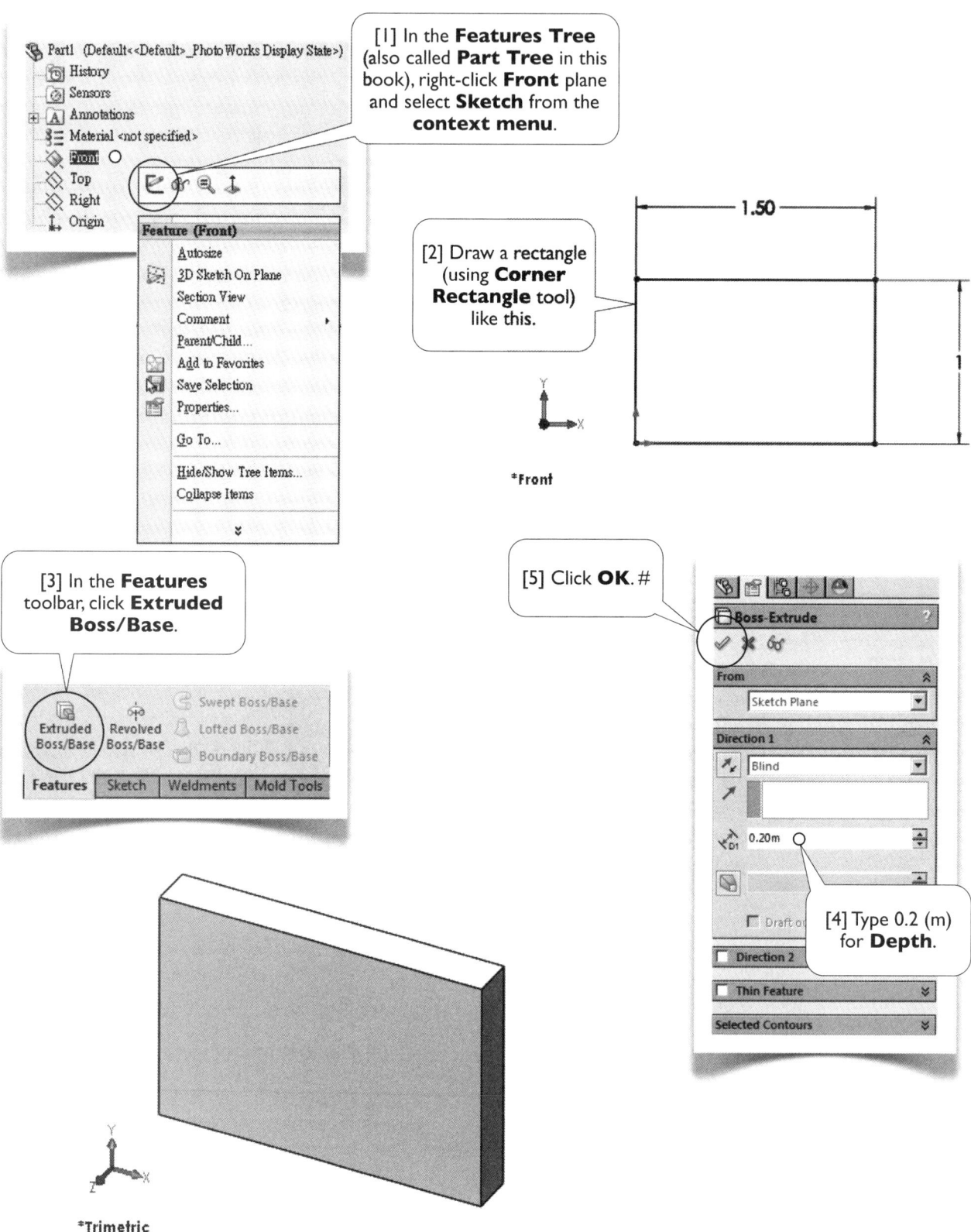

[1] In the **Features Tree** (also called **Part Tree** in this book), right-click **Front** plane and select **Sketch** from the **context menu**.

[2] Draw a rectangle (using **Corner Rectangle** tool) like this.

*Front

[3] In the **Features** toolbar, click **Extruded Boss/Base**.

[5] Click **OK**. #

[4] Type 0.2 (m) for **Depth**.

*Trimetric

1.1-5 Create Two **Points** on the **Block**

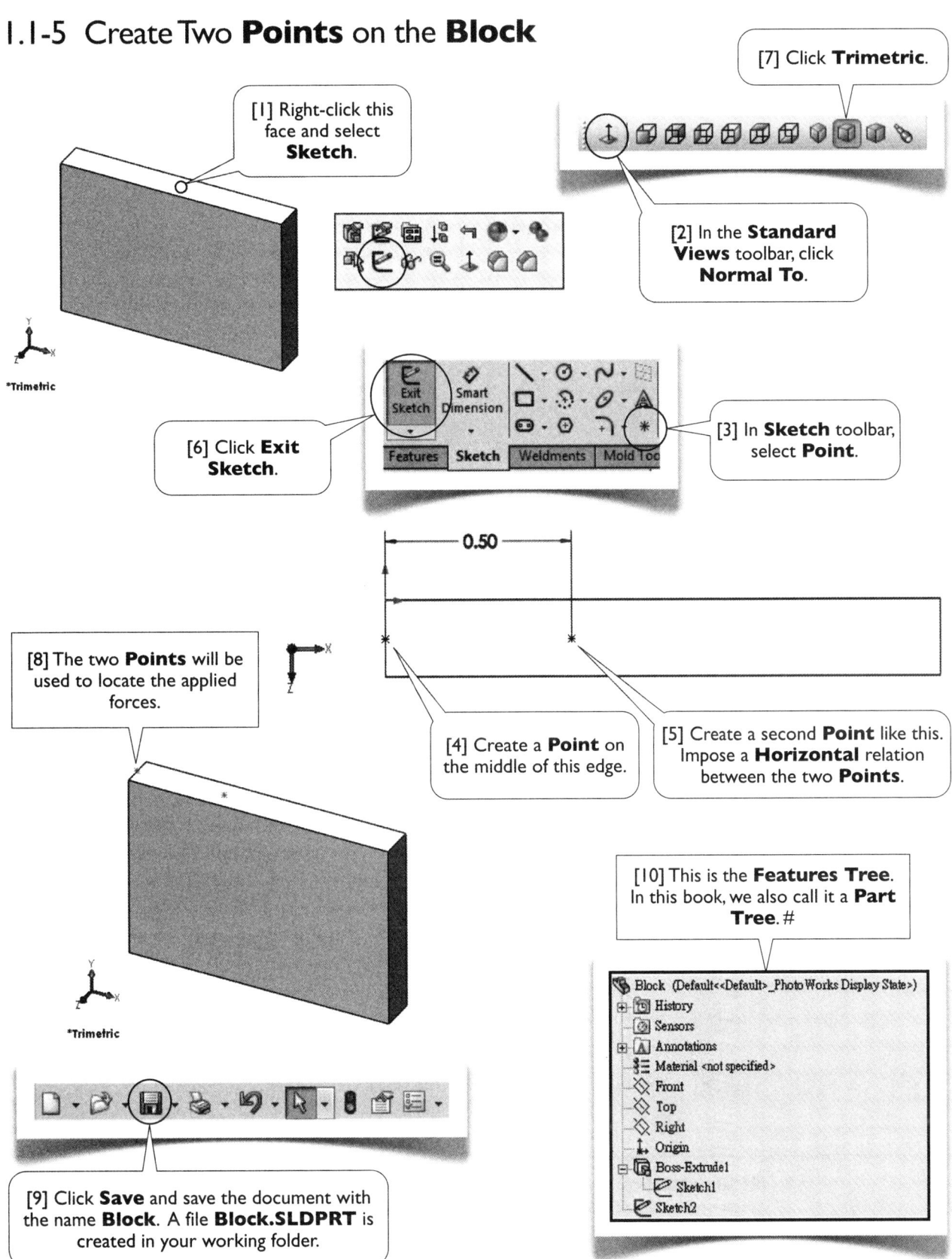

[7] Click **Trimetric**.

[1] Right-click this face and select **Sketch**.

[2] In the **Standard Views** toolbar, click **Normal To**.

*Trimetric

[6] Click **Exit Sketch**.

[3] In **Sketch** toolbar, select **Point**.

Exit Sketch   Smart Dimension

Features   **Sketch**   Weldments   Mold Too

0.50

[8] The two **Points** will be used to locate the applied forces.

[4] Create a **Point** on the middle of this edge.

[5] Create a second **Point** like this. Impose a **Horizontal** relation between the two **Points**.

*Trimetric

[10] This is the **Features Tree**. In this book, we also call it a **Part Tree**. #

[9] Click **Save** and save the document with the name **Block**. A file **Block.SLDPRT** is created in your working folder.

Block (Default<<Default>_PhotoWorks Display State>)
- History
- Sensors
- Annotations
- Material <not specified>
- Front
- Top
- Right
- Origin
- Boss-Extrude1
  - Sketch1
- Sketch2

1.1-6 Create an Assembly: **BlockSupported**

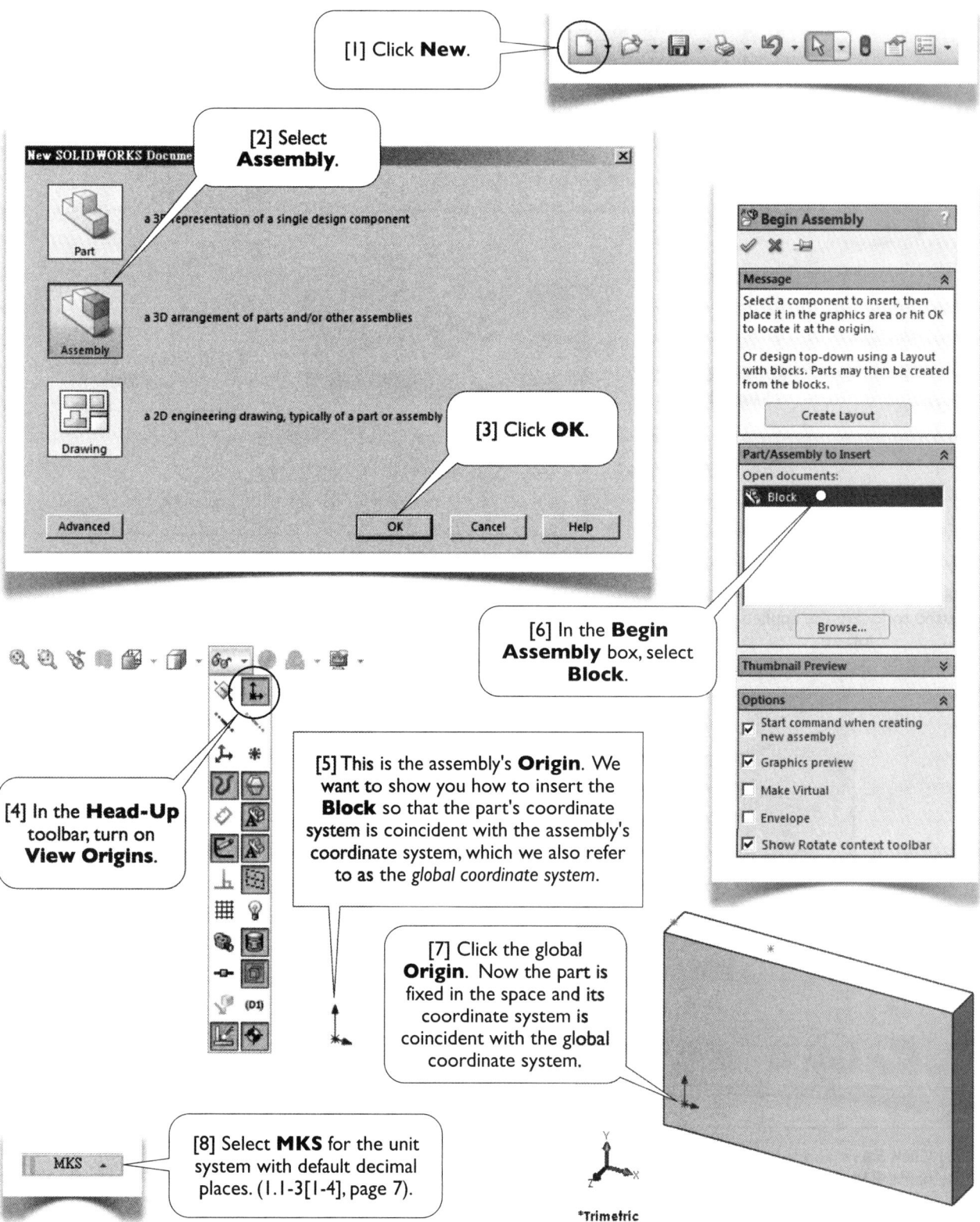

[1] Click **New**.

[2] Select **Assembly**.

New SOLIDWORKS Docume...

a 3D representation of a single design component

Part

a 3D arrangement of parts and/or other assemblies

Assembly

a 2D engineering drawing, typically of a part or assembly

Drawing

[3] Click **OK**.

Advanced   OK   Cancel   Help

Begin Assembly

Message

Select a component to insert, then place it in the graphics area or hit OK to locate it at the origin.

Or design top-down using a Layout with blocks. Parts may then be created from the blocks.

Create Layout

Part/Assembly to Insert

Open documents:

Block

Browse...

Thumbnail Preview

Options

☑ Start command when creating new assembly

☑ Graphics preview

☐ Make Virtual

☐ Envelope

☑ Show Rotate context toolbar

[6] In the **Begin Assembly** box, select **Block**.

[4] In the **Head-Up** toolbar, turn on **View Origins**.

[5] This is the assembly's **Origin**. We want to show you how to insert the **Block** so that the part's coordinate system is coincident with the assembly's coordinate system, which we also refer to as the *global coordinate system*.

[7] Click the global **Origin**. Now the part is fixed in the space and its coordinate system is coincident with the global coordinate system.

[8] Select **MKS** for the unit system with default decimal places. (1.1-3[1-4], page 7).

MKS

*Trimetric

[10] This is the **Features Tree** of the **Assembly**. In this book, we simply call it an **Assembly Tree**.

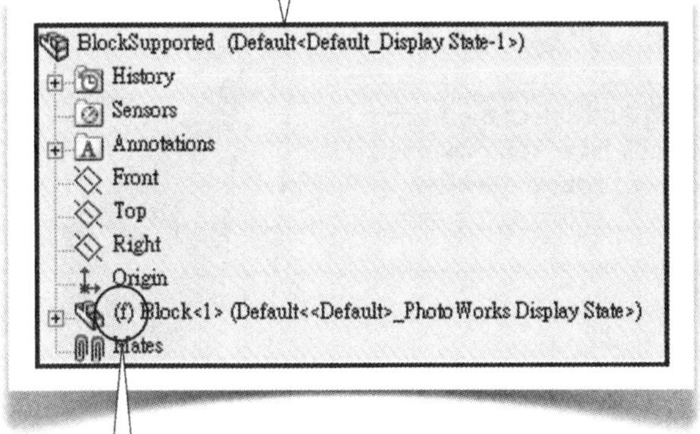

BlockSupported (Default<Default_Display State-1>)
History
Sensors
Annotations
Front
Top
Right
Origin
(f) Block<1> (Default<<Default>_PhotoWorks Display State>)
Mates

[11] An **(f)** sign indicates that the **Block** is fixed in the space. Now, right-click it and select **Float**.

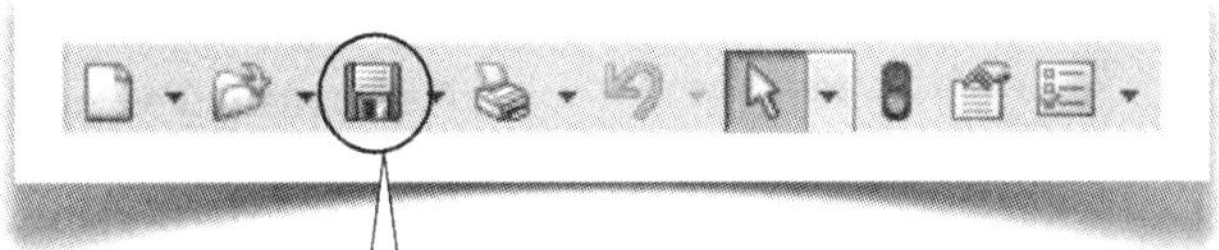

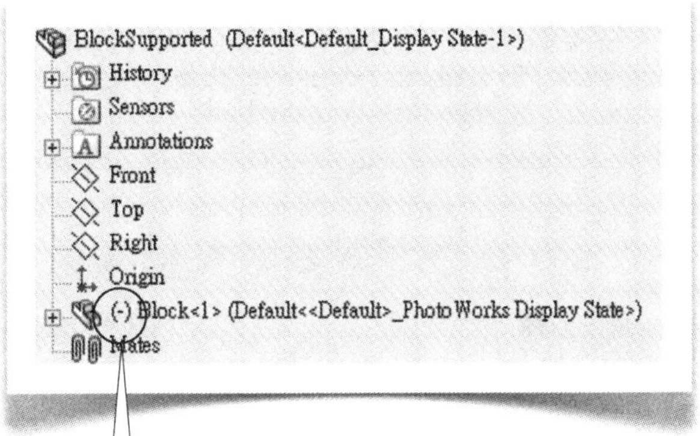

BlockSupported (Default<Default_Display State-1>)
History
Sensors
Annotations
Front
Top
Right
Origin
(-) Block<1> (Default<<Default>_PhotoWorks Display State>)
Mates

[12] The **(f)** sign changes to **(-)**, indicating that the **Block** is no more fixed. We'll show that it can be moved freely [13-16]. In 1.1-7 (next page), we'll set up constraints to fully support the **Block**.

[9] Click **Save** and save the document with the name **BlockSupported**. A file **BlockSupported.SLDASM** is created in your working folder.

[13] Left-click-drag the body to move it around the space.

*Trimetric

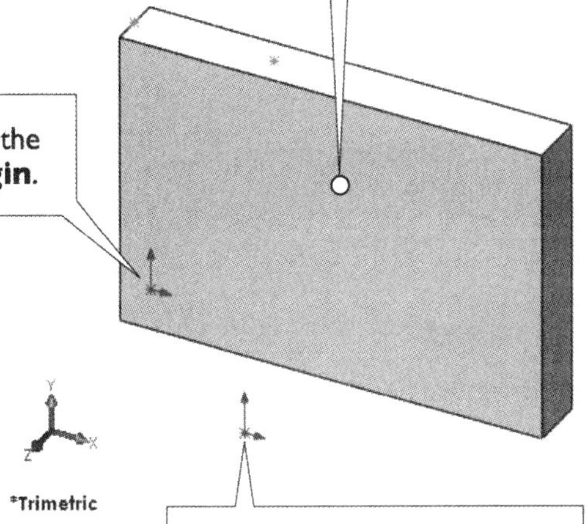

[14] This is the part's **Origin**.

[15] Now, the global **Origin** may not be coincident with the part's **Origin**.

Edit Component
Insert Components
Mate
Linear Compone...
Smart Fasteners
Move Component
Show Hidden Components
Assembly Features
Reference Geometry
Assembly
Layout
Sheet Metal
Sketch
Evaluate
Move Component
Rotate Component

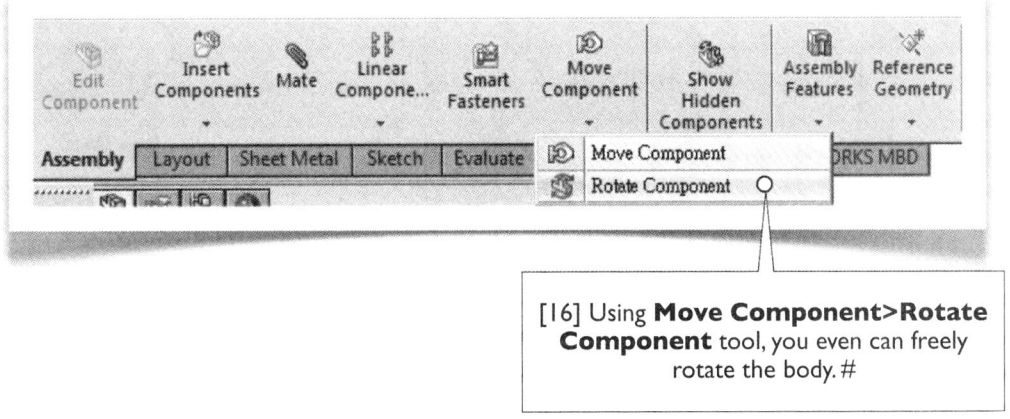

[16] Using **Move Component>Rotate Component** tool, you even can freely rotate the body. #

1.1-7 Set Up Supports

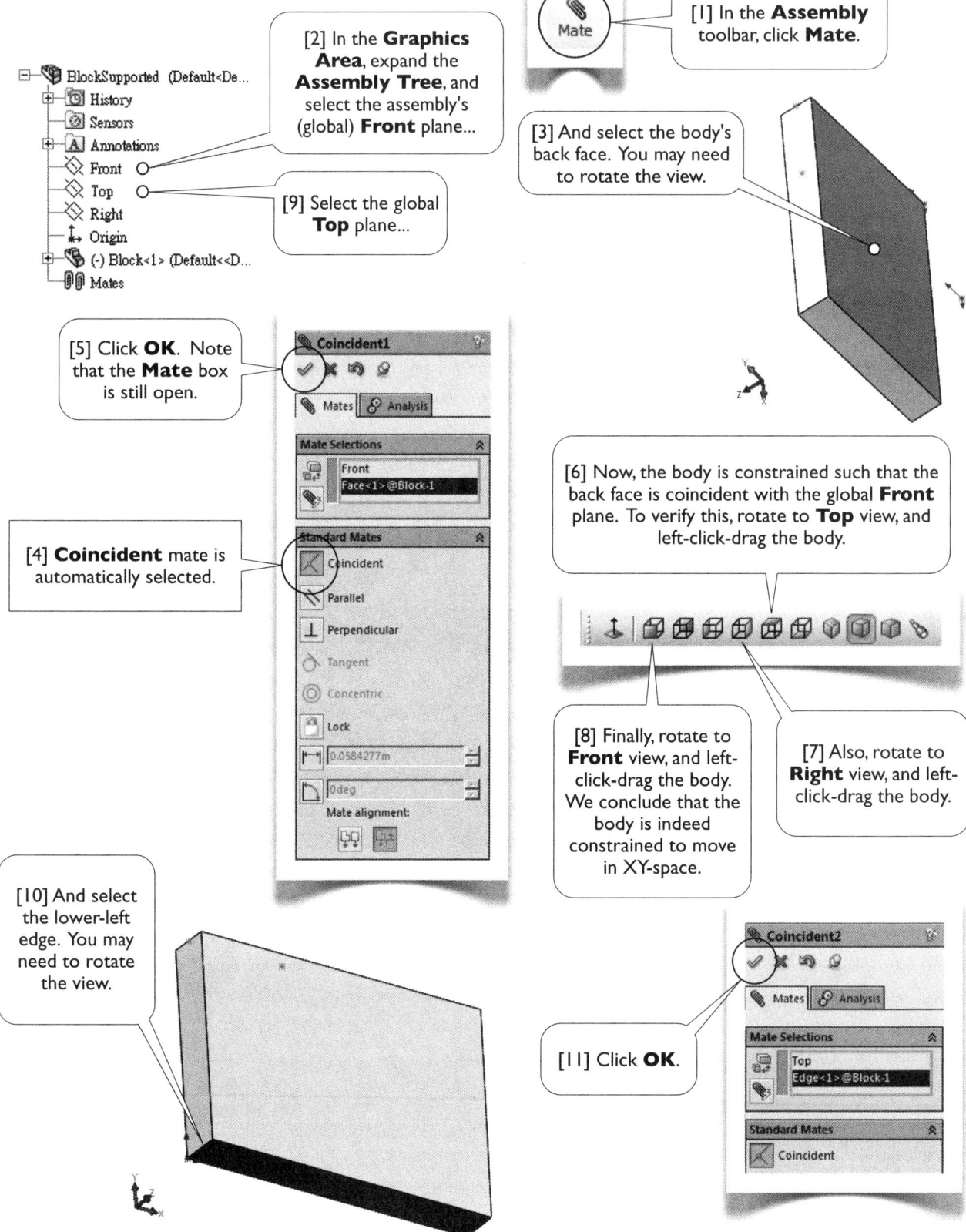

[1] In the **Assembly** toolbar, click **Mate**.

[2] In the **Graphics Area**, expand the **Assembly Tree**, and select the assembly's (global) **Front** plane...

[9] Select the global **Top** plane...

[3] And select the body's back face. You may need to rotate the view.

[5] Click **OK**. Note that the **Mate** box is still open.

[4] **Coincident** mate is automatically selected.

[6] Now, the body is constrained such that the back face is coincident with the global **Front** plane. To verify this, rotate to **Top** view, and left-click-drag the body.

[8] Finally, rotate to **Front** view, and left-click-drag the body. We conclude that the body is indeed constrained to move in XY-space.

[7] Also, rotate to **Right** view, and left-click-drag the body.

[10] And select the lower-left edge. You may need to rotate the view.

[11] Click **OK**.

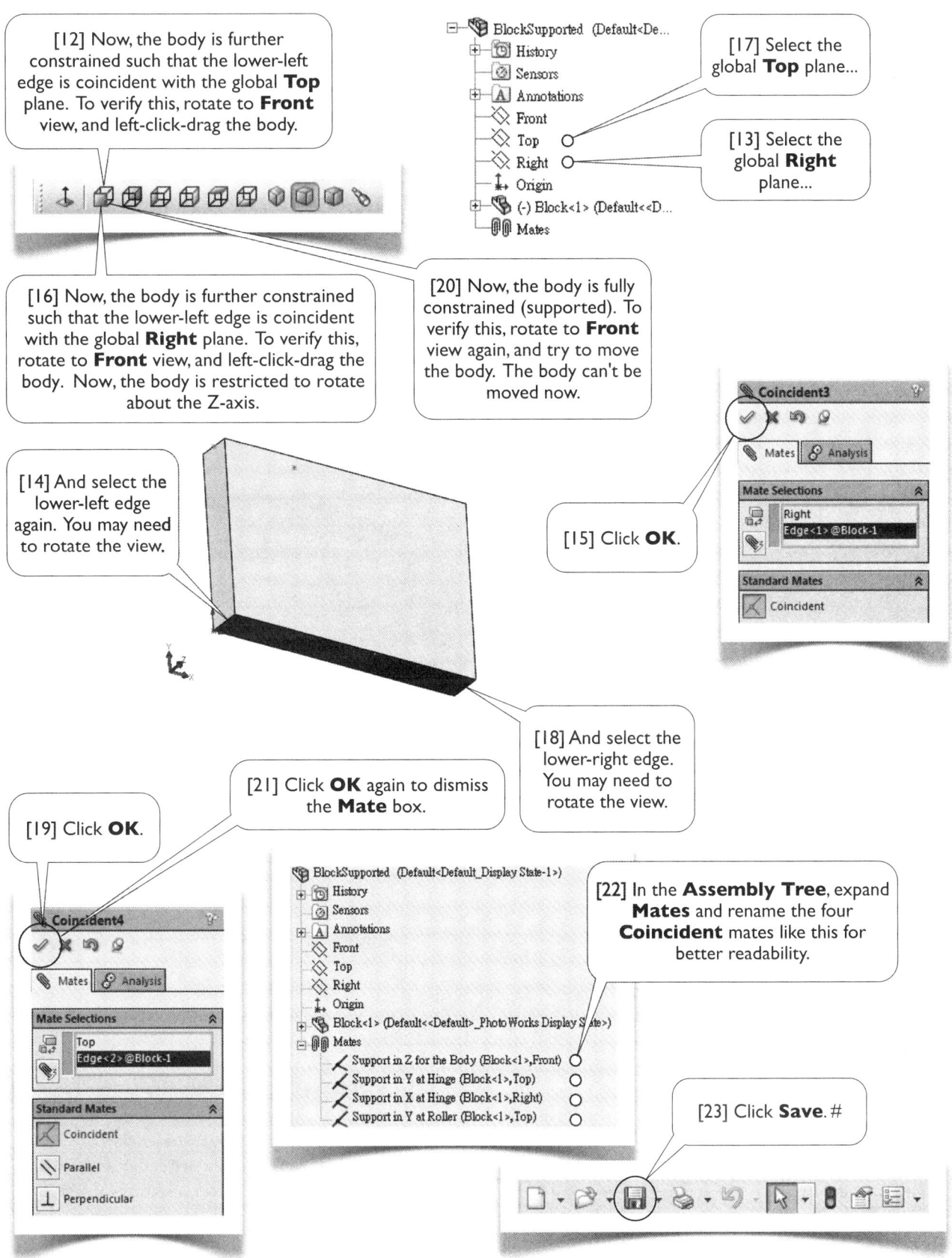

[12] Now, the body is further constrained such that the lower-left edge is coincident with the global **Top** plane. To verify this, rotate to **Front** view, and left-click-drag the body.

[16] Now, the body is further constrained such that the lower-left edge is coincident with the global **Right** plane. To verify this, rotate to **Front** view, and left-click-drag the body. Now, the body is restricted to rotate about the Z-axis.

[14] And select the lower-left edge again. You may need to rotate the view.

BlockSupported (Default<De...
History
Sensors
Annotations
Front
Top
Right
Origin
(-) Block<1> (Default<<D...
Mates

[17] Select the global **Top** plane...

[13] Select the global **Right** plane...

[20] Now, the body is fully constrained (supported). To verify this, rotate to **Front** view again, and try to move the body. The body can't be moved now.

Coincident3
Mates | Analysis
Mate Selections
Right
Edge<1> @Block-1
Standard Mates
Coincident

[15] Click **OK**.

[18] And select the lower-right edge. You may need to rotate the view.

[21] Click **OK** again to dismiss the **Mate** box.

[19] Click **OK**.

Coincident4
Mates | Analysis
Mate Selections
Top
Edge<2> @Block-1
Standard Mates
Coincident
Parallel
Perpendicular

BlockSupported (Default<Default_Display State-1>)
History
Sensors
Annotations
Front
Top
Right
Origin
Block<1> (Default<<Default>_PhotoWorks Display State>)
Mates
Support in Z for the Body (Block<1>,Front)
Support in Y at Hinge (Block<1>,Top)
Support in X at Hinge (Block<1>,Right)
Support in Y at Roller (Block<1>,Top)

[22] In the **Assembly Tree**, expand **Mates** and rename the four **Coincident** mates like this for better readability.

[23] Click **Save**. #

1.1-8 Load **SOLIDWORKS** Motion

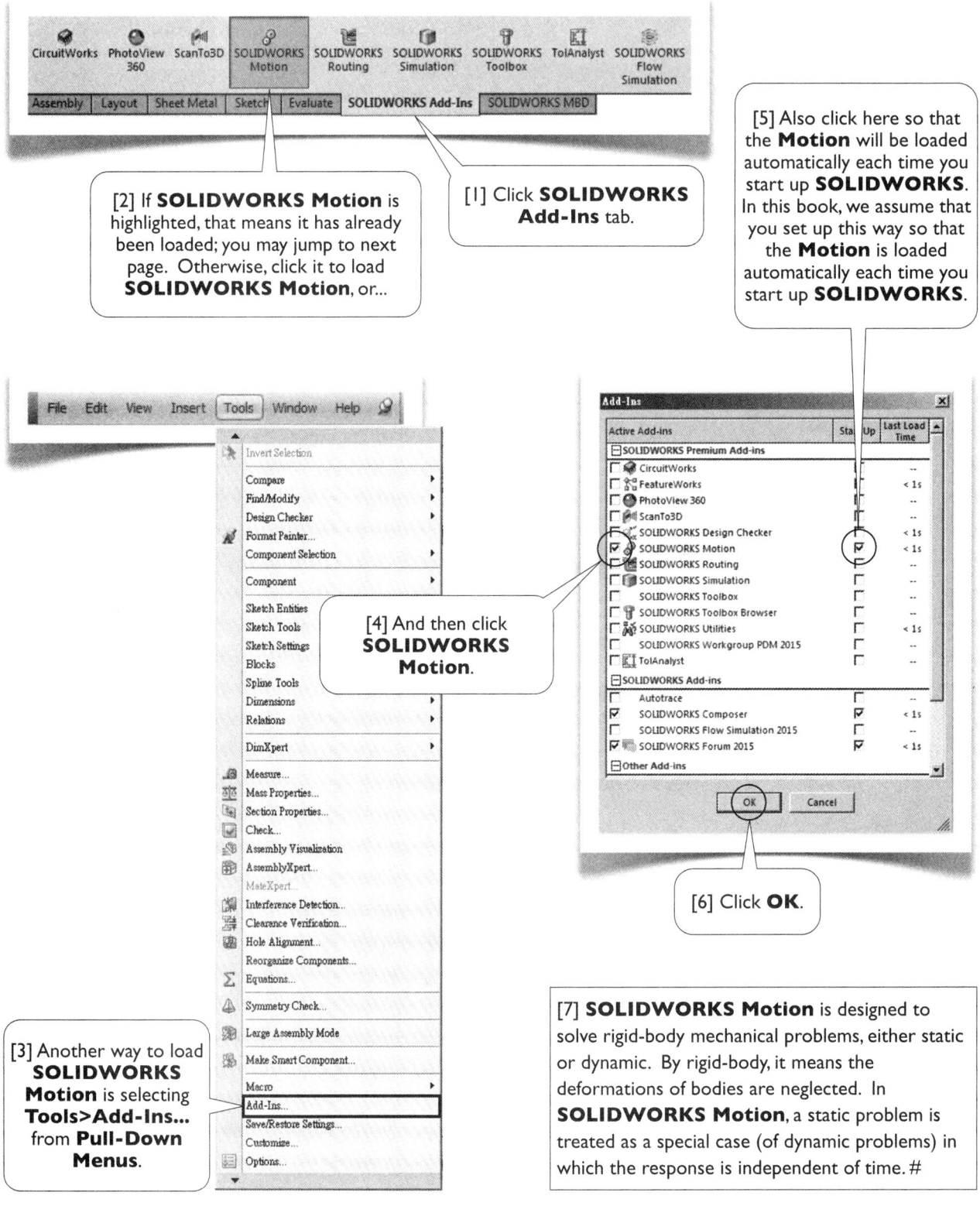

[2] If **SOLIDWORKS Motion** is highlighted, that means it has already been loaded; you may jump to next page. Otherwise, click it to load **SOLIDWORKS Motion**, or...

[1] Click **SOLIDWORKS Add-Ins** tab.

[5] Also click here so that the **Motion** will be loaded automatically each time you start up **SOLIDWORKS**. In this book, we assume that you set up this way so that the **Motion** is loaded automatically each time you start up **SOLIDWORKS**.

[4] And then click **SOLIDWORKS Motion**.

[6] Click **OK**.

[3] Another way to load **SOLIDWORKS Motion** is selecting **Tools>Add-Ins...** from **Pull-Down Menus**.

[7] **SOLIDWORKS Motion** is designed to solve rigid-body mechanical problems, either static or dynamic. By rigid-body, it means the deformations of bodies are neglected. In **SOLIDWORKS Motion**, a static problem is treated as a special case (of dynamic problems) in which the response is independent of time. #

1.1-9 Create a **Motion Study**

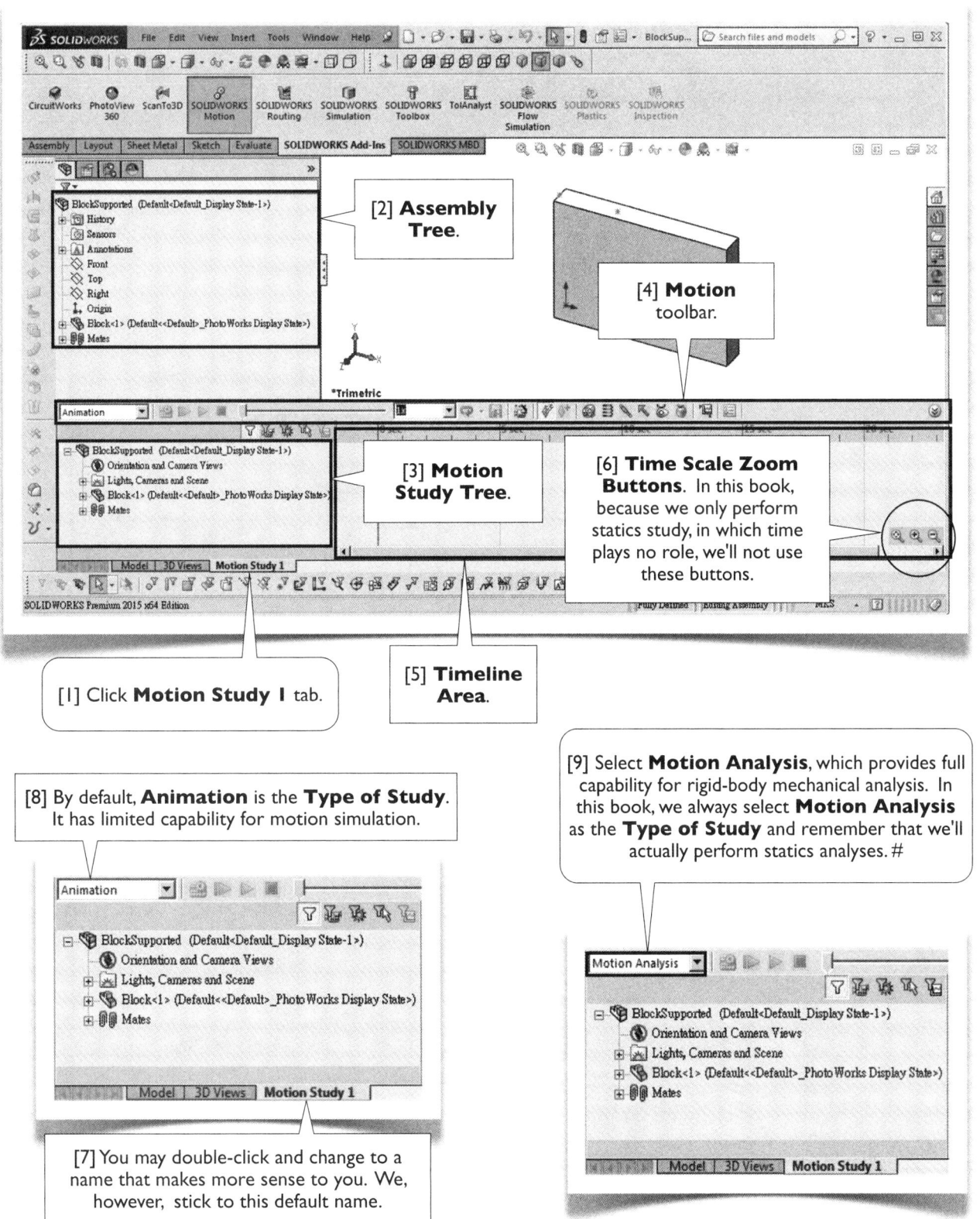

[1] Click **Motion Study 1** tab.

[2] **Assembly Tree**.

[3] **Motion Study Tree**.

[4] **Motion** toolbar.

[5] **Timeline Area**.

[6] **Time Scale Zoom Buttons**. In this book, because we only perform statics study, in which time plays no role, we'll not use these buttons.

[8] By default, **Animation** is the **Type of Study**. It has limited capability for motion simulation.

[7] You may double-click and change to a name that makes more sense to you. We, however, stick to this default name.

[9] Select **Motion Analysis**, which provides full capability for rigid-body mechanical analysis. In this book, we always select **Motion Analysis** as the **Type of Study** and remember that we'll actually perform statics analyses. #

1.1-10 Set Up Forces

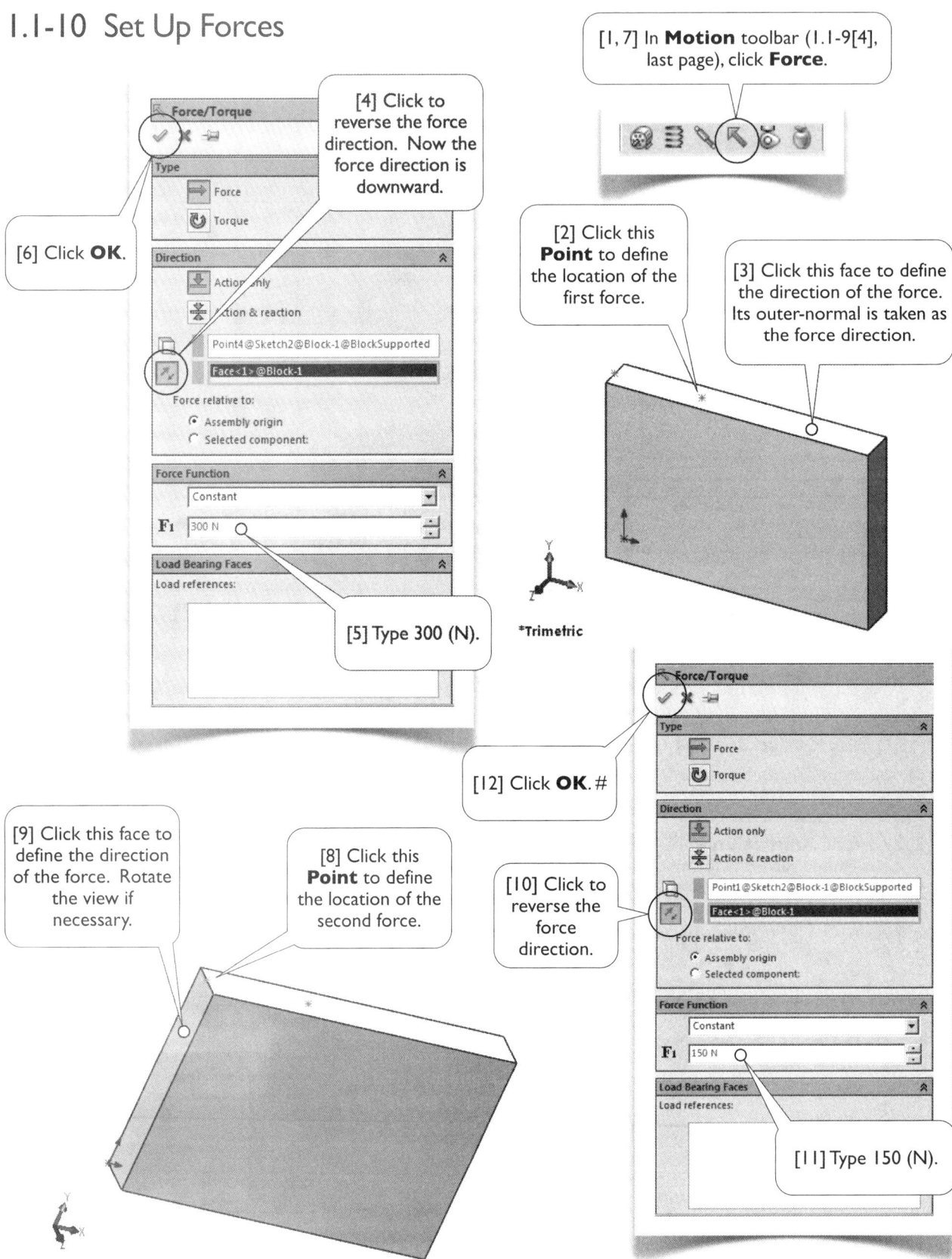

[1, 7] In **Motion** toolbar (1.1-9[4], last page), click **Force**.

[4] Click to reverse the force direction. Now the force direction is downward.

[6] Click **OK**.

[2] Click this **Point** to define the location of the first force.

[3] Click this face to define the direction of the force. Its outer-normal is taken as the force direction.

[5] Type 300 (N).

*Trimetric

[12] Click **OK**. #

[9] Click this face to define the direction of the force. Rotate the view if necessary.

[8] Click this **Point** to define the location of the second force.

[10] Click to reverse the force direction.

[11] Type 150 (N).

1.1-11 Calculate Results

[3] The calculation completes in a few seconds, and the **Time Slider** proceeds to right-most position. By default, the simulation time is 5 seconds, which is an arbitrarily chosen time. Since the body is fully supported and the forces are constant, the results are time independent. We're in effect performing a static analysis. One thing you need to remember about this **Time Slider** is that whenever you want to re-define the forces, make sure the **Time Slider** is at the beginning (the left-most position). #

[1] Click **Calculate**.

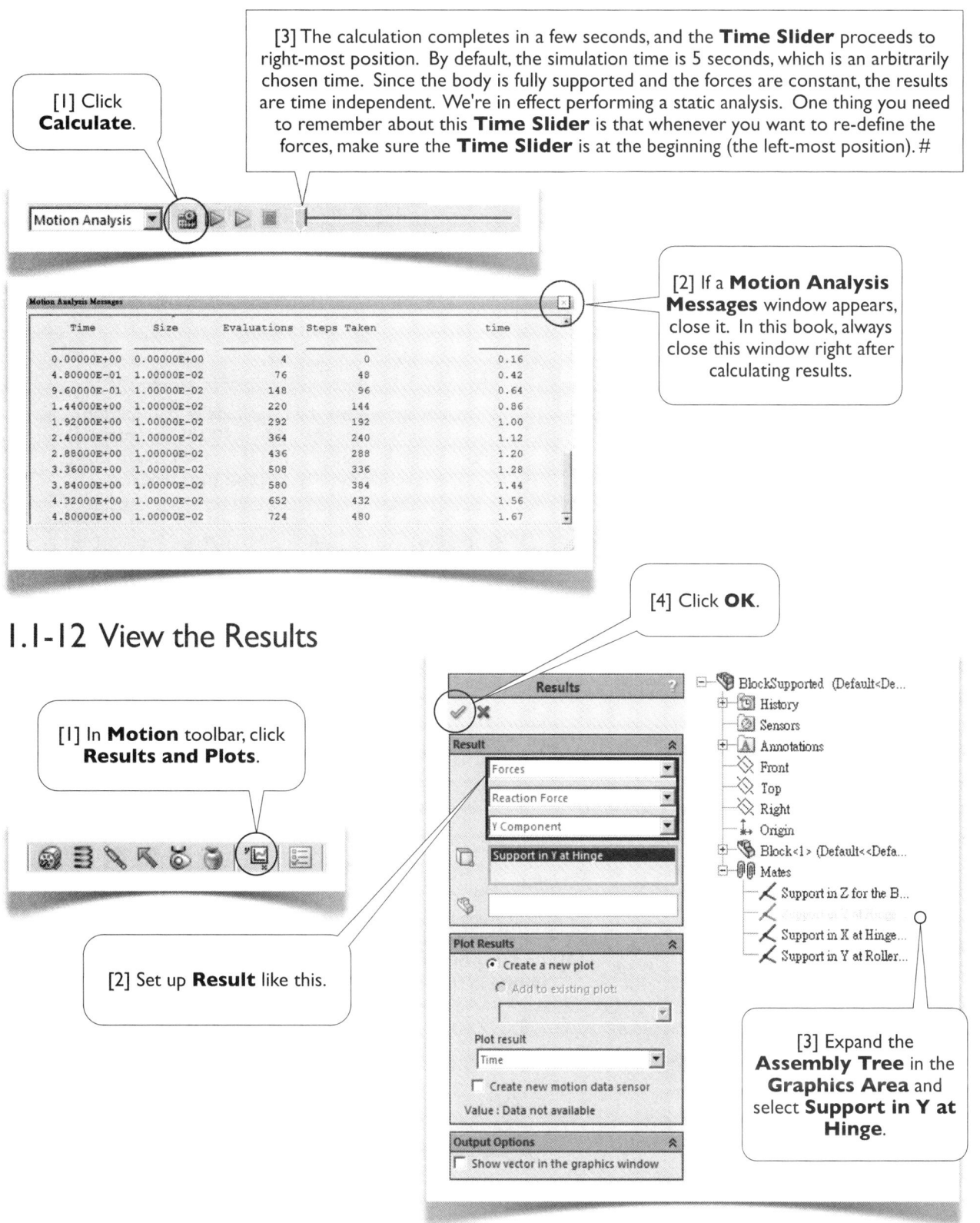

[2] If a **Motion Analysis Messages** window appears, close it. In this book, always close this window right after calculating results.

[4] Click **OK**.

1.1-12 View the Results

[1] In **Motion** toolbar, click **Results and Plots**.

[2] Set up **Result** like this.

[3] Expand the **Assembly Tree** in the **Graphics Area** and select **Support in Y at Hinge**.

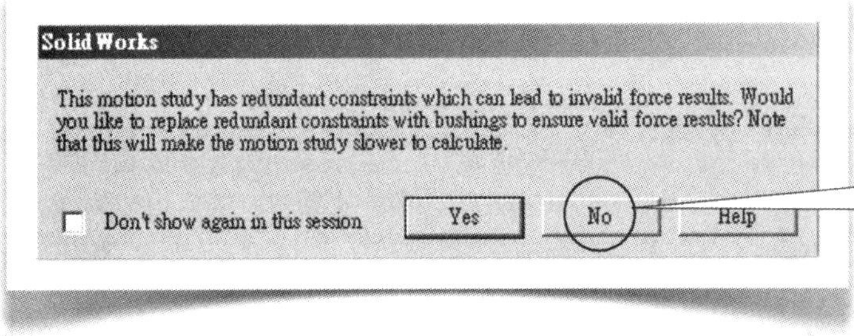

SolidWorks

This motion study has redundant constraints which can lead to invalid force results. Would you like to replace redundant constraints with bushings to ensure valid force results? Note that this will make the motion study slower to calculate.

☐ Don't show again in this session   [Yes]   (No)   [Help]

[5] Click **No**. In this book, always click **No** if this message appears.

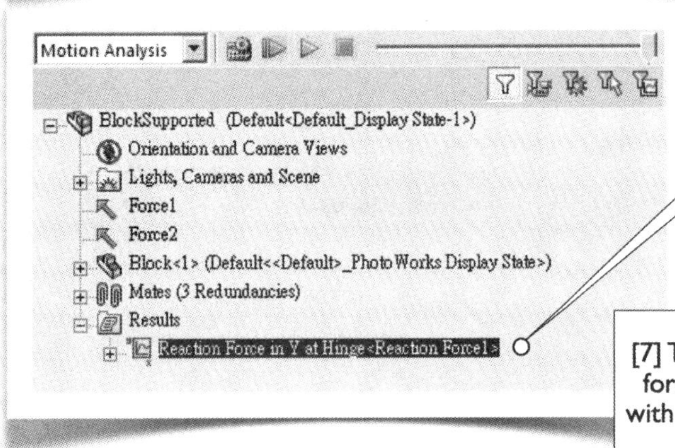

Motion Analysis ▼ | 🖳 ▷ ▷ ■ ―――――――― |

▽ ⯅ ⯆ ⯆ ⯆

⊟ 🖿 BlockSupported (Default<Default_Display State-1>)
　　⊙ Orientation and Camera Views
　⊞ 🔦 Lights, Cameras and Scene
　　⚲ Force1
　　⚲ Force2
　⊞ 🖿 Block<1> (Default<<Default>_PhotoWorks Display State>)
　⊞ 🔗 Mates (3 Redundancies)
　⊟ 📂 Results
　　⊞ 📈 Reaction Force in Y at Hinge<Reaction Force1> ○

[6] In the **Motion Study Tree**, expand **Results**, click **Plot1** twice (not double-click) and change name to **Reaction Force in Y at Hinge**.

[7] The plot shows that the magnitude of the reaction force in Y-direction at the hinge is 100 N, consistent with the one calculated in 1.1-1[1] (page 5). Note that the reaction force is constant over time.

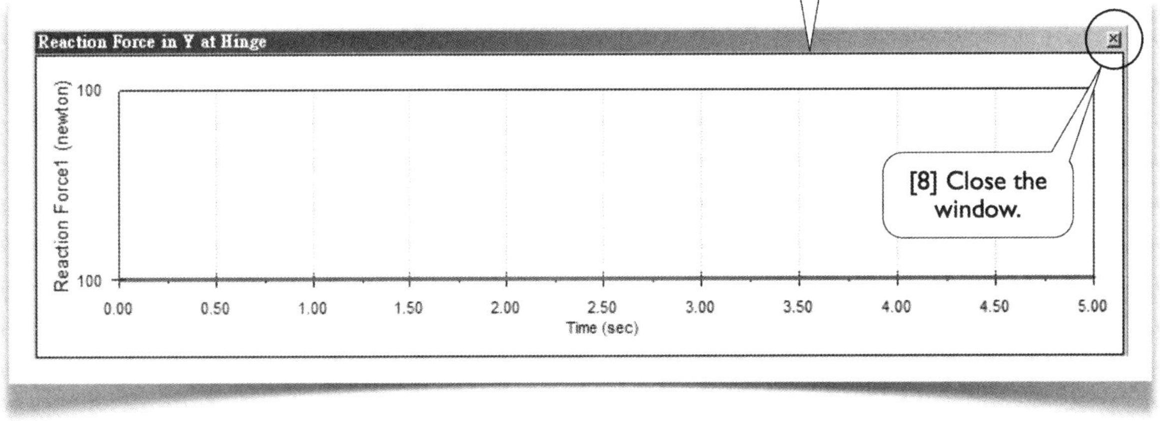

Reaction Force in Y at Hinge   [×]

[8] Close the window.

(plot with Y-axis "Reaction Force1 (newton)" marked 100, 100 and X-axis "Time (sec)" 0.00 to 5.00)

Results

✓ ✗

Result   ⯅
　Forces   ○ ▼
　Reaction Force   ○ ▼
　X Component   ○ ▼
　Support in X at Hinge   ●

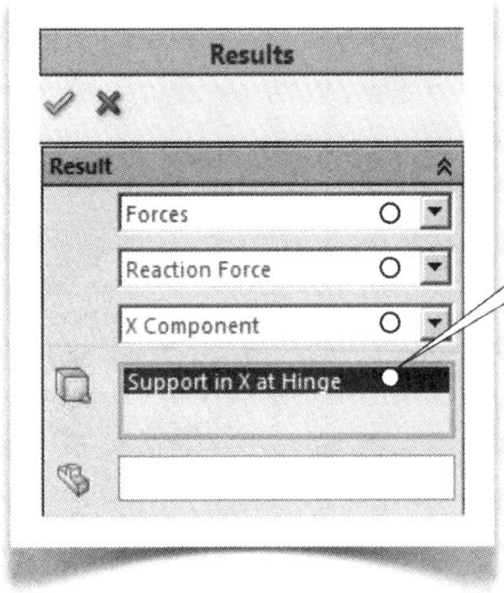

[9] Repeat steps [1-5] except selecting **X Component** in step [2] and **Support in X at Hinge** in step [3],

[10] In the **Motion Study Tree**, click **Plot2** twice and change name to **Reaction Force in X at Hinge**.

⊟ 📂 Results
　⊞ 📈 Reaction Force in Y at Hinge<Reaction Force1>
　⊞ 📈 Reaction Force in X at Hinge<Reaction Force2> ○

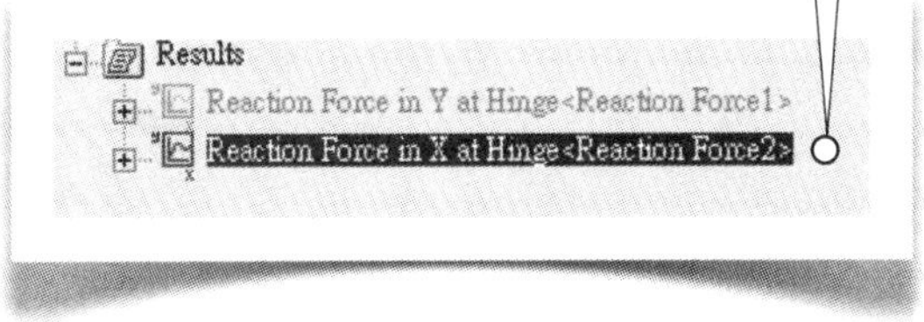

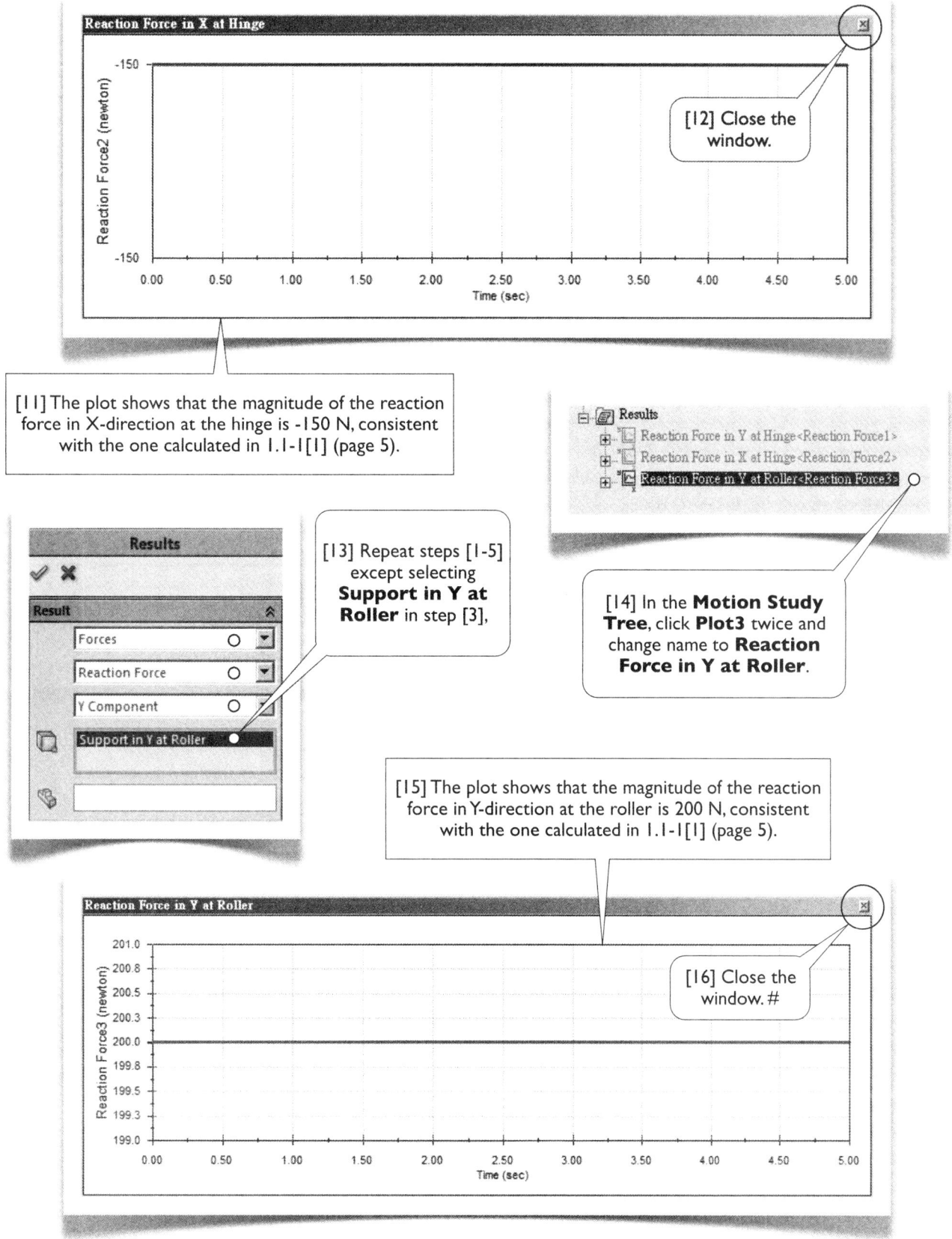

Reaction Force in X at Hinge

[12] Close the window.

[11] The plot shows that the magnitude of the reaction force in X-direction at the hinge is -150 N, consistent with the one calculated in 1.1-1[1] (page 5).

Results
  Reaction Force in Y at Hinge<Reaction Force1>
  Reaction Force in X at Hinge<Reaction Force2>
  Reaction Force in Y at Roller<Reaction Force3>

[13] Repeat steps [1-5] except selecting **Support in Y at Roller** in step [3],

Results
Result
  Forces
  Reaction Force
  Y Component
  Support in Y at Roller

[14] In the **Motion Study Tree**, click **Plot3** twice and change name to **Reaction Force in Y at Roller**.

[15] The plot shows that the magnitude of the reaction force in Y-direction at the roller is 200 N, consistent with the one calculated in 1.1-1[1] (page 5).

[16] Close the window. #

Reaction Force in Y at Roller

1.1-13 Do It Yourself: Validation of the Results

Do It Yourself

[1] To verify the validity of the results, you may check the force and moment equilibria for the body. In a 2D problem like this, you need to check 3 equilibrium equations to conclude the validity of the results. Of course, the 3 equilibrium equations must be independent. #

1.1-14 Wrap Up

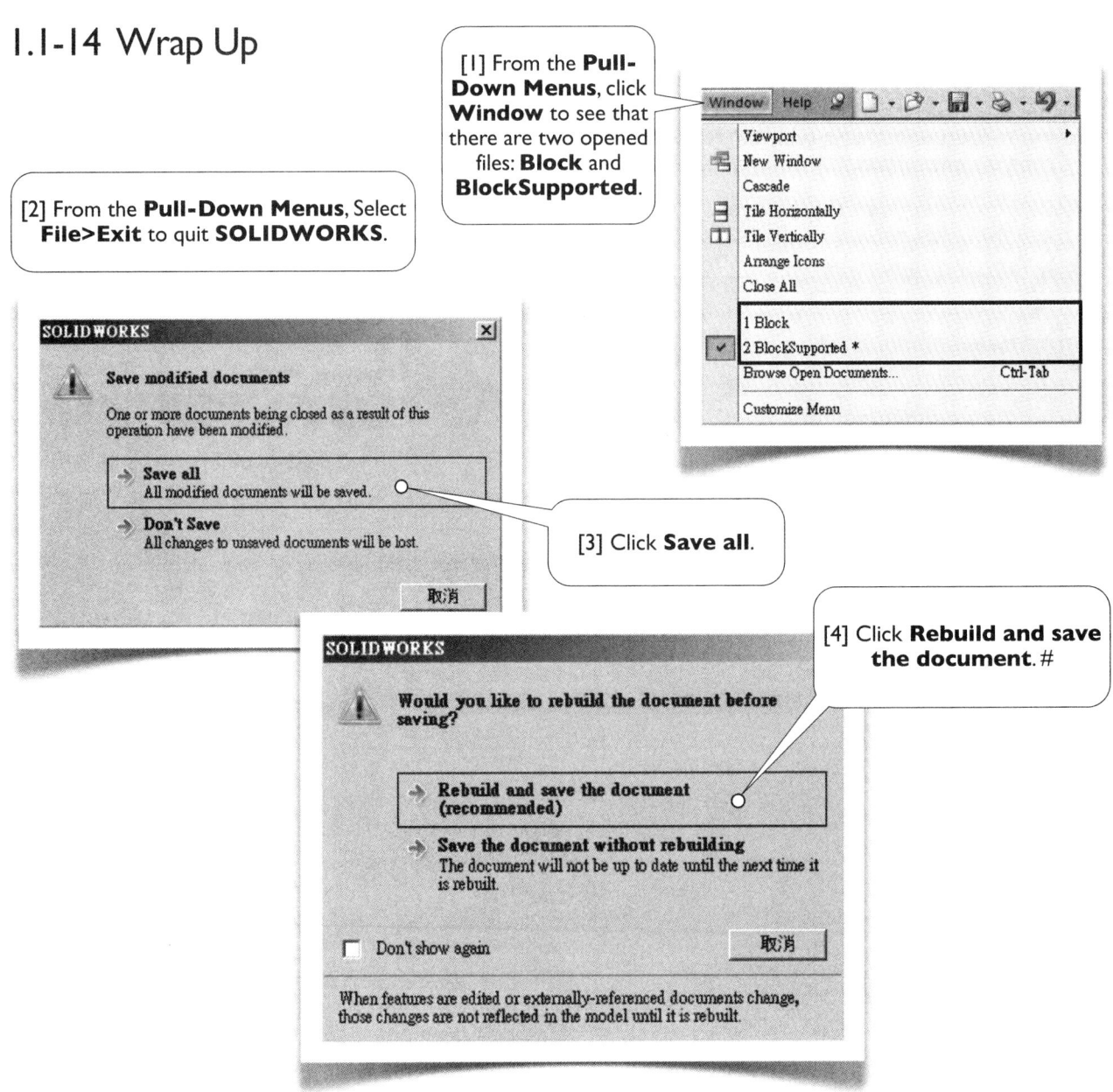

[1] From the **Pull-Down Menus**, click **Window** to see that there are two opened files: **Block** and **BlockSupported**.

[2] From the **Pull-Down Menus**, Select **File>Exit** to quit **SOLIDWORKS**.

[3] Click **Save all**.

[4] Click **Rebuild and save the document**. #

Section 1.2

Supported L-Plate: A 3D Case

1.2-1 Introduction

[1] Consider an L-shaped plate of thickness 5 mm [2] supported at three corners [3-5] and subject to a force P [6]. We want to find the reaction forces at the supports.

There are six reaction force components in this problem, namely D_x, D_y, D_z, B_x, B_z, and C_y. It is possible to establish six equations, according to force and moment equilibria, and solve these six reaction forces. However, we may calculate C_y directly by considering the moment equilibrium about the axis passing through B and D,

$$\sum M_{BD} = 0$$

$$\vec{\lambda}_{BD} \cdot (\vec{r}_{A/B} \times \vec{P}) + \vec{\lambda}_{BD} \cdot (\vec{r}_{C/D} \times \vec{C}_Y) = 0$$

where the unit vector along BD,

$$\vec{\lambda}_{BD} = \frac{(-8 \text{ cm})\vec{i} + (-9 \text{ cm})\vec{j} + (12 \text{ cm})\vec{k}}{\sqrt{(6^2 + 9^2 + 12^2)} \text{ cm}}$$

the position vectors,

$$\vec{r}_{A/B} = (-6 \text{ cm})\vec{i}, \ \vec{r}_{C/D} = (8 \text{ cm})\vec{i}$$

and the forces,

$$\vec{P} = (200 \text{ N})\vec{k}, \ \vec{C}_Y = (C_Y)\vec{j}$$

Substituting these into the moment equilibrium equation, and after mild calculation, we have the reaction force in Y at C,

$$C_Y = 112.5 \text{ N} \tag{1}$$

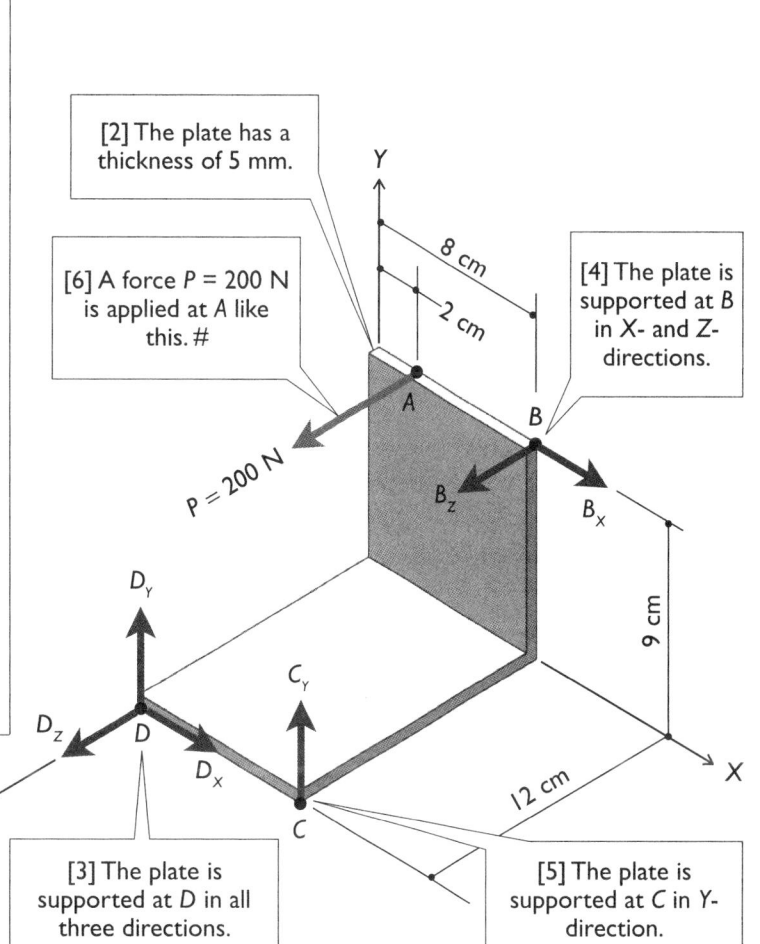

[2] The plate has a thickness of 5 mm.

[6] A force $P = 200$ N is applied at A like this. #

[4] The plate is supported at B in X- and Z- directions.

8 cm
2 cm
9 cm
12 cm

A
B
B_z
B_x
D_Y
D_Z
D
D_X
C_Y
C
$P = 200$ N
Y
Z
X

[3] The plate is supported at D in all three directions.

[5] The plate is supported at C in Y- direction.

1.2-2 Start Up and Create a Part: **Plate**

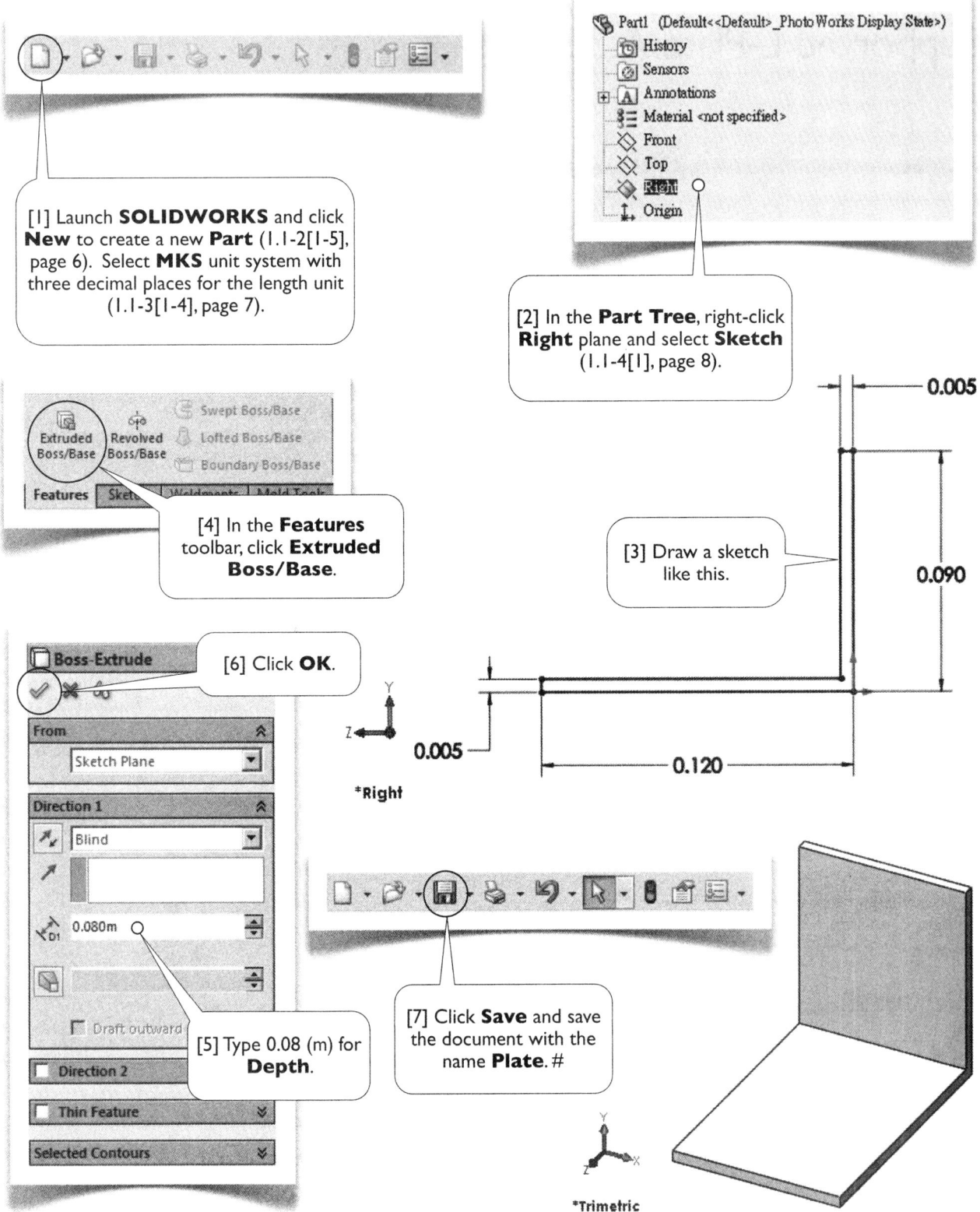

[1] Launch **SOLIDWORKS** and click **New** to create a new **Part** (1.1-2[1-5], page 6). Select **MKS** unit system with three decimal places for the length unit (1.1-3[1-4], page 7).

[2] In the **Part Tree**, right-click **Right** plane and select **Sketch** (1.1-4[1], page 8).

[3] Draw a sketch like this.

[4] In the **Features** toolbar, click **Extruded Boss/Base**.

[6] Click **OK**.

[5] Type 0.08 (m) for **Depth**.

[7] Click **Save** and save the document with the name **Plate**. #

1.2-3 Create a **Point** on the **Plate**

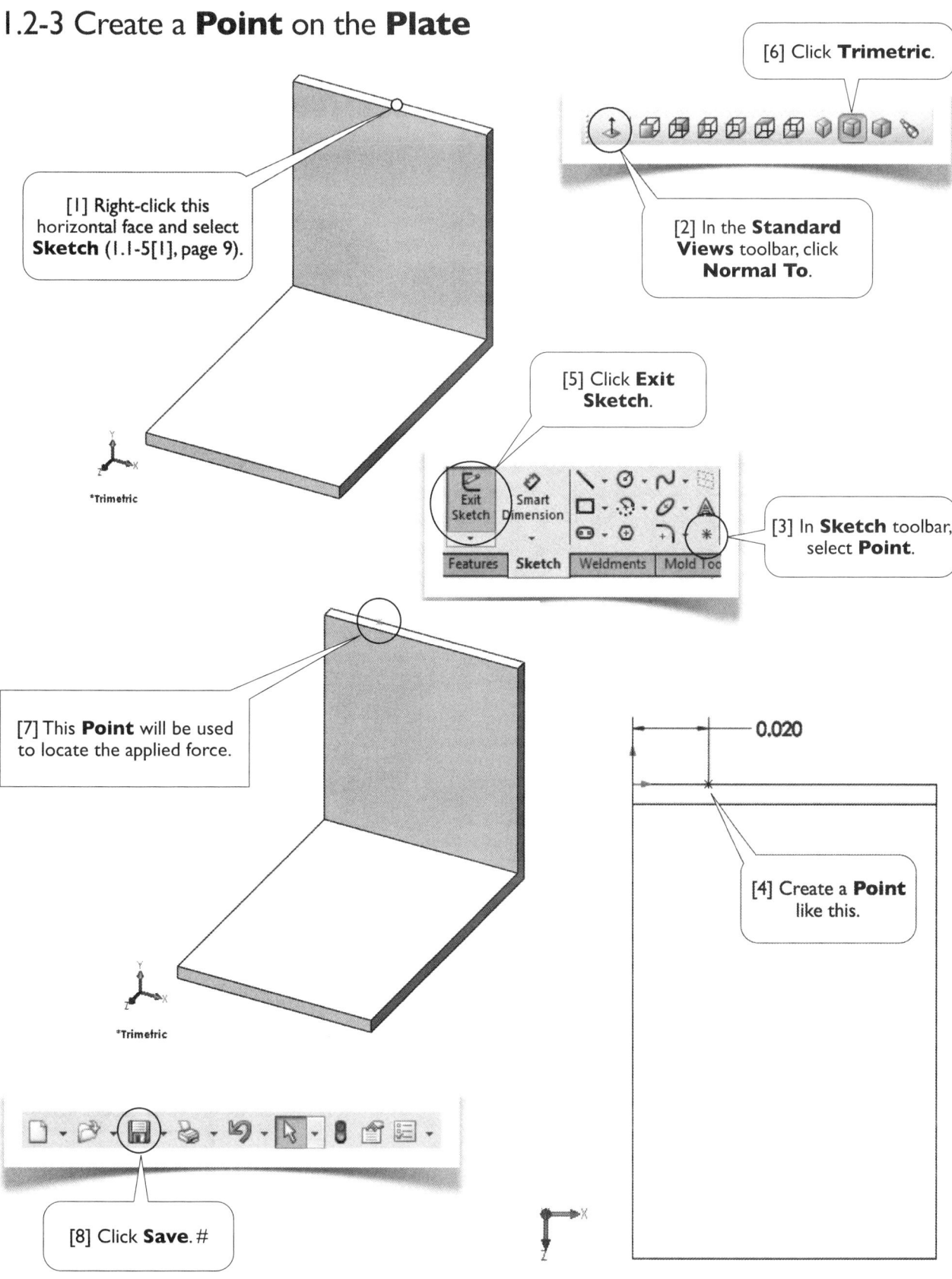

[6] Click **Trimetric**.

[1] Right-click this horizontal face and select **Sketch** (1.1-5[1], page 9).

[2] In the **Standard Views** toolbar, click **Normal To**.

*Trimetric

[5] Click **Exit Sketch**.

Exit Sketch   Smart Dimension

Features   **Sketch**   Weldments   Mold Too

[3] In **Sketch** toolbar, select **Point**.

[7] This **Point** will be used to locate the applied force.

*Trimetric

0.020

[4] Create a **Point** like this.

[8] Click **Save**. #

1.2-4 Create an Assembly: **PlateSupported**

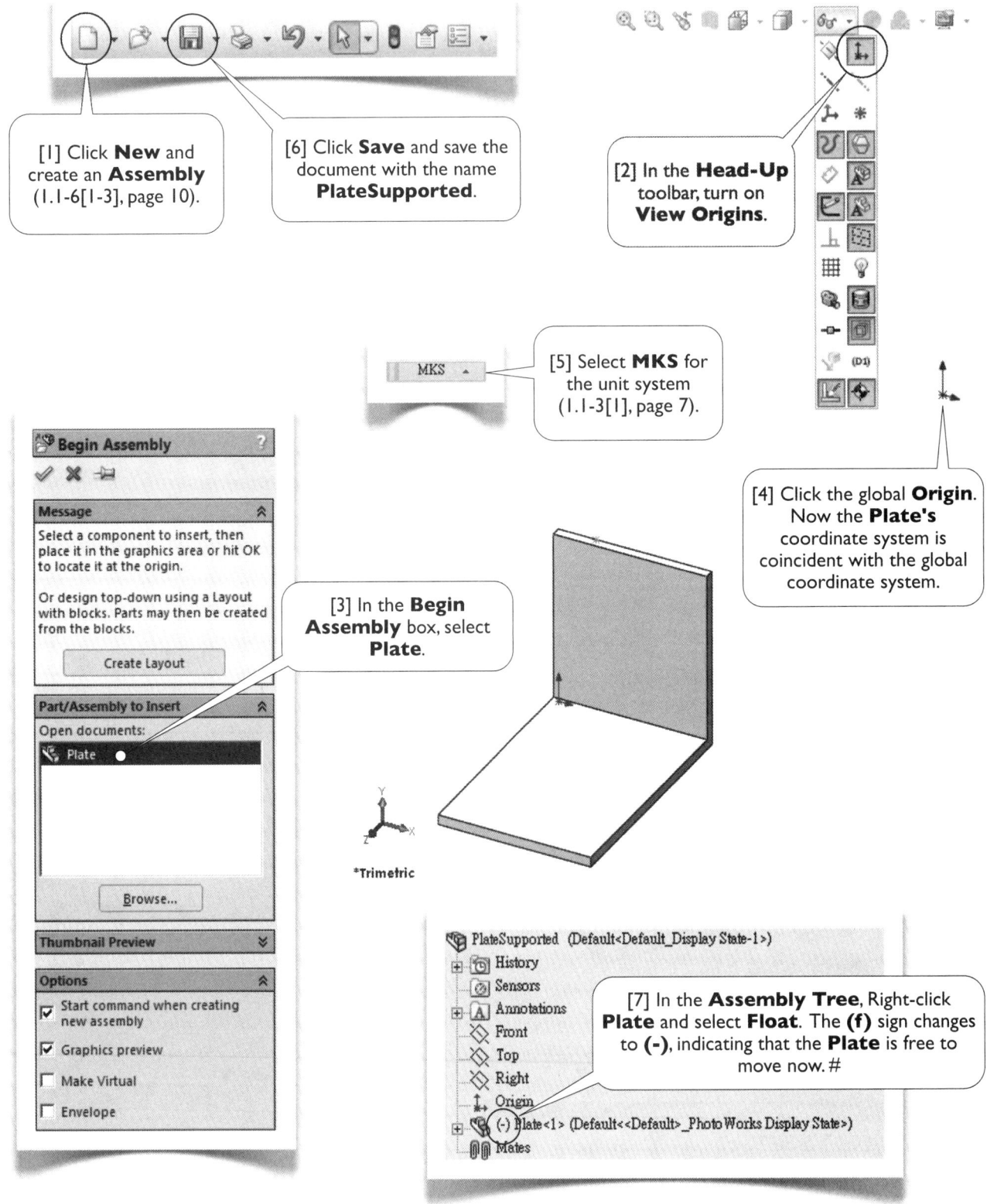

[1] Click **New** and create an **Assembly** (1.1-6[1-3], page 10).

[6] Click **Save** and save the document with the name **PlateSupported**.

[2] In the **Head-Up** toolbar, turn on **View Origins**.

[5] Select **MKS** for the unit system (1.1-3[1], page 7).

[4] Click the global **Origin**. Now the **Plate's** coordinate system is coincident with the global coordinate system.

Begin Assembly

Message

Select a component to insert, then place it in the graphics area or hit OK to locate it at the origin.

Or design top-down using a Layout with blocks. Parts may then be created from the blocks.

Create Layout

[3] In the **Begin Assembly** box, select **Plate**.

Part/Assembly to Insert

Open documents:

Plate

Browse...

Thumbnail Preview

Options

☑ Start command when creating new assembly

☑ Graphics preview

☐ Make Virtual

☐ Envelope

*Trimetric

PlateSupported (Default<Default_Display State-1>)
  History
  Sensors
  Annotations
  Front
  Top
  Right
  Origin
  (-) Plate<1> (Default<<Default>_PhotoWorks Display State>)
  Mates

[7] In the **Assembly Tree**, Right-click **Plate** and select **Float**. The **(f)** sign changes to **(-)**, indicating that the **Plate** is free to move now. #

1.2-5 Create Planes

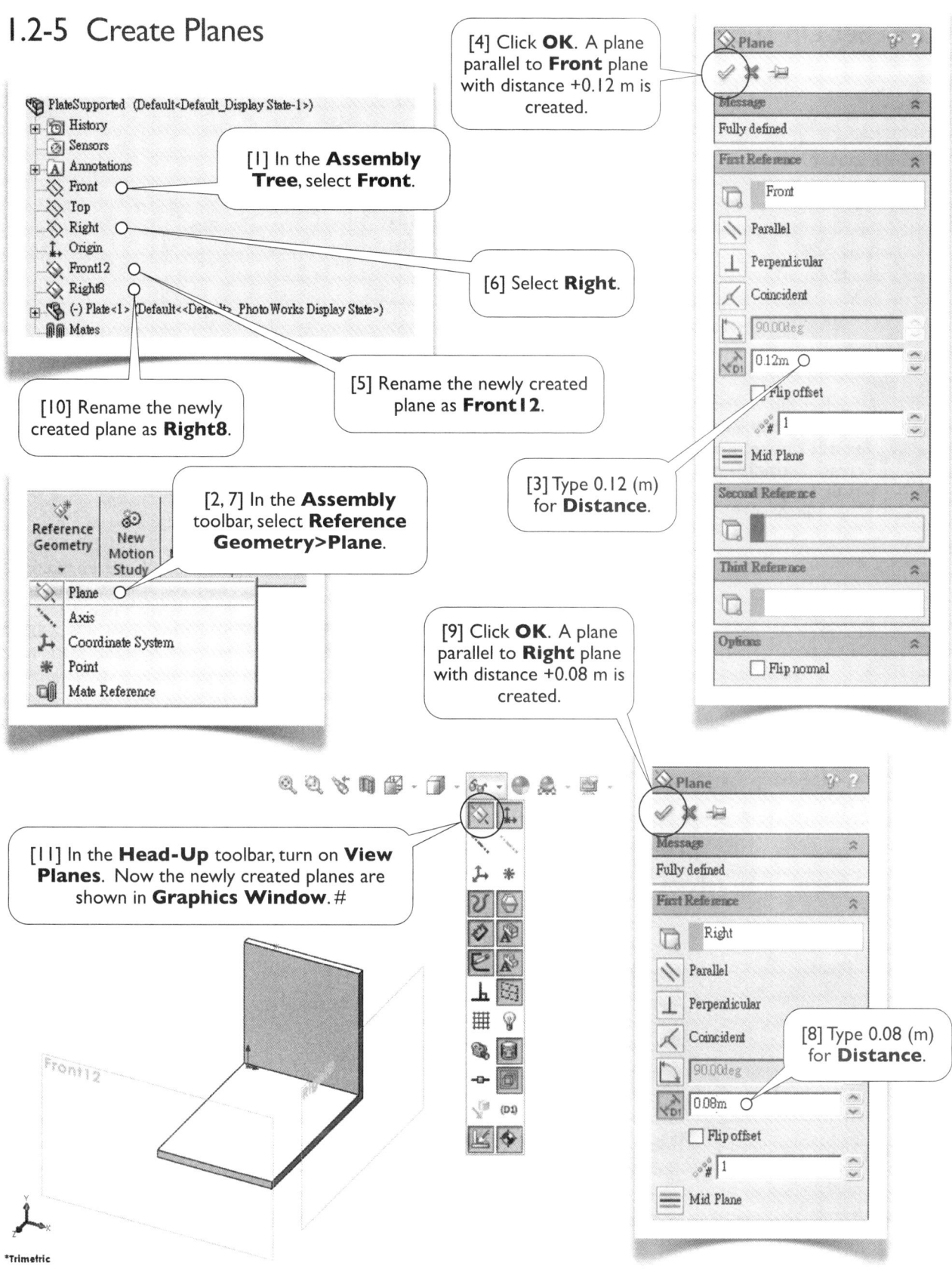

[4] Click **OK**. A plane parallel to **Front** plane with distance +0.12 m is created.

[1] In the **Assembly Tree**, select **Front**.

[6] Select **Right**.

[5] Rename the newly created plane as **Front12**.

[10] Rename the newly created plane as **Right8**.

[2, 7] In the **Assembly** toolbar, select **Reference Geometry>Plane**.

[3] Type 0.12 (m) for **Distance**.

[9] Click **OK**. A plane parallel to **Right** plane with distance +0.08 m is created.

[11] In the **Head-Up** toolbar, turn on **View Planes**. Now the newly created planes are shown in **Graphics Window**. #

[8] Type 0.08 (m) for **Distance**.

1.2-6 Set Up Supports

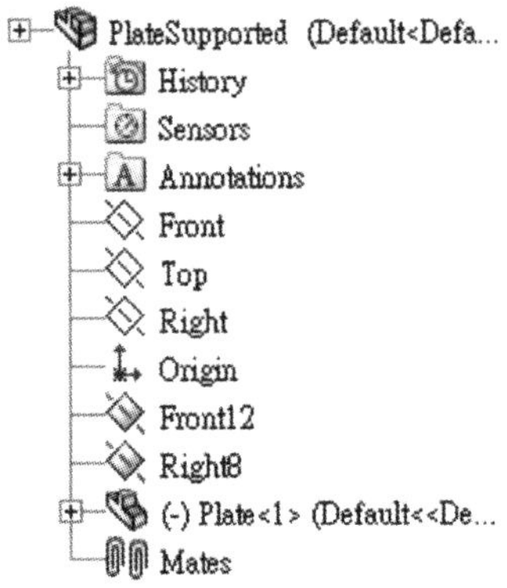

PlateSupported (Default<Defa...
  History
  Sensors
  Annotations
  Front
  Top
  Right
  Origin
  Front12
  Right8
  (-) Plate<1> (Default<<De...
  Mates

[1] In **Assembly** toolbar, click **Mate**.

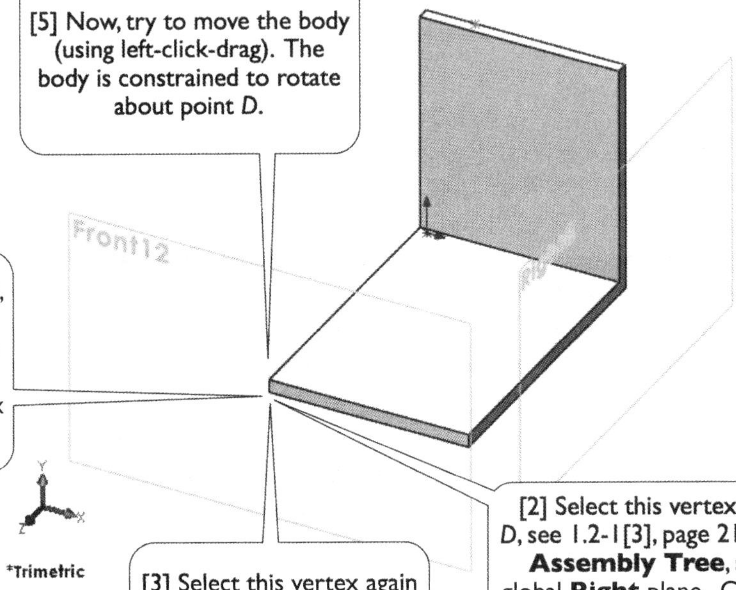

[5] Now, try to move the body (using left-click-drag). The body is constrained to rotate about point D.

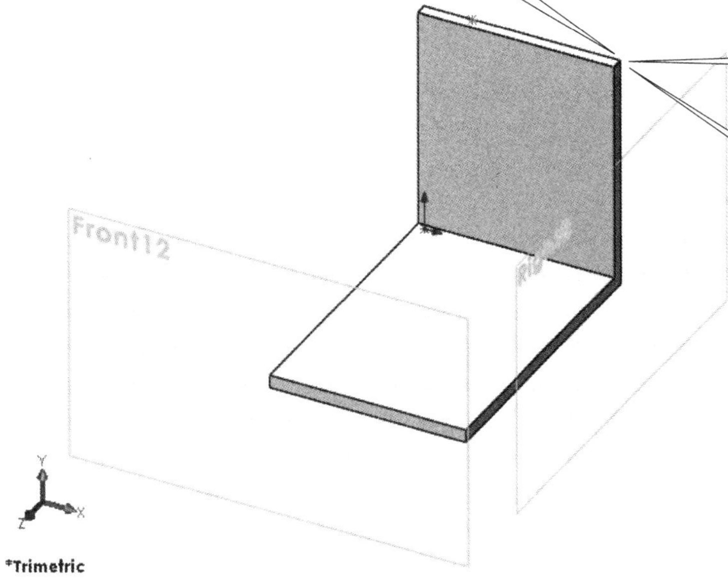

[4] Select this vertex again and, in **Assembly Tree**, select **Front12** plane. Click **OK** to create **Coincident3**. This sets up a support at this vertex in Z-direction.

*Trimetric

[3] Select this vertex again and, in **Assembly Tree**, select global **Top** plane. Click **OK** to create **Coincident2**. This sets up a support at this vertex in Y-direction.

[2] Select this vertex (point D, see 1.2-1[3], page 21) and, in **Assembly Tree**, select global **Right** plane. Click **OK** to create **Coincident1**. This sets up a support at this vertex in X-direction.

[8] Now, try to move the body (using left-click-drag). The body is constrained to rotate about the axis passing through BD.

[6] Select this vertex (point B, see 1.2-1[4], page 21) and, in **Assembly Tree**, select **Right8** plane. Click **OK** to create **Coincident4**. This sets up a support at this vertex in X-direction.

[7] Select this vertex again and, in **Assembly Tree**, select global **Front** plane. Click **OK** to create **Coincident5**. This sets up a support at this vertex in Z-direction.

*Trimetric

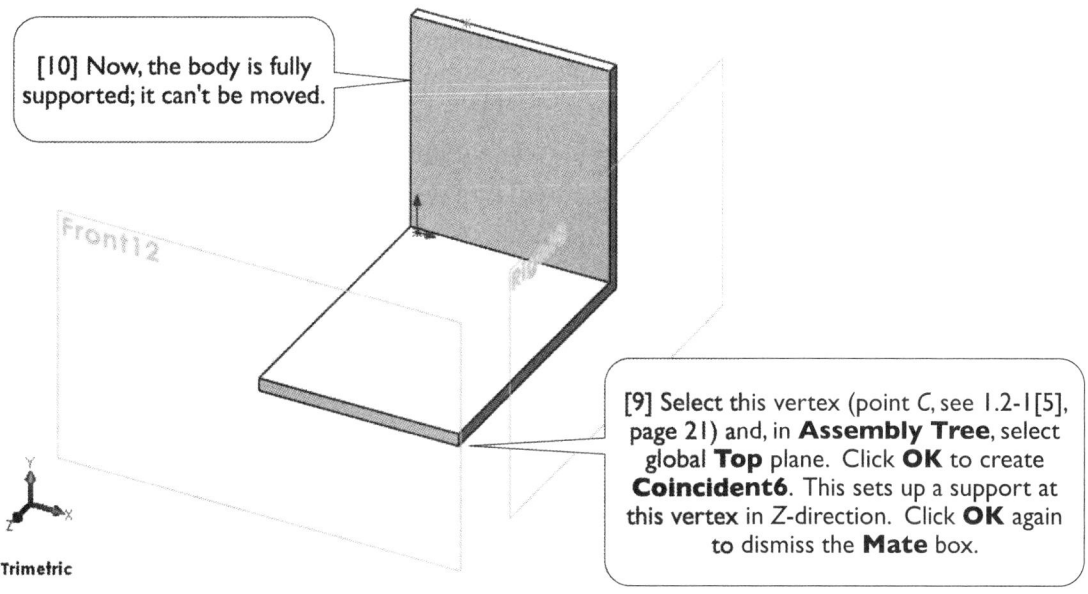

[10] Now, the body is fully supported; it can't be moved.

Front12

*Trimetric

[9] Select this vertex (point C, see 1.2-1[5], page 21) and, in **Assembly Tree**, select global **Top** plane. Click **OK** to create **Coincident6**. This sets up a support at this vertex in Z-direction. Click **OK** again to dismiss the **Mate** box.

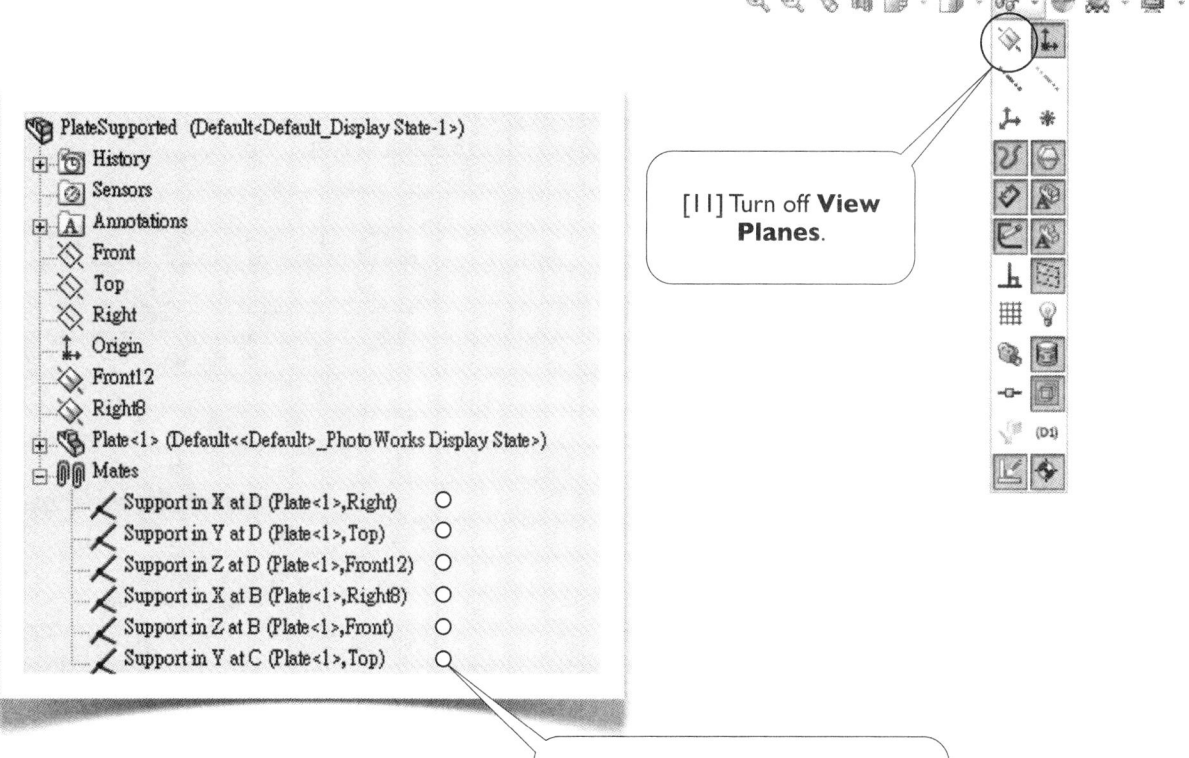

[11] Turn off **View Planes**.

[12] In the **Assembly Tree**, expand **Mates** and rename the six **Coincident** mates like this (for better readability). #

1.2-7 Create a **Study**

[1] Load **SOLIDWORKS Motion** if it is not loaded yet (1.1-8, page 14).

[3] Select **Motion Analysis** (1.1-9[8, 9], page 15). #

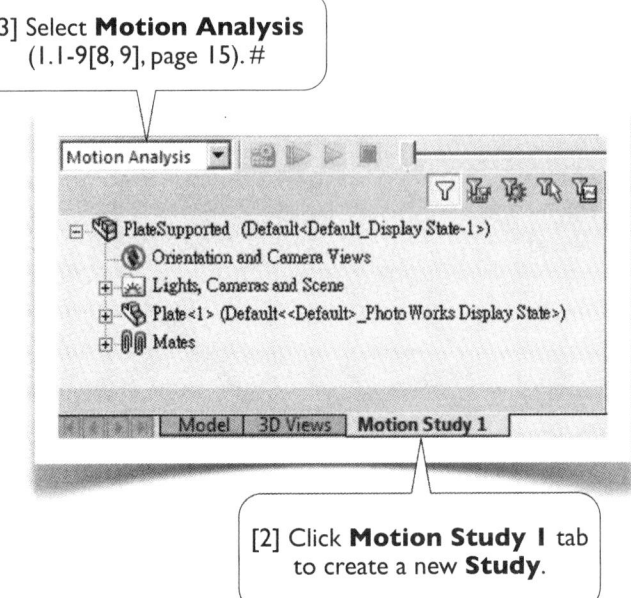

[2] Click **Motion Study 1** tab to create a new **Study**.

1.2-8 Set Up Forces

[1] In **Motion** toolbar, click **Force**.

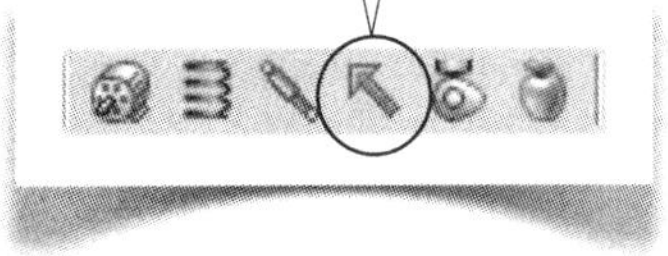

[5] Click **OK**. #

[2] Click this **Point** to define the location of the force.

[3] Click this face to define the direction of the force. Its outer-normal is the force direction.

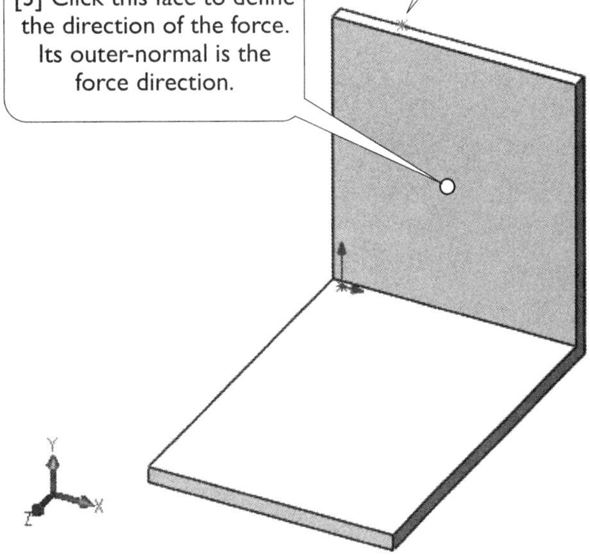

Force/Torque

Direction

Action only

Action & reaction

Point1@Sketch2@Plate-1@PlateSupported

Face<1>@Plate-1

Force relative to:
- Assembly origin
- Selected component:

Force Function

Constant

F_1 200 N

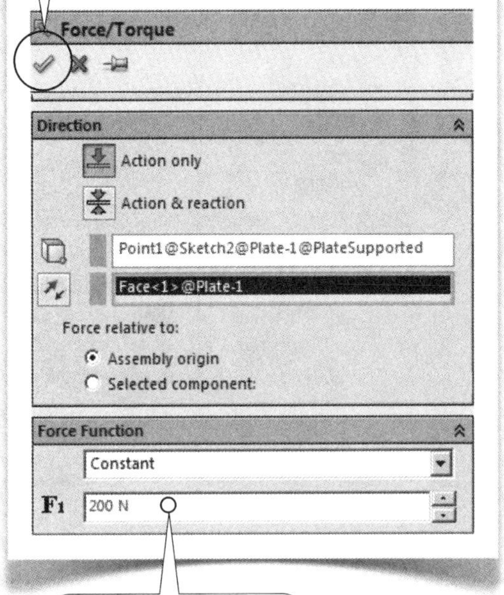

[4] Type 200 (N).

*Trimetric

1.2-9 Calculate Results

[1] Click **Calculate**. Remember that, If a **Motion Analysis Messages** window appears, close it. #

1.2-10 Retrieve the Reaction Force at C

[4] Click **OK**.

[1] In **Motion** toolbar, click **Results and Plots**.

[2] Set up **Result** like this.

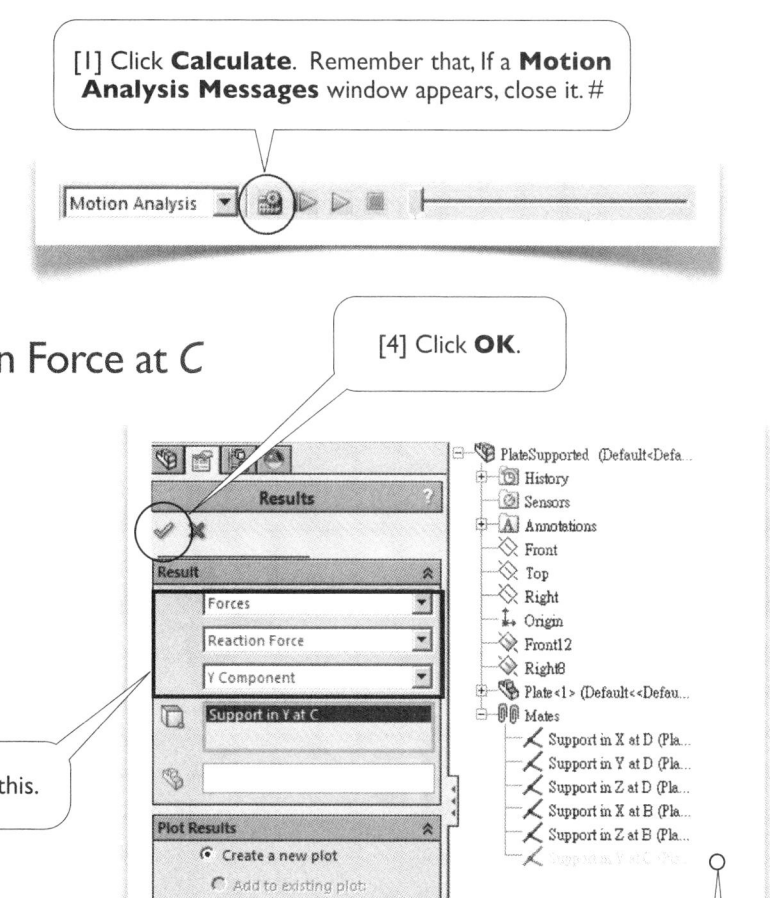

[3] Expand the **Assembly Tree** in the **Graphics Area** and select **Support in Y at C**.

Results
    Reaction Force in Y at C <Reaction Force1>

[5] In the **Motion Study Tree**, Rename **Plot1** as **Reaction Force in Y at C**.

[7] Close the window. #

[6] The plot shows that the reaction force in Y-direction at C is +112.5 N, consistent with the value in Eq. 1.2-1(1) (page 21).

1.2-11 Do It Yourself: Other Reaction Forces and Validation of the Results

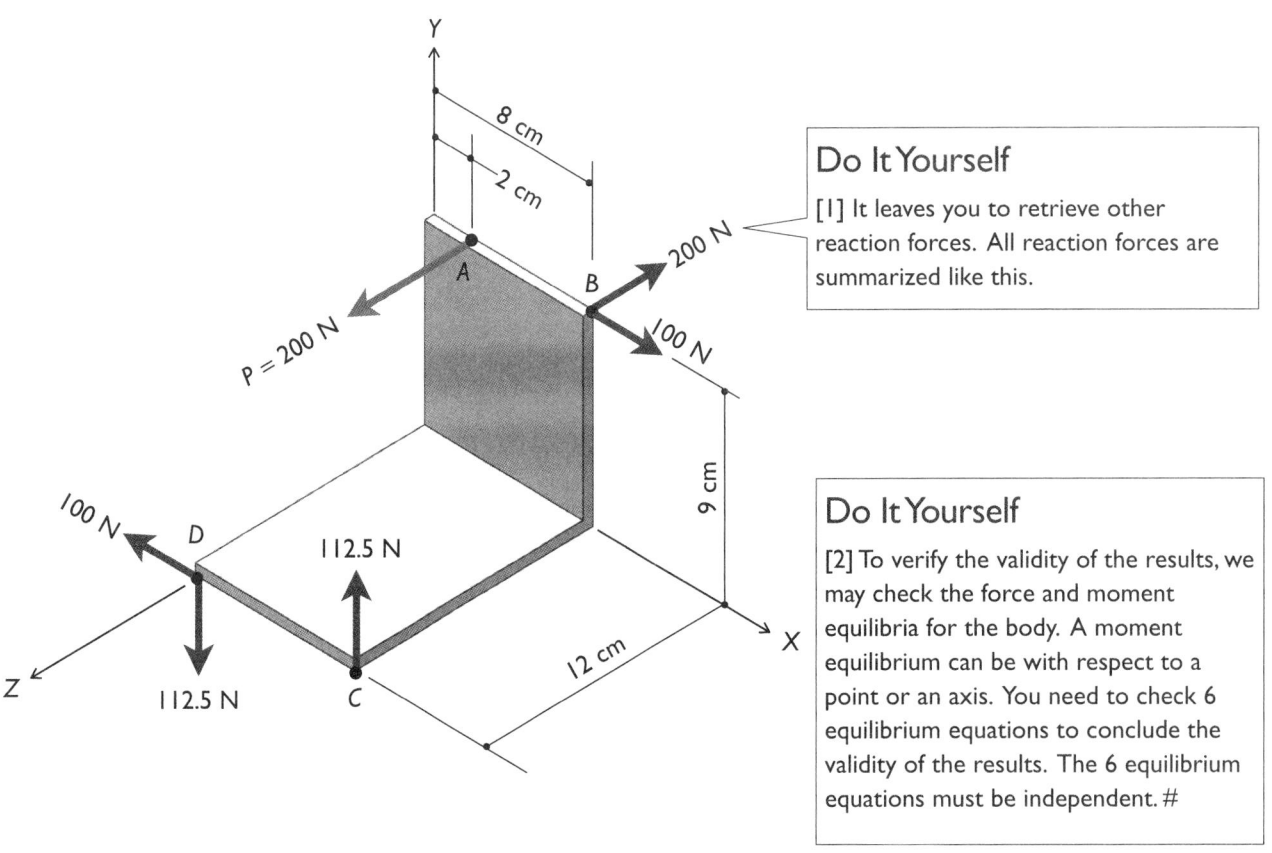

Do It Yourself

[1] It leaves you to retrieve other reaction forces. All reaction forces are summarized like this.

Do It Yourself

[2] To verify the validity of the results, we may check the force and moment equilibria for the body. A moment equilibrium can be with respect to a point or an axis. You need to check 6 equilibrium equations to conclude the validity of the results. The 6 equilibrium equations must be independent. #

1.2-12 Wrap Up

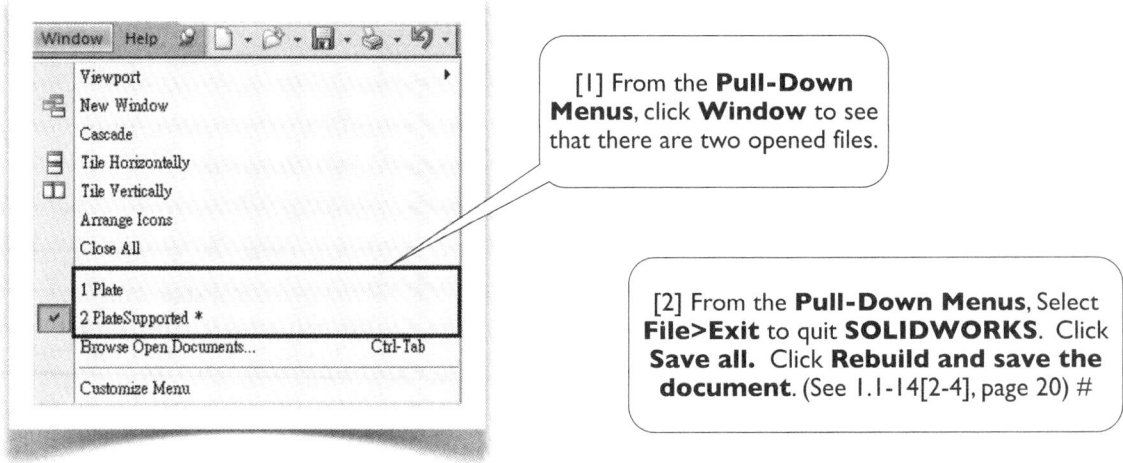

[1] From the **Pull-Down Menus**, click **Window** to see that there are two opened files.

[2] From the **Pull-Down Menus**, Select **File>Exit** to quit **SOLIDWORKS**. Click **Save all**. Click **Rebuild and save the document**. (See 1.1-14[2-4], page 20) #

Chapter 2

Trusses

A truss is a structural system that uses slender members to achieve material-efficiency. Each member is designed mainly to withstand tension or compression. Therefore, truss members are also called two-force members. For a truss member to be a two-force member, it must connect to other members with joints that allow free rotations and the external forces can only be applied at joints.

To achieve free-rotation, pin joints would be used in 2D trusses, and ball-and-socket joints would be used in 3D trusses. In reality, pin joints or ball-and-socket joints are seldom used. Modern-day trusses are constructed with joints using multiple bolts or welding. The members are rigid-jointed, rather than pin-jointed or ball-and-socket-jointed.

Historically, we assume truss members as two-force members, because the assumption largely reduces computational efforts. You may raise a question: how many numerical errors are caused by the assumption. The answer is: as far as the truss members are slender enough, the errors can be neglected. Of course, the next question would be: how slender is "slender enough?" It is a good exercise problem. After completing the exercises in this chapter and the following chapters and learning the analysis techniques, you may be able to answer this question yourself, by doing some extra exercises.

In Section 2.1, we'll show you a way to solve a pin-jointed 2D truss with **SOLIDWORKS** and, in Section 2.2, solve a ball-and-socket-jointed 3D truss.

As mentioned in the opening of Chapter 1, this book discusses only statically determinate structures. To solve a statically indeterminate structure, deformation of the structural members must be taken into account. In such cases, the rigid-body assumption must be lifted and **SOLIDWORKS Motion**, which implements rigid-body mechanics, can no longer be used to solve such problems. A more general way to solve a truss problem (including statically indeterminate problems) is to use software that implements finite element methods, such as **SOLIDWORKS Simulation**. This is demonstrated in Section 11.1, *Mechanics of Materials Labs with SOLIDWORKS Simulation 2015*, by Huei-Huang Lee.

Section 2.1

Plane Truss

2.1-1 Introduction

[1] In this section, we consider a plane truss supported by a hinge at the left and a roller at the right and subject to a downward force of 6000 N at node C [2, 3]. We want to find the reaction forces and the member forces.

From the free-body diagram of the entire truss [3], taking the moment equilibrium about G, we may calculate A_Y,

$$\sum M_G = 0, \quad (A_Y)(18 \text{ m}) - (6000 \text{ N})(12 \text{ m}) = 0,$$

$$A_Y = 4000 \text{ N} \qquad (1)$$

From the free-body diagram shown in [4], taking the moment equilibrium about C, we may calculate F_{BD},

$$\sum M_C = 0, \quad (A_Y)(6 \text{ m}) + (F_{BD})(4 \text{ m}) = 0,$$

$$F_{BD} = -6000 \text{ N} \qquad (2)$$

The negative sign indicates that it is a compressive force.

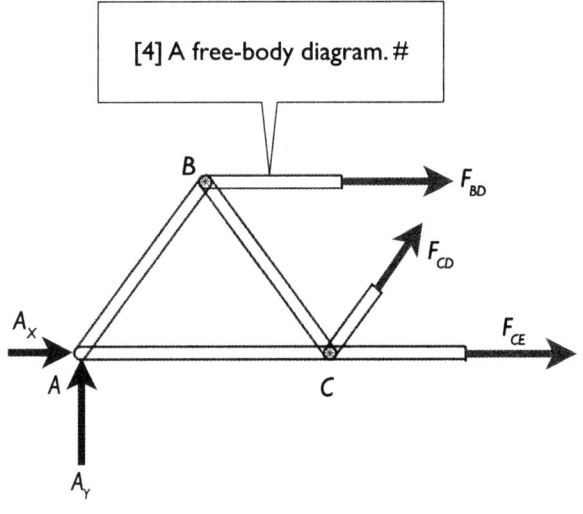

[4] A free-body diagram. #

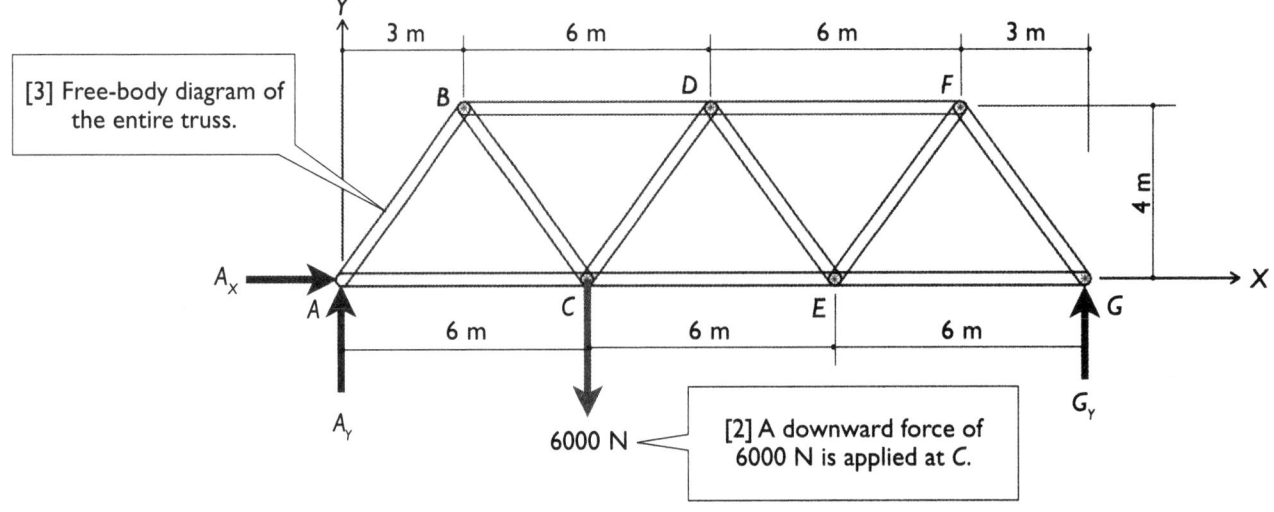

[3] Free-body diagram of the entire truss.

[2] A downward force of 6000 N is applied at C.

2.1-2 Start Up and Create a Part: **Diagonal**

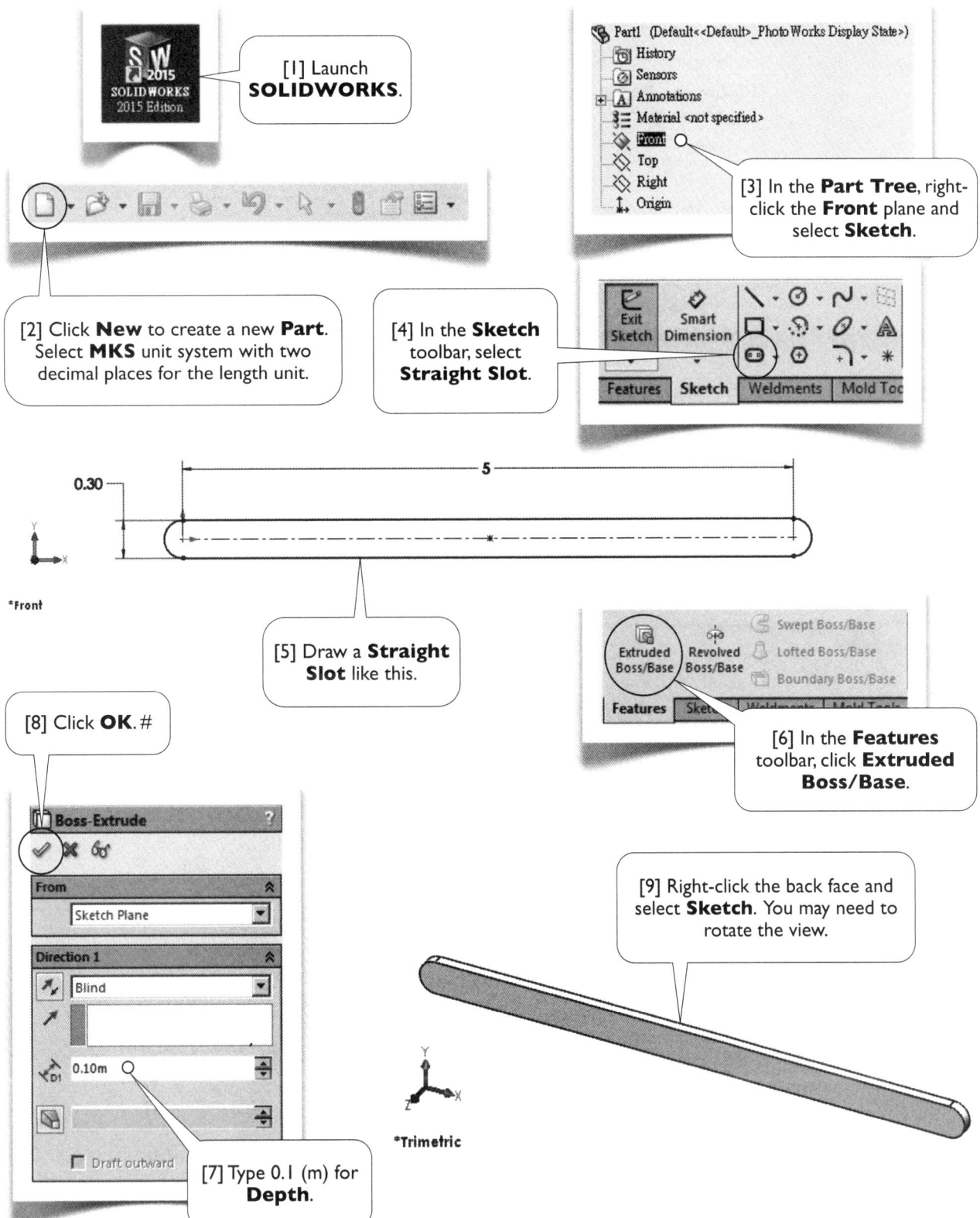

[1] Launch **SOLIDWORKS**.

Part1 (Default<<Default>_PhotoWorks Display State>)
- History
- Sensors
- Annotations
- Material <not specified>
- Front
- Top
- Right
- Origin

[3] In the **Part Tree**, right-click the **Front** plane and select **Sketch**.

[2] Click **New** to create a new **Part**. Select **MKS** unit system with two decimal places for the length unit.

[4] In the **Sketch** toolbar, select **Straight Slot**.

Exit Sketch Smart Dimension

Features Sketch Weldments Mold Too

0.30

5

*Front

[5] Draw a **Straight Slot** like this.

Extruded Boss/Base   Revolved Boss/Base   Swept Boss/Base   Lofted Boss/Base   Boundary Boss/Base

Features Sket Weldments Mold Tools

[6] In the **Features** toolbar, click **Extruded Boss/Base**.

[8] Click **OK**. #

Boss-Extrude   ?

From   ⌃
Sketch Plane

Direction 1   ⌃
Blind

0.10m

☐ Draft outward

[7] Type 0.1 (m) for **Depth**.

[9] Right-click the back face and select **Sketch**. You may need to rotate the view.

*Trimetric

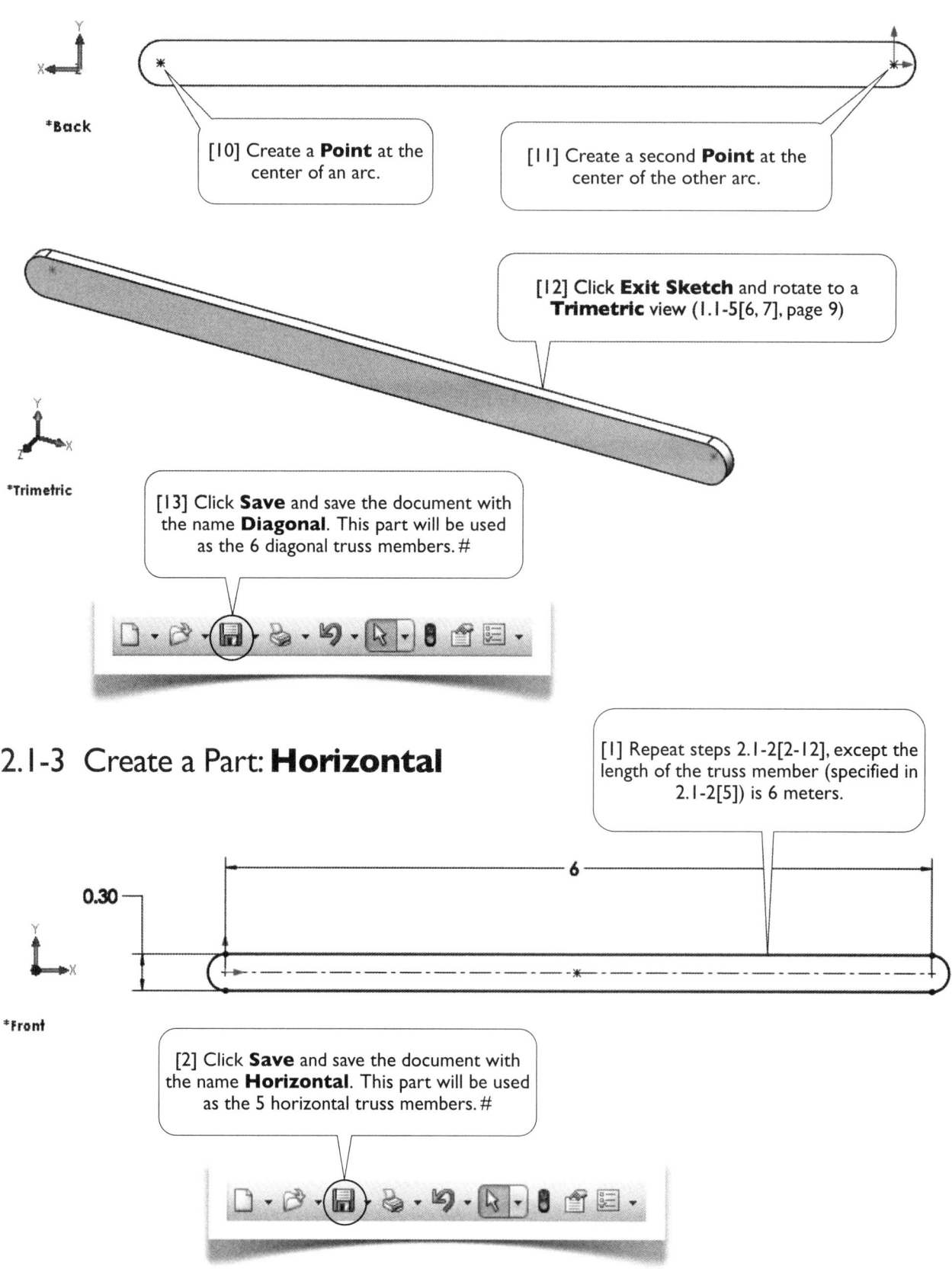

*Back

[10] Create a **Point** at the center of an arc.

[11] Create a second **Point** at the center of the other arc.

[12] Click **Exit Sketch** and rotate to a **Trimetric** view (1.1-5[6, 7], page 9)

*Trimetric

[13] Click **Save** and save the document with the name **Diagonal**. This part will be used as the 6 diagonal truss members. #

2.1-3 Create a Part: **Horizontal**

[1] Repeat steps 2.1-2[2-12], except the length of the truss member (specified in 2.1-2[5]) is 6 meters.

0.30

6

*Front

[2] Click **Save** and save the document with the name **Horizontal**. This part will be used as the 5 horizontal truss members. #

2.1-4 Create an Assembly: **PlaneTruss**

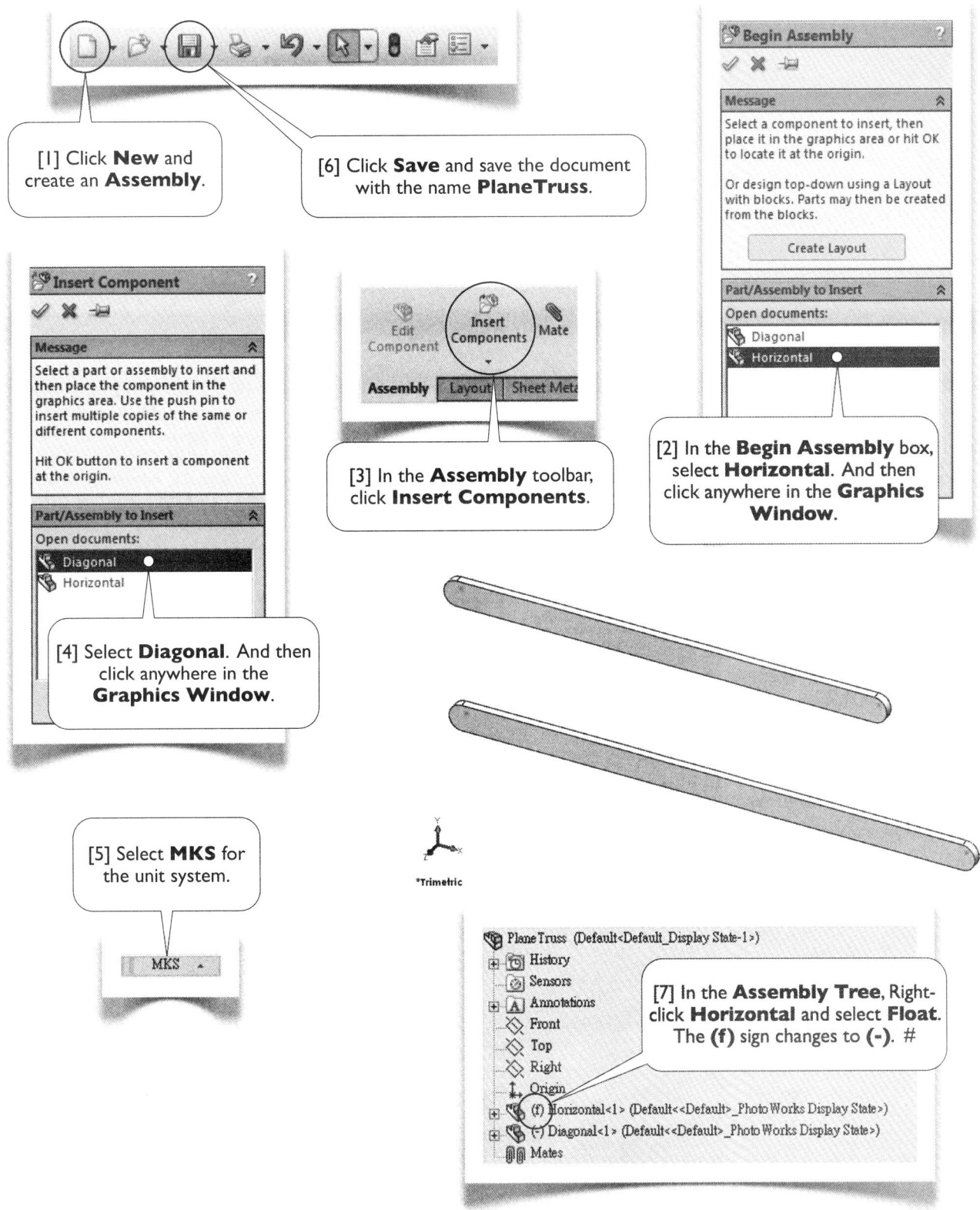

[1] Click **New** and create an **Assembly**.

[6] Click **Save** and save the document with the name **PlaneTruss**.

Begin Assembly

Message

Select a component to insert, then place it in the graphics area or hit OK to locate it at the origin.

Or design top-down using a Layout with blocks. Parts may then be created from the blocks.

Create Layout

Part/Assembly to Insert

Open documents:

Diagonal
Horizontal

Insert Component

Message

Select a part or assembly to insert and then place the component in the graphics area. Use the push pin to insert multiple copies of the same or different components.

Hit OK button to insert a component at the origin.

Part/Assembly to Insert

Open documents:

Diagonal
Horizontal

Edit Component   Insert Components   Mate

Assembly Layout Sheet Meta

[3] In the **Assembly** toolbar, click **Insert Components**.

[2] In the **Begin Assembly** box, select **Horizontal**. And then click anywhere in the **Graphics Window**.

[4] Select **Diagonal**. And then click anywhere in the **Graphics Window**.

[5] Select **MKS** for the unit system.

MKS

*Trimetric

PlaneTruss (Default<Default_Display State-1>)
- History
- Sensors
- Annotations
- Front
- Top
- Right
- Origin
- (f) Horizontal<1> (Default<<Default>_PhotoWorks Display State>)
- (-) Diagonal<1> (Default<<Default>_PhotoWorks Display State>)
- Mates

[7] In the **Assembly Tree**, Right-click **Horizontal** and select **Float**. The **(f)** sign changes to **(-)**. #

2.1-5 Duplicate **Parts** and Constrain all Members in XY-Space

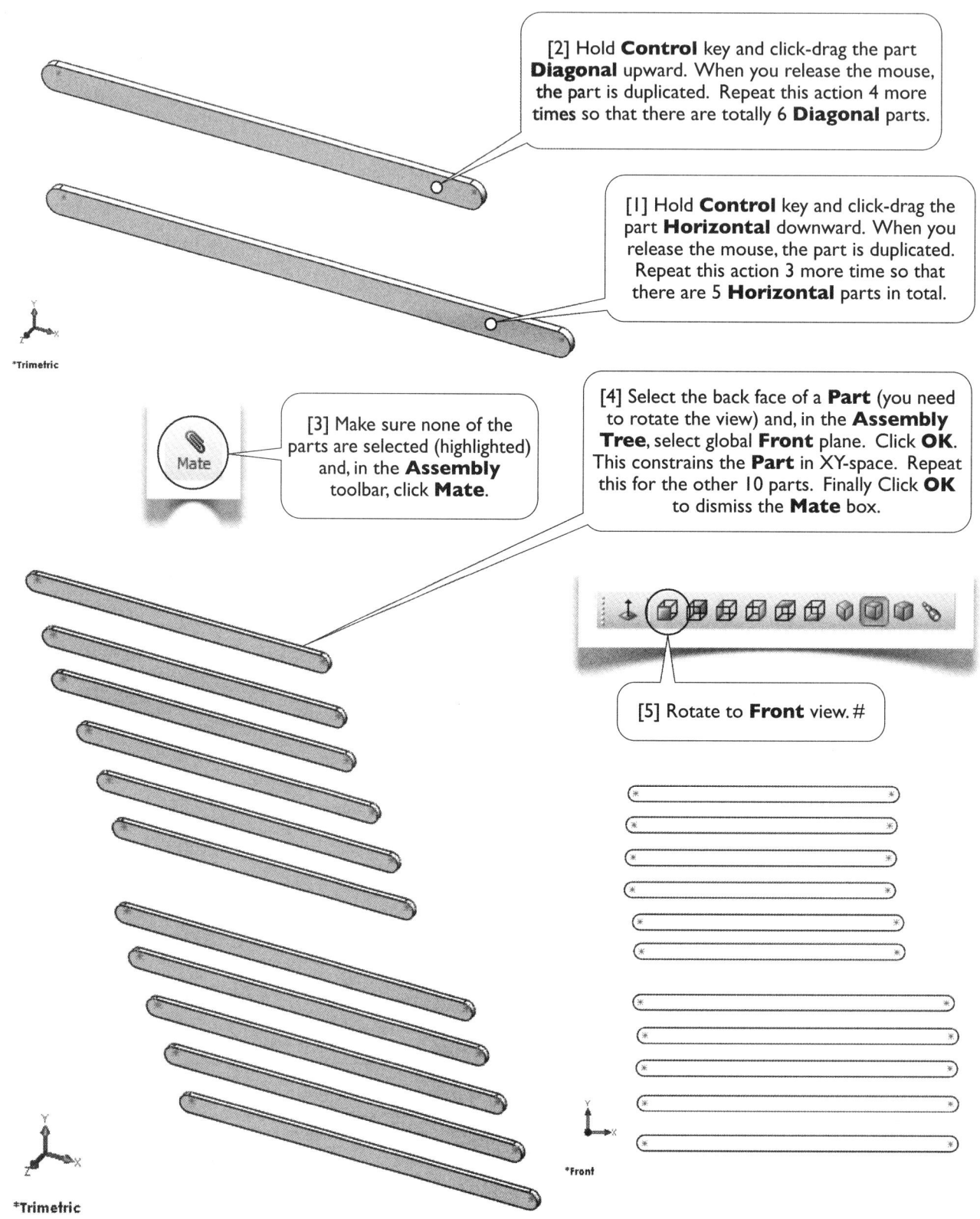

[2] Hold **Control** key and click-drag the part **Diagonal** upward. When you release the mouse, the part is duplicated. Repeat this action 4 more times so that there are totally 6 **Diagonal** parts.

[1] Hold **Control** key and click-drag the part **Horizontal** downward. When you release the mouse, the part is duplicated. Repeat this action 3 more time so that there are 5 **Horizontal** parts in total.

*Trimetric

[3] Make sure none of the parts are selected (highlighted) and, in the **Assembly** toolbar, click **Mate**.

Mate

[4] Select the back face of a **Part** (you need to rotate the view) and, in the **Assembly Tree**, select global **Front** plane. Click **OK**. This constrains the **Part** in XY-space. Repeat this for the other 10 parts. Finally Click **OK** to dismiss the **Mate** box.

[5] Rotate to **Front** view. #

*Front

*Trimetric

2.1-6 Set Up Lower Horizontal Members and Supports

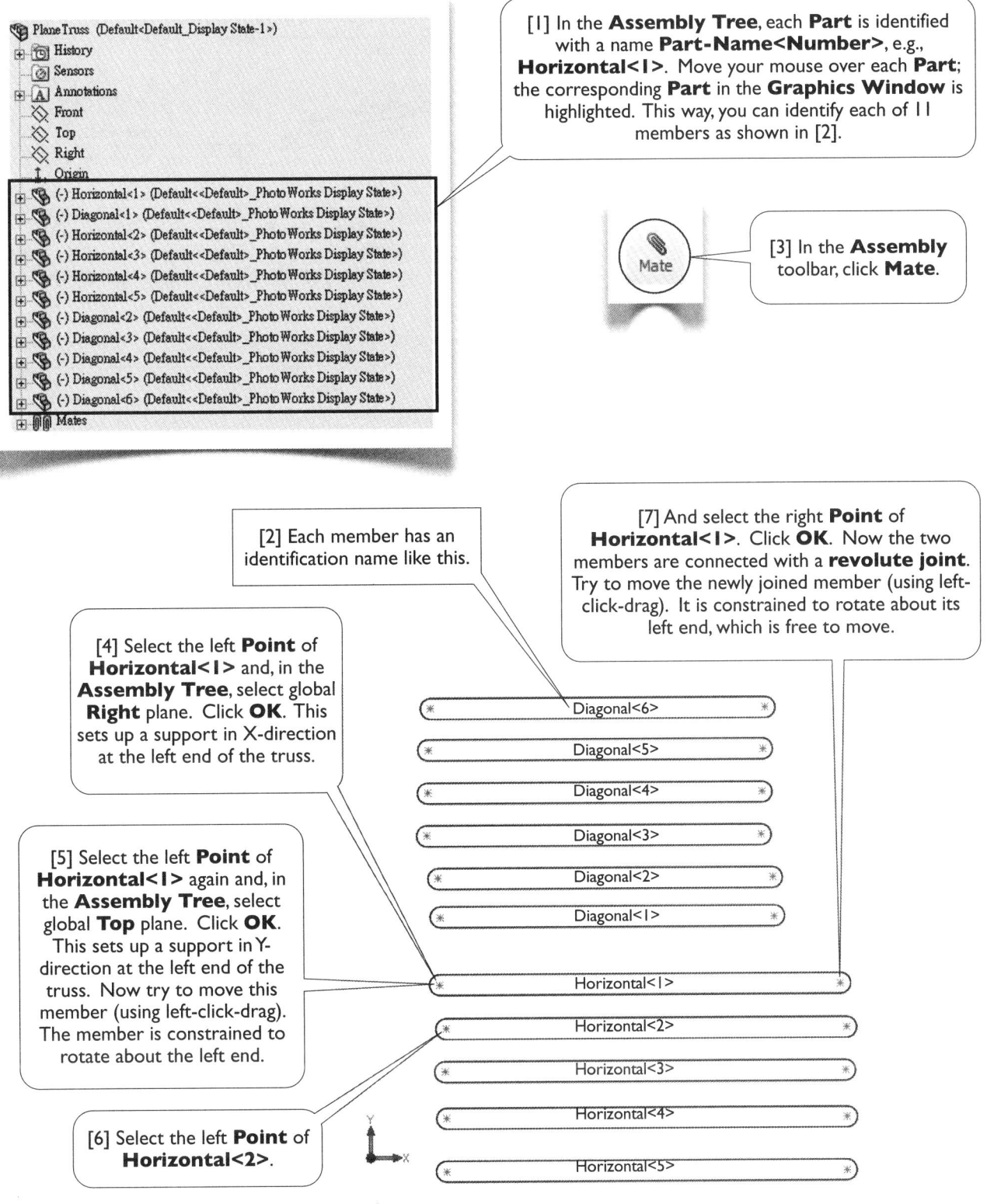

[1] In the **Assembly Tree**, each **Part** is identified with a name **Part-Name<Number>**, e.g., **Horizontal<1>**. Move your mouse over each **Part**; the corresponding **Part** in the **Graphics Window** is highlighted. This way, you can identify each of 11 members as shown in [2].

[3] In the **Assembly** toolbar, click **Mate**.

[2] Each member has an identification name like this.

[7] And select the right **Point** of **Horizontal<1>**. Click **OK**. Now the two members are connected with a **revolute joint**. Try to move the newly joined member (using left-click-drag). It is constrained to rotate about its left end, which is free to move.

[4] Select the left **Point** of **Horizontal<1>** and, in the **Assembly Tree**, select global **Right** plane. Click **OK**. This sets up a support in X-direction at the left end of the truss.

[5] Select the left **Point** of **Horizontal<1>** again and, in the **Assembly Tree**, select global **Top** plane. Click **OK**. This sets up a support in Y-direction at the left end of the truss. Now try to move this member (using left-click-drag). The member is constrained to rotate about the left end.

[6] Select the left **Point** of **Horizontal<2>**.

Diagonal<6>
Diagonal<5>
Diagonal<4>
Diagonal<3>
Diagonal<2>
Diagonal<1>
Horizontal<1>
Horizontal<2>
Horizontal<3>
Horizontal<4>
Horizontal<5>

*Front

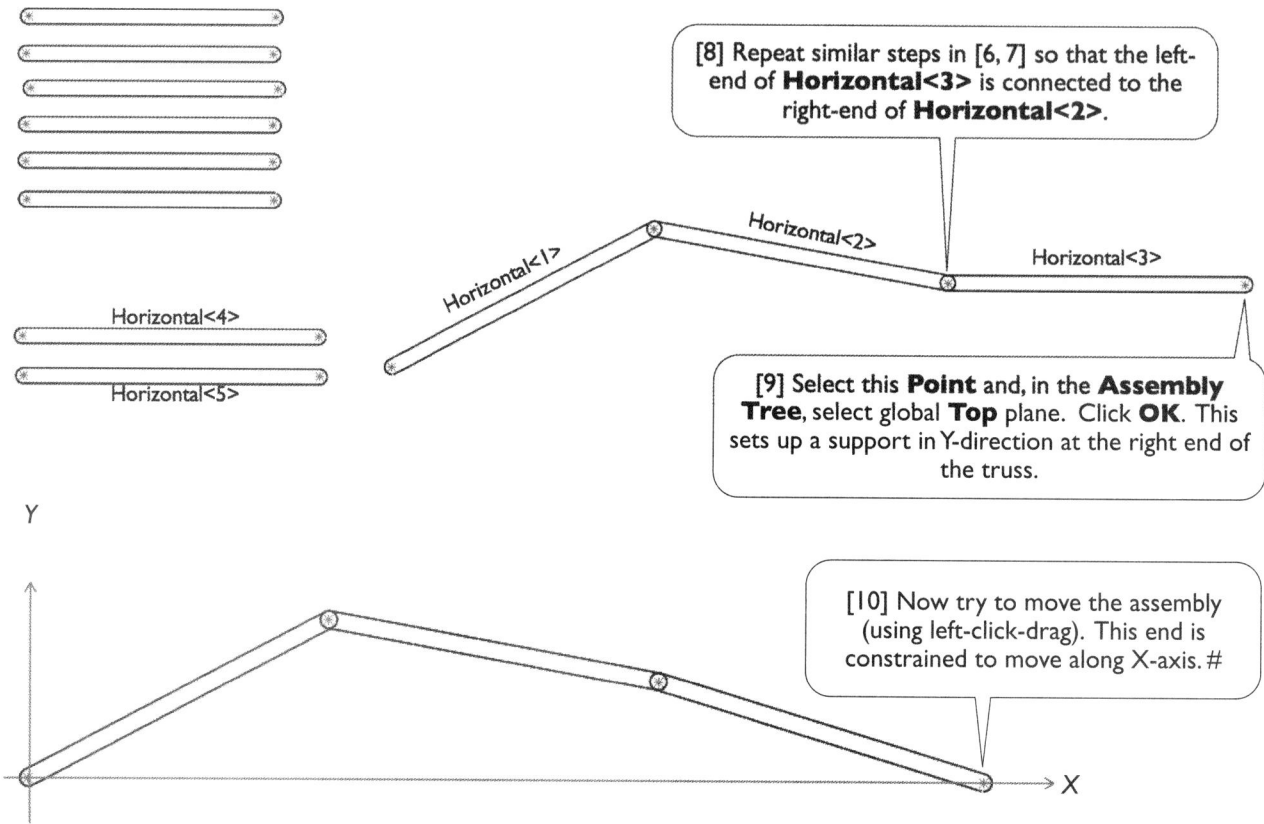

[8] Repeat similar steps in [6, 7] so that the left-end of **Horizontal<3>** is connected to the right-end of **Horizontal<2>**.

[9] Select this **Point** and, in the **Assembly Tree**, select global **Top** plane. Click **OK**. This sets up a support in Y-direction at the right end of the truss.

[10] Now try to move the assembly (using left-click-drag). This end is constrained to move along X-axis. #

2.1-7 Assemble other Truss Members

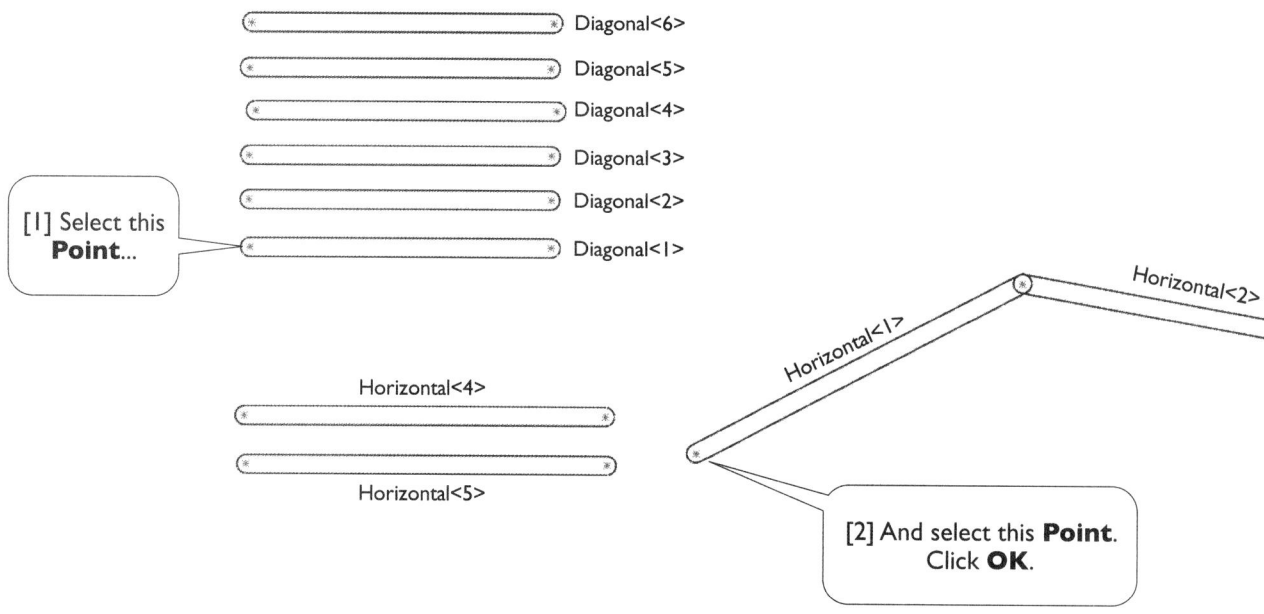

[1] Select this **Point**...

[2] And select this **Point**. Click **OK**.

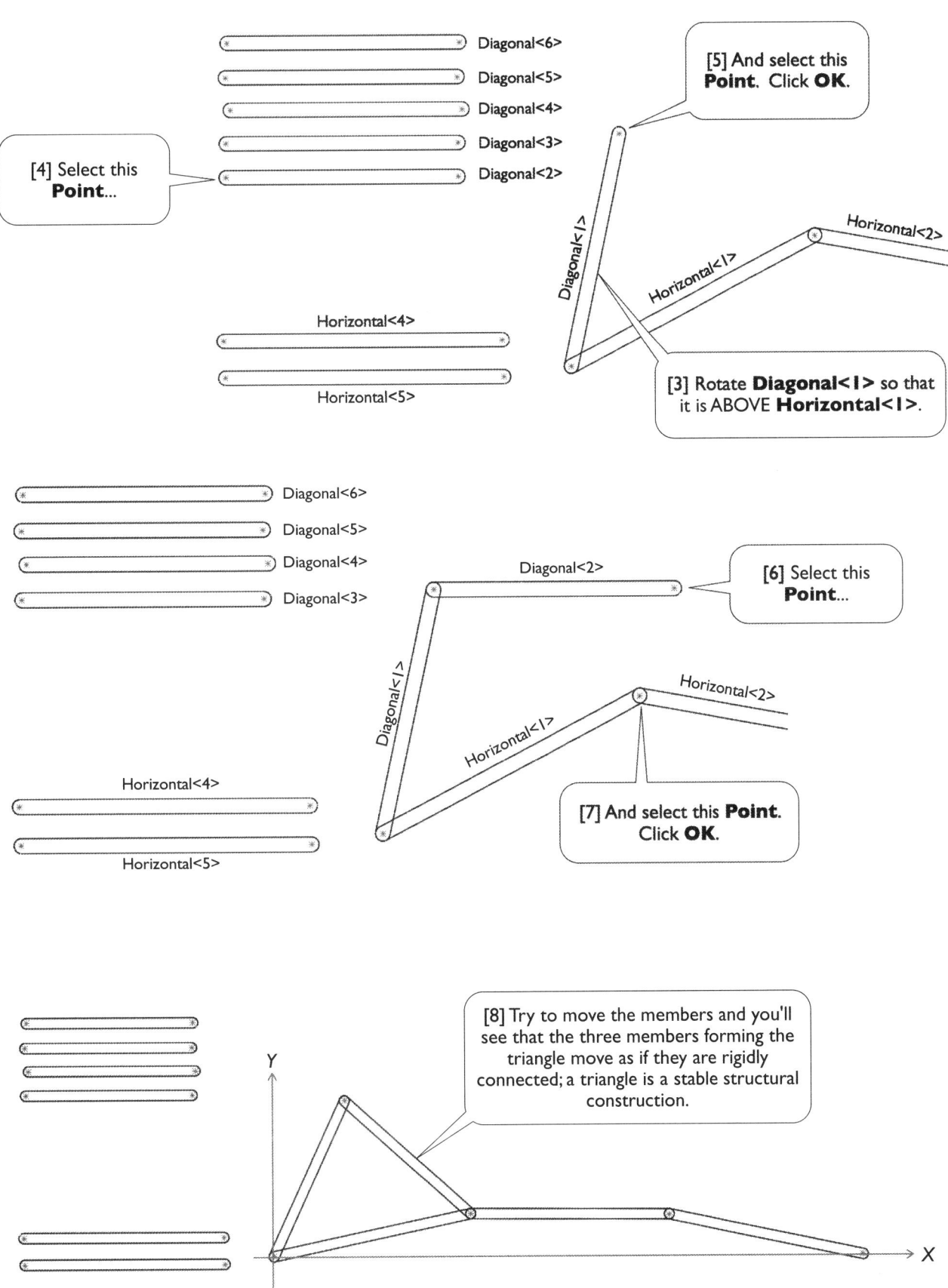

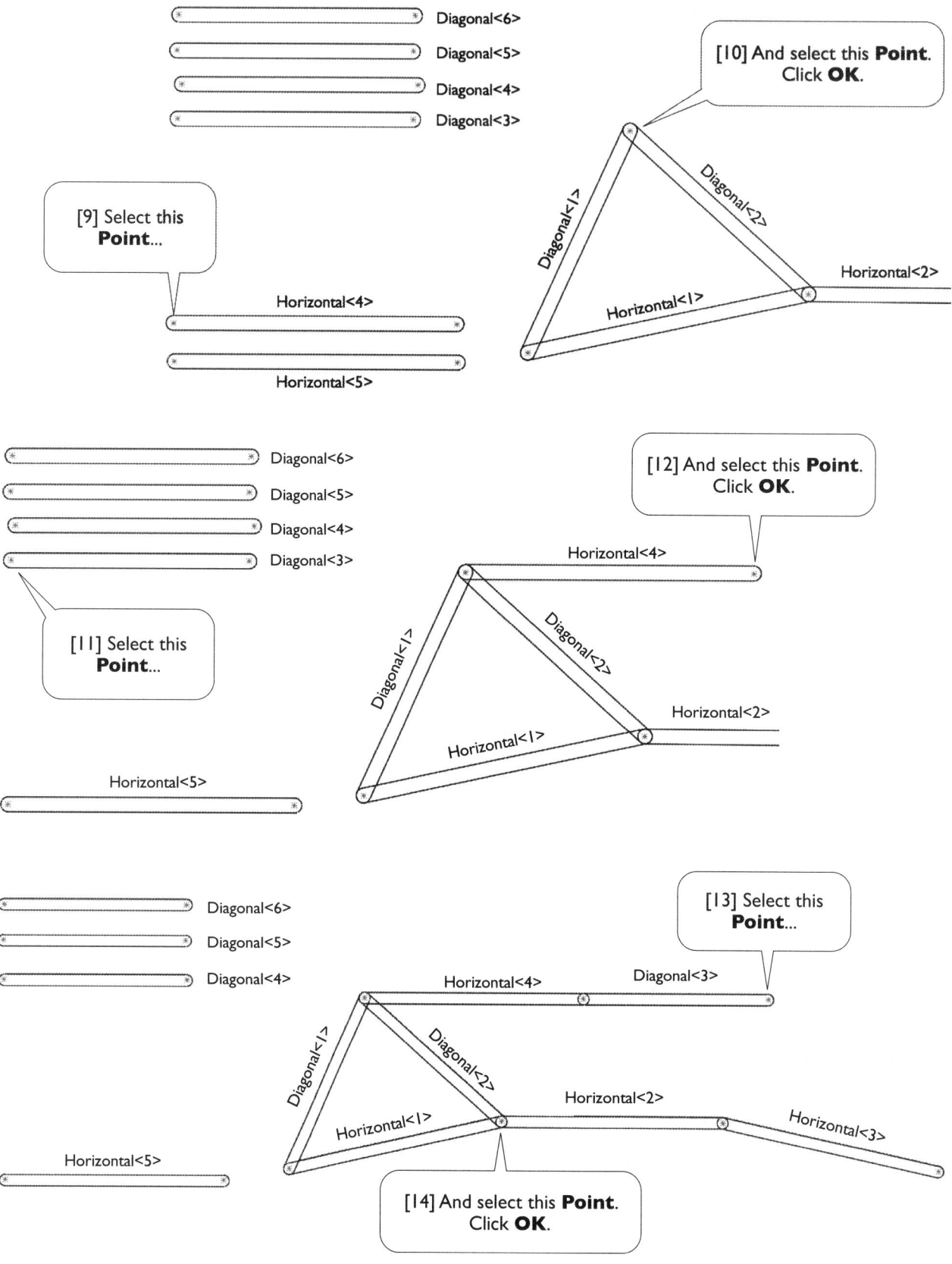

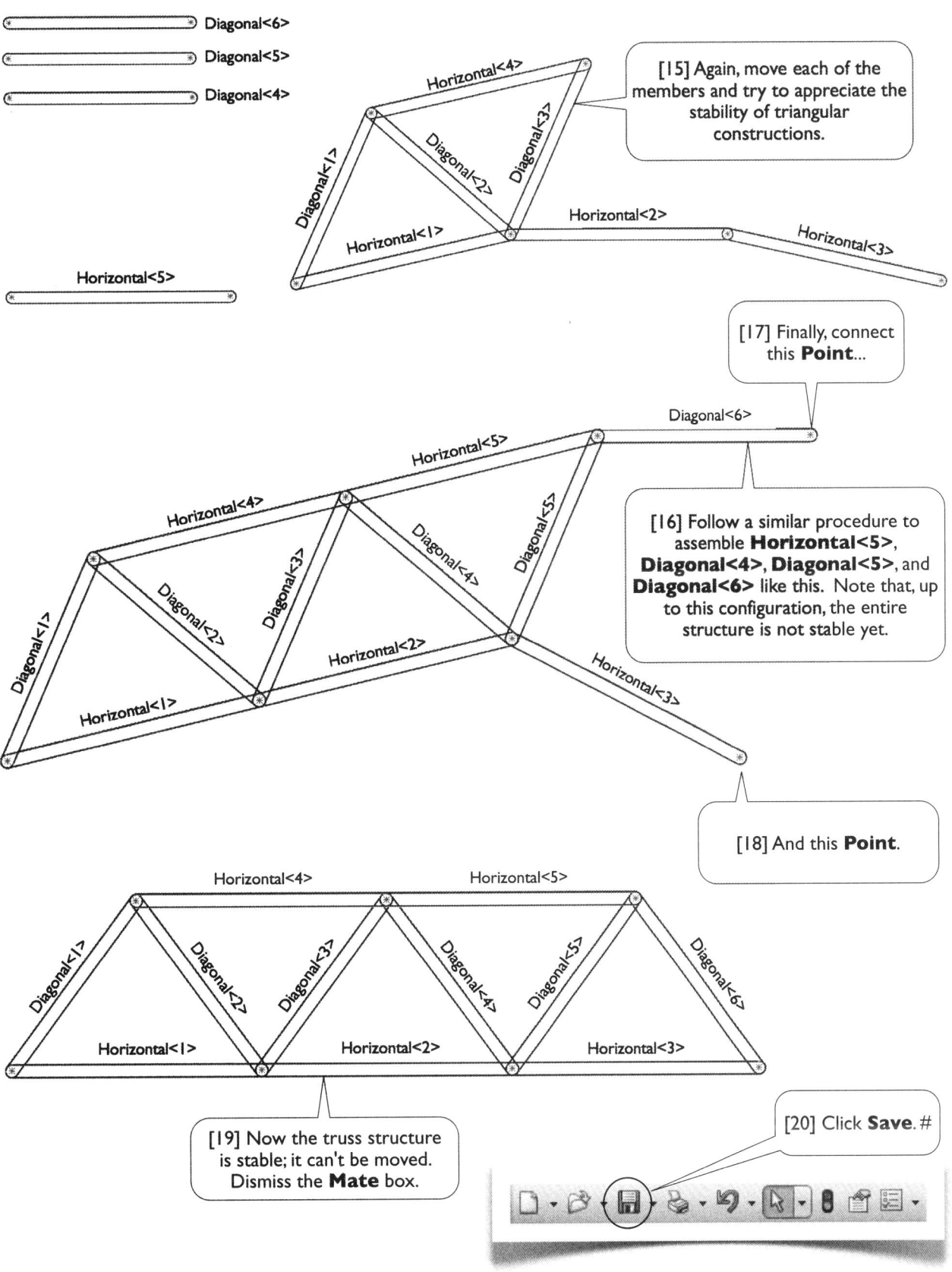

Diagonal<6>

Diagonal<5>

Diagonal<4>

[15] Again, move each of the members and try to appreciate the stability of triangular constructions.

Horizontal<4>

Diagonal<1>

Diagonal<2>

Diagonal<3>

Horizontal<1>

Horizontal<2>

Horizontal<3>

Horizontal<5>

[17] Finally, connect this **Point**...

Diagonal<6>

Horizontal<5>

Horizontal<4>

Diagonal<5>

Diagonal<4>

[16] Follow a similar procedure to assemble **Horizontal<5>**, **Diagonal<4>**, **Diagonal<5>**, and **Diagonal<6>** like this. Note that, up to this configuration, the entire structure is not stable yet.

Diagonal<1>

Diagonal<2>

Diagonal<3>

Horizontal<1>

Horizontal<2>

Horizontal<3>

[18] And this **Point**.

Horizontal<4>

Horizontal<5>

Diagonal<1>

Diagonal<2>

Diagonal<3>

Diagonal<4>

Diagonal<5>

Diagonal<6>

Horizontal<1>

Horizontal<2>

Horizontal<3>

[19] Now the truss structure is stable; it can't be moved. Dismiss the **Mate** box.

[20] Click **Save**. #

2.1-8 Create a **Study**

[2] Select **Motion Analysis**. #

[1] Click **Motion Study 1** tab to create a new **Study**.

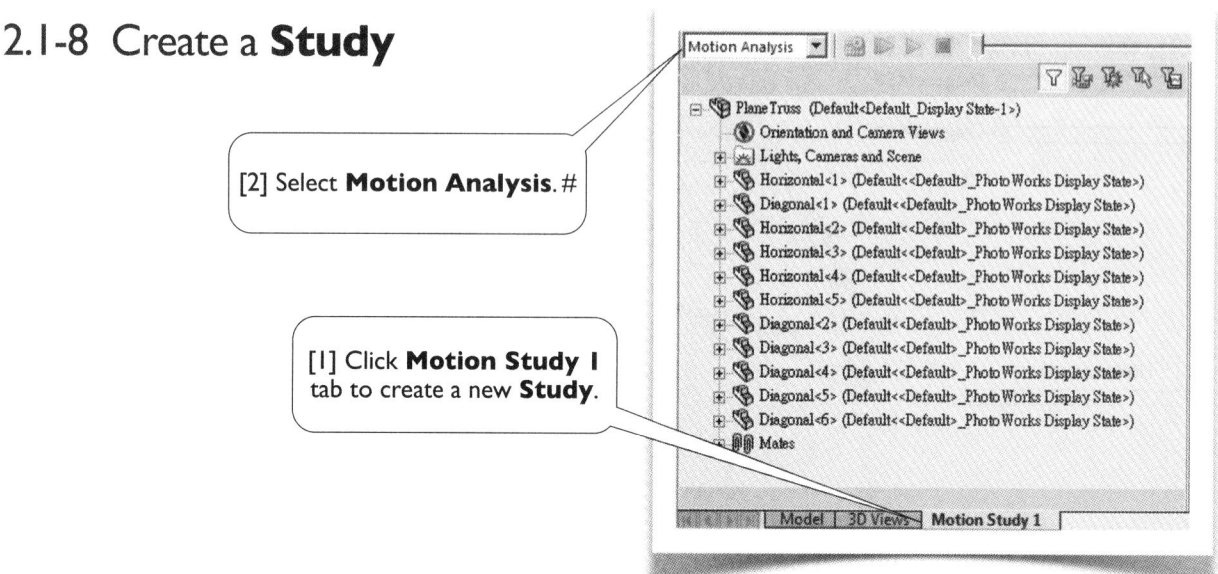

2.1-9 Set Up Forces and Calculate Results

[1] In **Motion** toolbar, click **Force**.

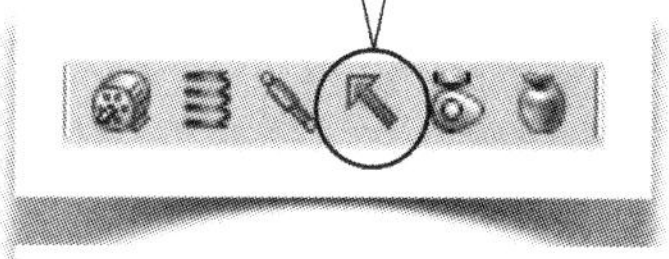

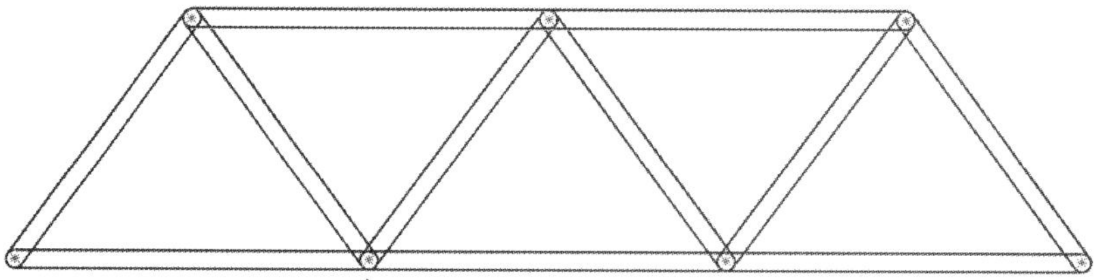

[6] Click **OK**.

Type

→ Force

↻ Torque

[4] Click **Reverse Direction** so that the force is downward.

Direction

⬇ Action only

⬇ Action & reaction

Point@Sketch2@Horizontal-2@PlaneTruss

Top

Force relative to:
○ Assembly origin
○ Selected component:

Force Function

Constant

F_1 6000 N

[5] Type 6000 (N).

[2] Click this **Point** to define the location of the force.

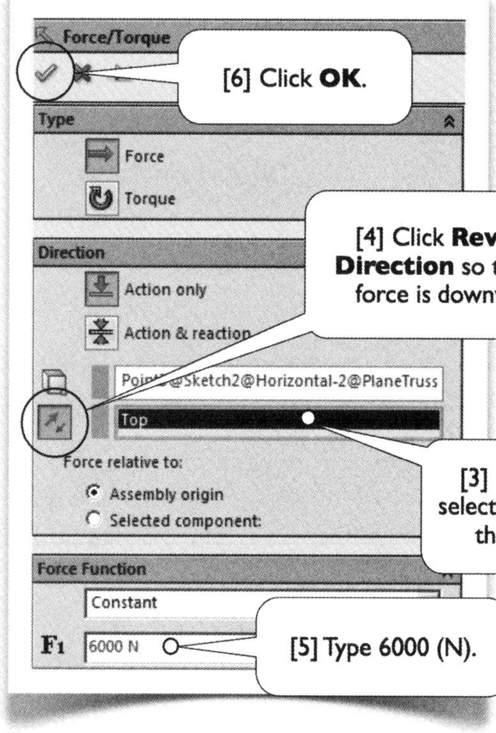

[3] In the **Assembly Tree**, select global **Top** plane to define the direction of the force.

[7] Click **Save**. #

[8] Click **Calculate**. #

Motion Analysis

2.1-10 View the Reaction Forces

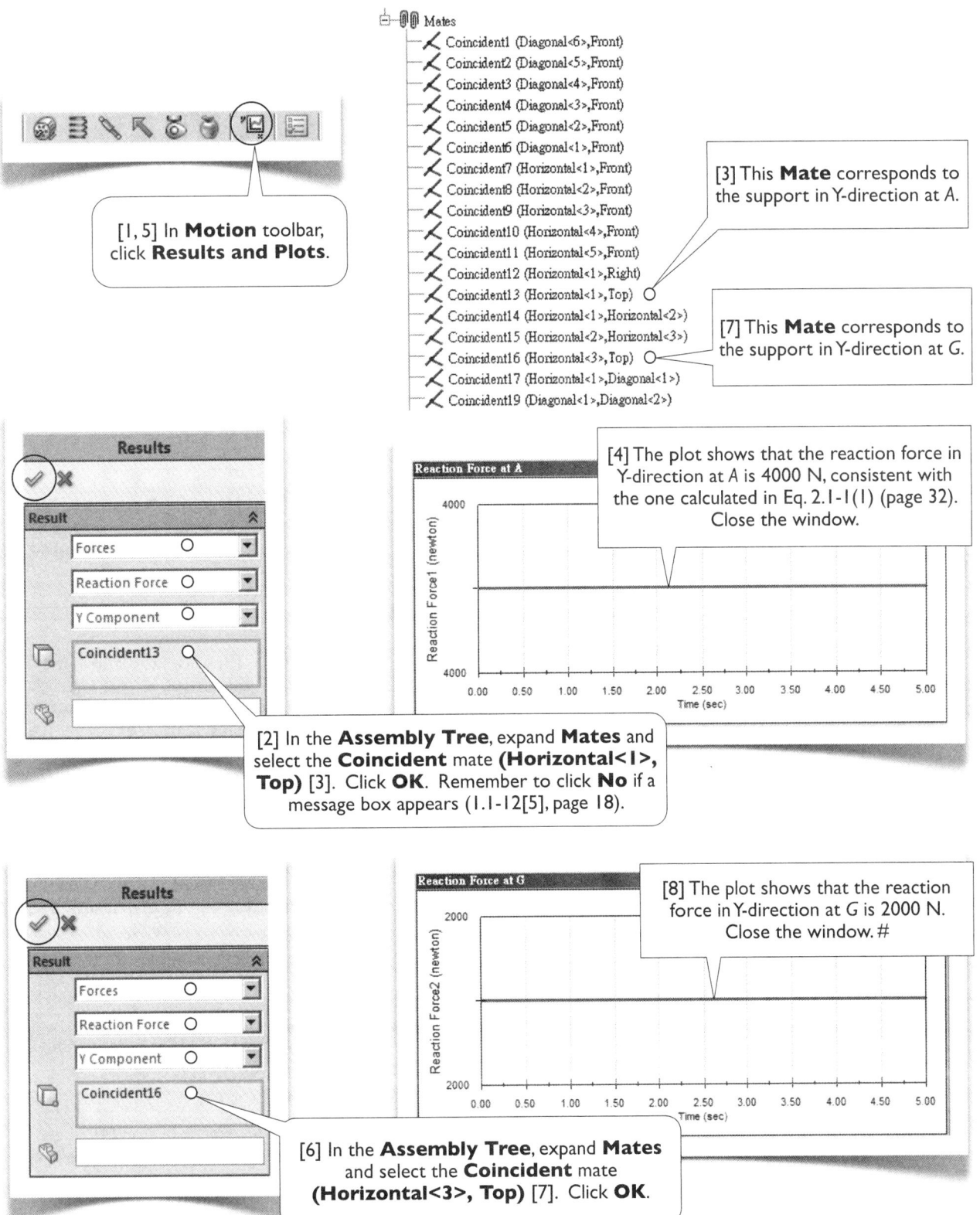

[1, 5] In **Motion** toolbar, click **Results and Plots**.

[3] This **Mate** corresponds to the support in Y-direction at A.

[7] This **Mate** corresponds to the support in Y-direction at G.

[4] The plot shows that the reaction force in Y-direction at A is 4000 N, consistent with the one calculated in Eq. 2.1-1(1) (page 32). Close the window.

[2] In the **Assembly Tree**, expand **Mates** and select the **Coincident** mate (**Horizontal<1>, Top**) [3]. Click **OK**. Remember to click **No** if a message box appears (1.1-12[5], page 18).

[8] The plot shows that the reaction force in Y-direction at G is 2000 N. Close the window. #

[6] In the **Assembly Tree**, expand **Mates** and select the **Coincident** mate (**Horizontal<3>, Top**) [7]. Click **OK**.

2.1-11 View the Member Force of **Horizontal<4>**

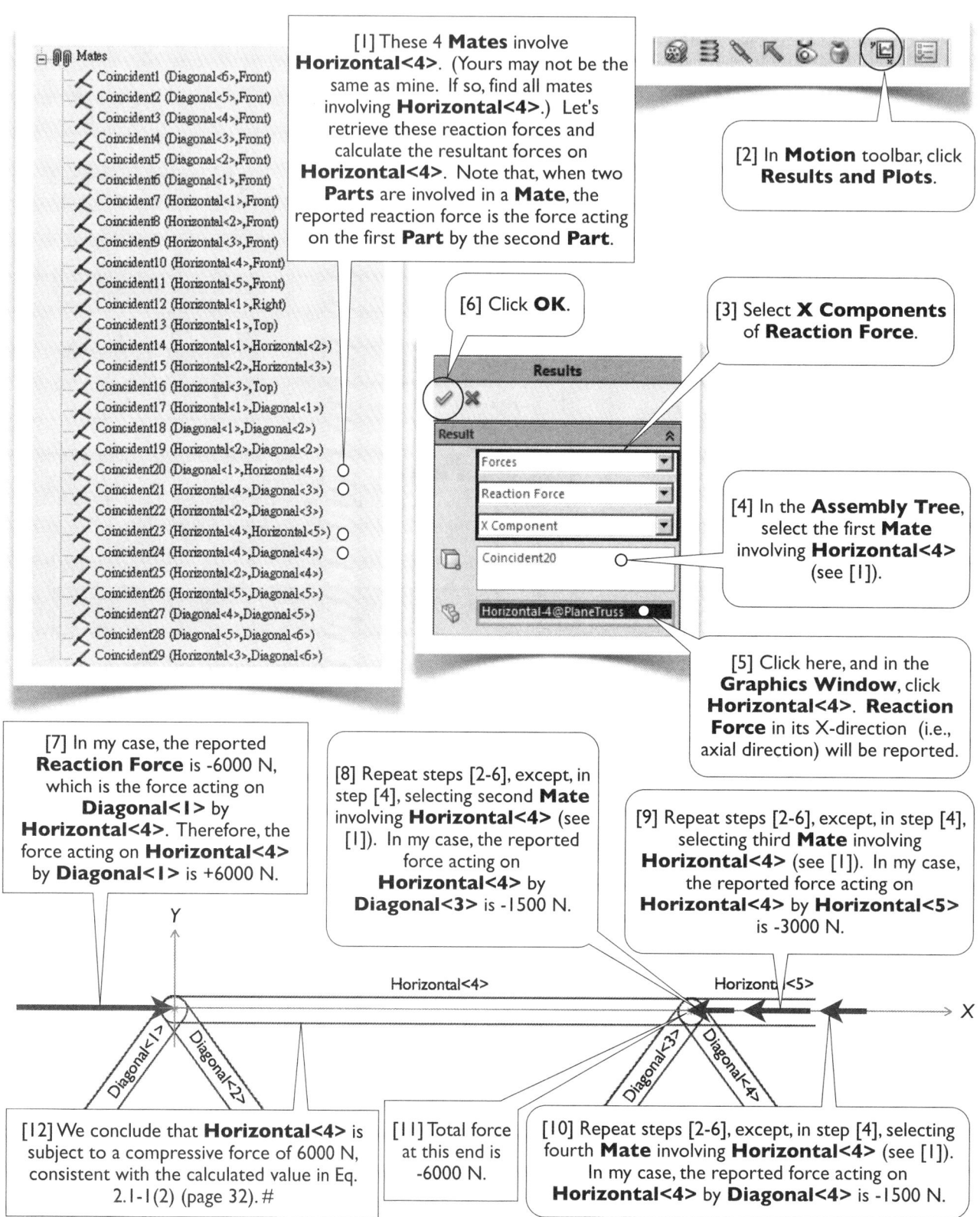

[1] These 4 **Mates** involve **Horizontal<4>**. (Yours may not be the same as mine. If so, find all mates involving **Horizontal<4>**.) Let's retrieve these reaction forces and calculate the resultant forces on **Horizontal<4>**. Note that, when two **Parts** are involved in a **Mate**, the reported reaction force is the force acting on the first **Part** by the second **Part**.

[2] In **Motion** toolbar, click **Results and Plots**.

[6] Click **OK**.

[3] Select **X Components** of **Reaction Force**.

Results

Result

Forces

Reaction Force

X Component

Coincident20

Horizontal-4@PlaneTruss

[4] In the **Assembly Tree**, select the first **Mate** involving **Horizontal<4>** (see [1]).

[5] Click here, and in the **Graphics Window**, click **Horizontal<4>**. **Reaction Force** in its X-direction (i.e., axial direction) will be reported.

[7] In my case, the reported **Reaction Force** is -6000 N, which is the force acting on **Diagonal<1>** by **Horizontal<4>**. Therefore, the force acting on **Horizontal<4>** by **Diagonal<1>** is +6000 N.

[8] Repeat steps [2-6], except, in step [4], selecting second **Mate** involving **Horizontal<4>** (see [1]). In my case, the reported force acting on **Horizontal<4>** by **Diagonal<3>** is -1500 N.

[9] Repeat steps [2-6], except, in step [4], selecting third **Mate** involving **Horizontal<4>** (see [1]). In my case, the reported force acting on **Horizontal<4>** by **Horizontal<5>** is -3000 N.

[12] We conclude that **Horizontal<4>** is subject to a compressive force of 6000 N, consistent with the calculated value in Eq. 2.1-1(2) (page 32). #

[11] Total force at this end is -6000 N.

[10] Repeat steps [2-6], except, in step [4], selecting fourth **Mate** involving **Horizontal<4>** (see [1]). In my case, the reported force acting on **Horizontal<4>** by **Diagonal<4>** is -1500 N.

2.1-12 View the Member Force of **Diagonal<2>**

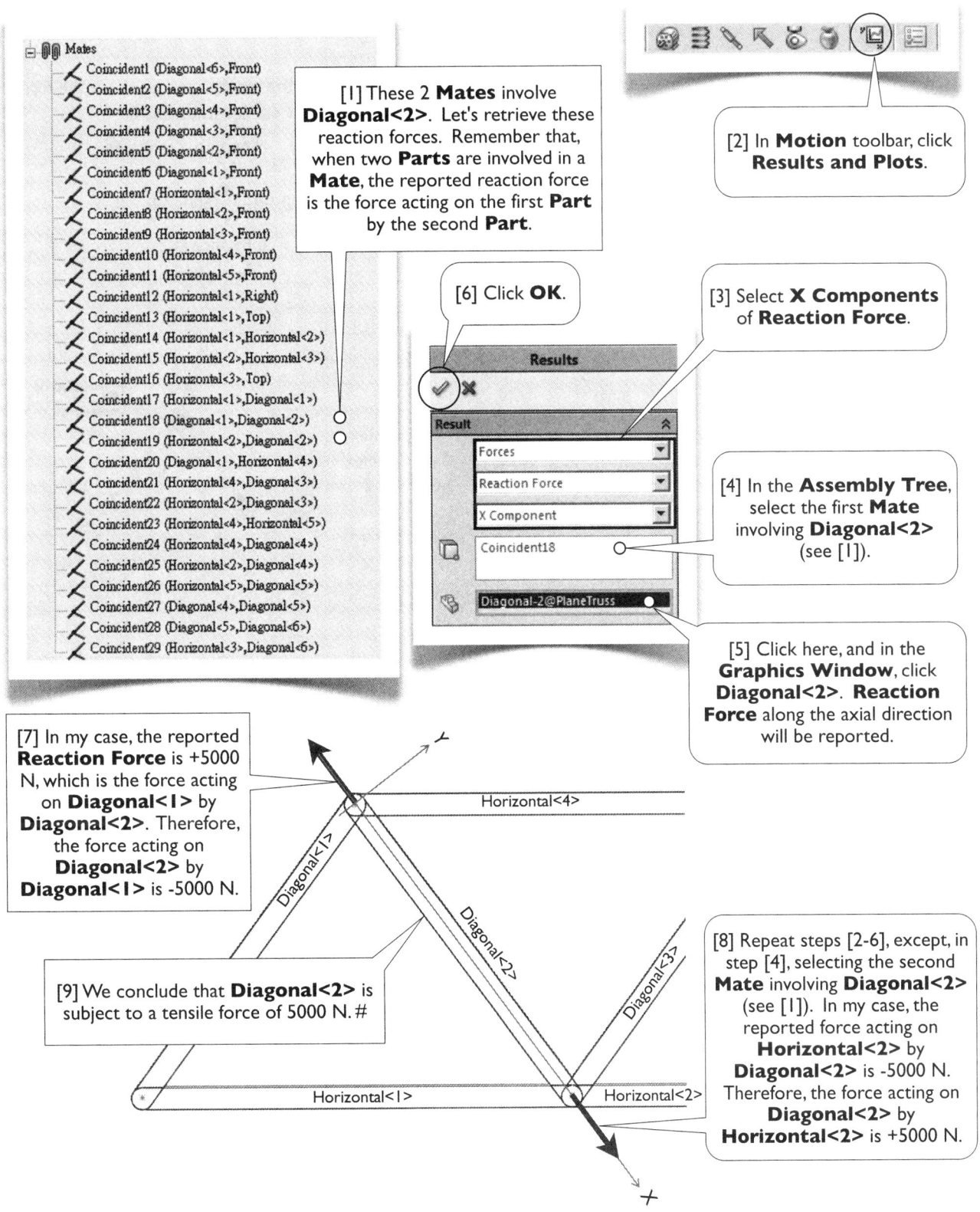

[1] These 2 **Mates** involve **Diagonal<2>**. Let's retrieve these reaction forces. Remember that, when two **Parts** are involved in a **Mate**, the reported reaction force is the force acting on the first **Part** by the second **Part**.

[2] In **Motion** toolbar, click **Results and Plots**.

[6] Click **OK**.

[3] Select **X Components** of **Reaction Force**.

Results

Result

Forces

Reaction Force

X Component

Coincident18

Diagonal-2@PlaneTruss

[4] In the **Assembly Tree**, select the first **Mate** involving **Diagonal<2>** (see [1]).

[5] Click here, and in the **Graphics Window**, click **Diagonal<2>**. **Reaction Force** along the axial direction will be reported.

[7] In my case, the reported **Reaction Force** is +5000 N, which is the force acting on **Diagonal<1>** by **Diagonal<2>**. Therefore, the force acting on **Diagonal<2>** by **Diagonal<1>** is -5000 N.

[9] We conclude that **Diagonal<2>** is subject to a tensile force of 5000 N. #

[8] Repeat steps [2-6], except, in step [4], selecting the second **Mate** involving **Diagonal<2>** (see [1]). In my case, the reported force acting on **Horizontal<2>** by **Diagonal<2>** is -5000 N. Therefore, the force acting on **Diagonal<2>** by **Horizontal<2>** is +5000 N.

2.1-13 Do It Yourself: Other Member Forces and Validation of the Results

Do It Yourself

[1] It leaves you to explore the other member forces. All member forces, along with the reaction forces at supports, are summarized like this.

Do It Yourself

[2] To verify the validity of the results, we may check the force equilibria at each node. For example, at node B,

$$\sum F_X = \frac{3}{5}(5000 \text{ N}) + \frac{3}{5}(5000 \text{ N}) - (6000 \text{ N}) = 0,$$

$$\sum F_Y = \frac{4}{5}(5000 \text{ N}) - \frac{4}{5}(5000 \text{ N}) = 0.$$

It leaves you to check the force balance at other nodes. #

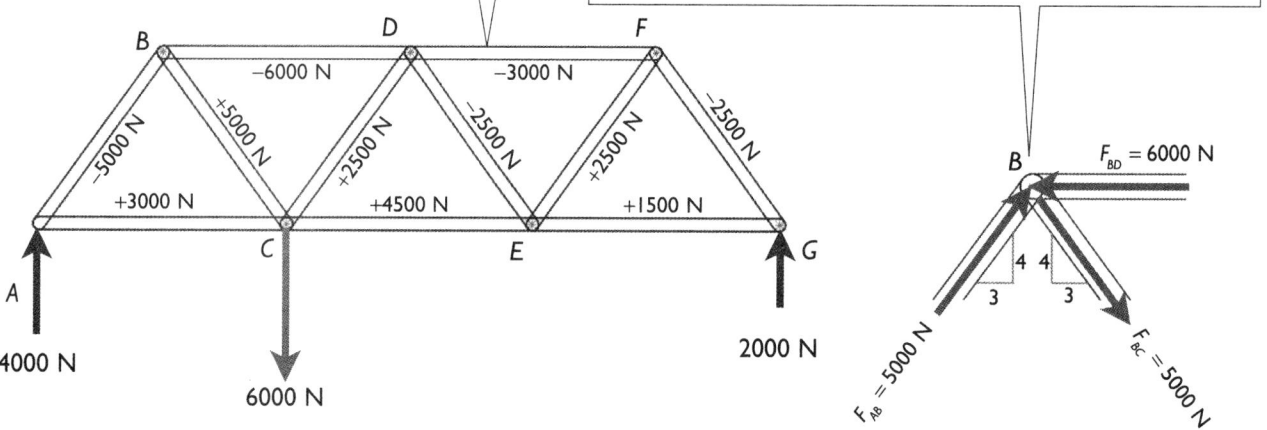

2.1-14 Wrap Up

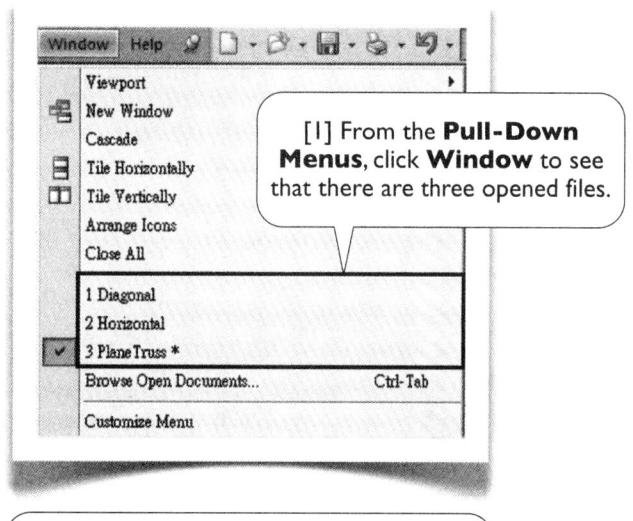

[1] From the **Pull-Down Menus**, click **Window** to see that there are three opened files.

[2] From the **Pull-Down Menus**, Select **File>Exit** to quit **SOLIDWORKS**. Click **Save all**. Click **Rebuild and save the document**.

Remark

[3] It seems, from 2.1-11[8-11] (page 44), that the compressive force 6000 N acting on the right-end of **Horizontal<4>** consists of three forces (1500 N, 3000 N, and 1500 N) which come from three members respectively. That may not be true. There is no way to know how much portion of force is from a specific member.

There are more than one way to "mate" the members at a specific joint. And they end up with different force components. However, the total force acting on the end of a member is always the same. #

Section 2.2

Space Truss

2.2-1 Introduction

[1] Consider a space truss subject to supports and loads as shown [2-5]. All members are connected with spherical (ball-and-socket) joints. We want to find the reaction forces at the supports and the member forces.

The truss is a *statically determinate structure*, that is, the reaction forces and the member forces can be solved using static equilibrium equations without any cross-sectional information. Here, we arbitrarily assume that all members have a circular cross-section of diameter 50 mm. There are four kinds of members, different in their lengths, which are respectively 1 m, $\sqrt{5}/2$ m (approximately 1118 mm), $\sqrt{2}$ m (approximately 1414 mm), and 1.5 m.

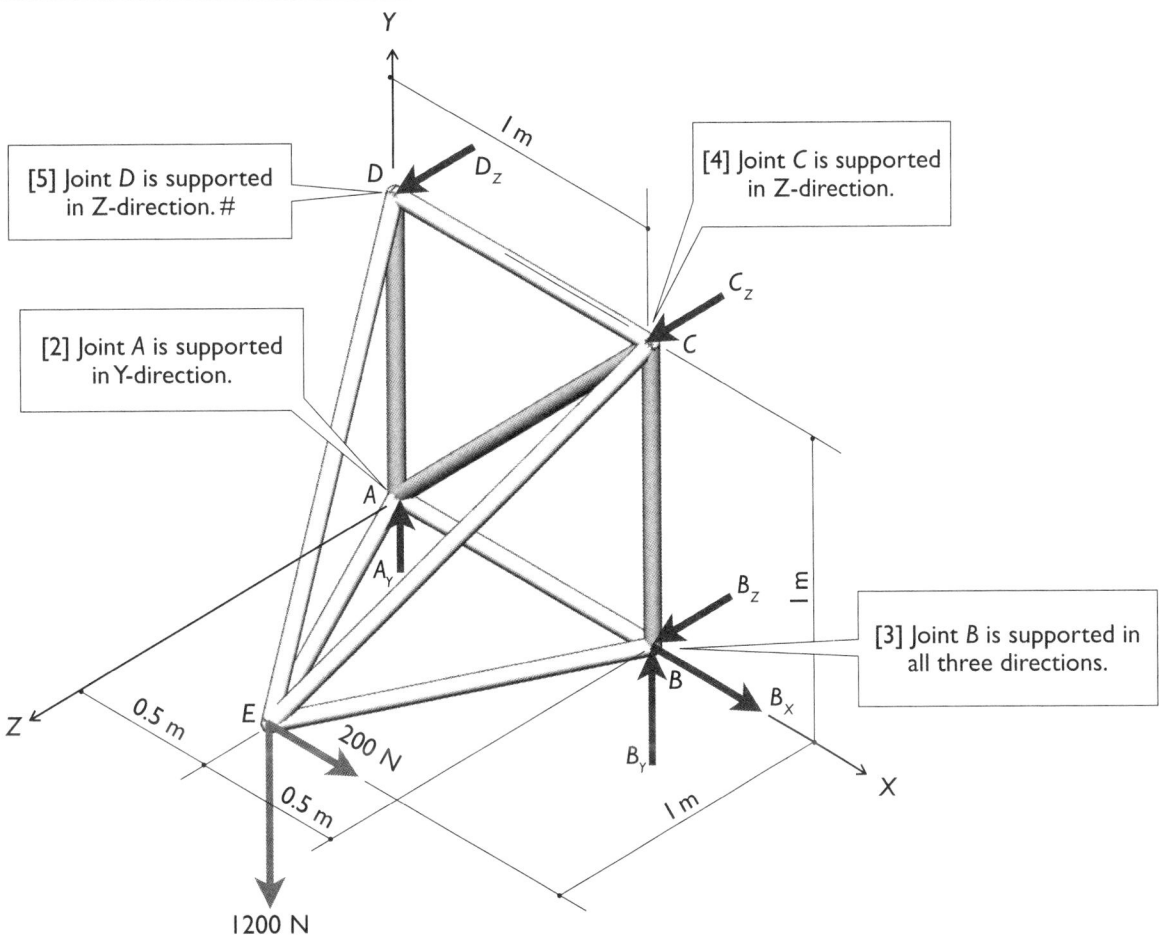

2.2-2 Start Up and Create a Part: **Member1000**

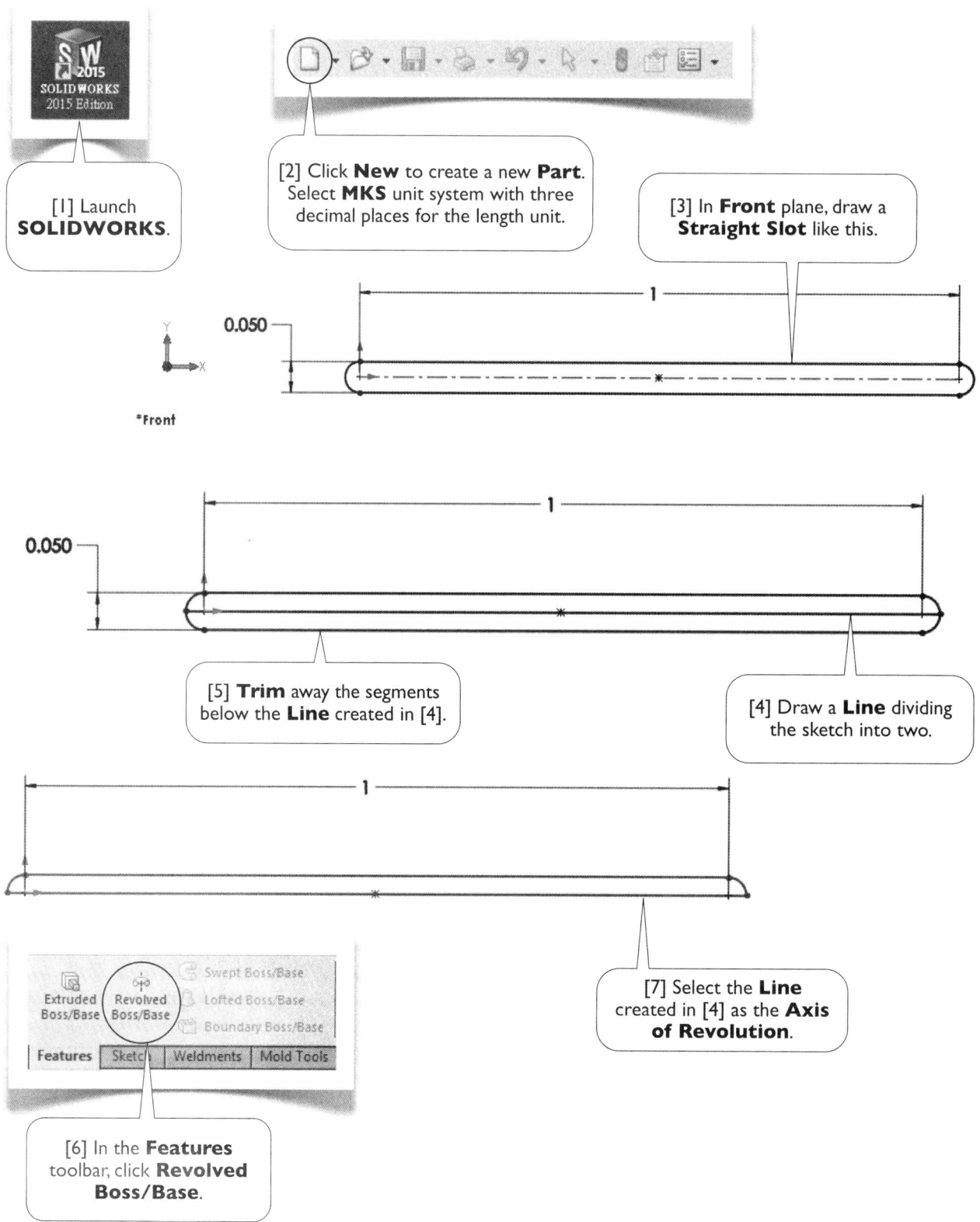

[1] Launch **SOLIDWORKS**.

[2] Click **New** to create a new **Part**. Select **MKS** unit system with three decimal places for the length unit.

[3] In **Front** plane, draw a **Straight Slot** like this.

0.050

1

*Front

0.050

1

[5] **Trim** away the segments below the **Line** created in [4].

[4] Draw a **Line** dividing the sketch into two.

1

[7] Select the **Line** created in [4] as the **Axis of Revolution**.

Extruded Boss/Base   Revolved Boss/Base   Swept Boss/Base   Lofted Boss/Base   Boundary Boss/Base

Features   Sketch   Weldments   Mold Tools

[6] In the **Features** toolbar, click **Revolved Boss/Base**.

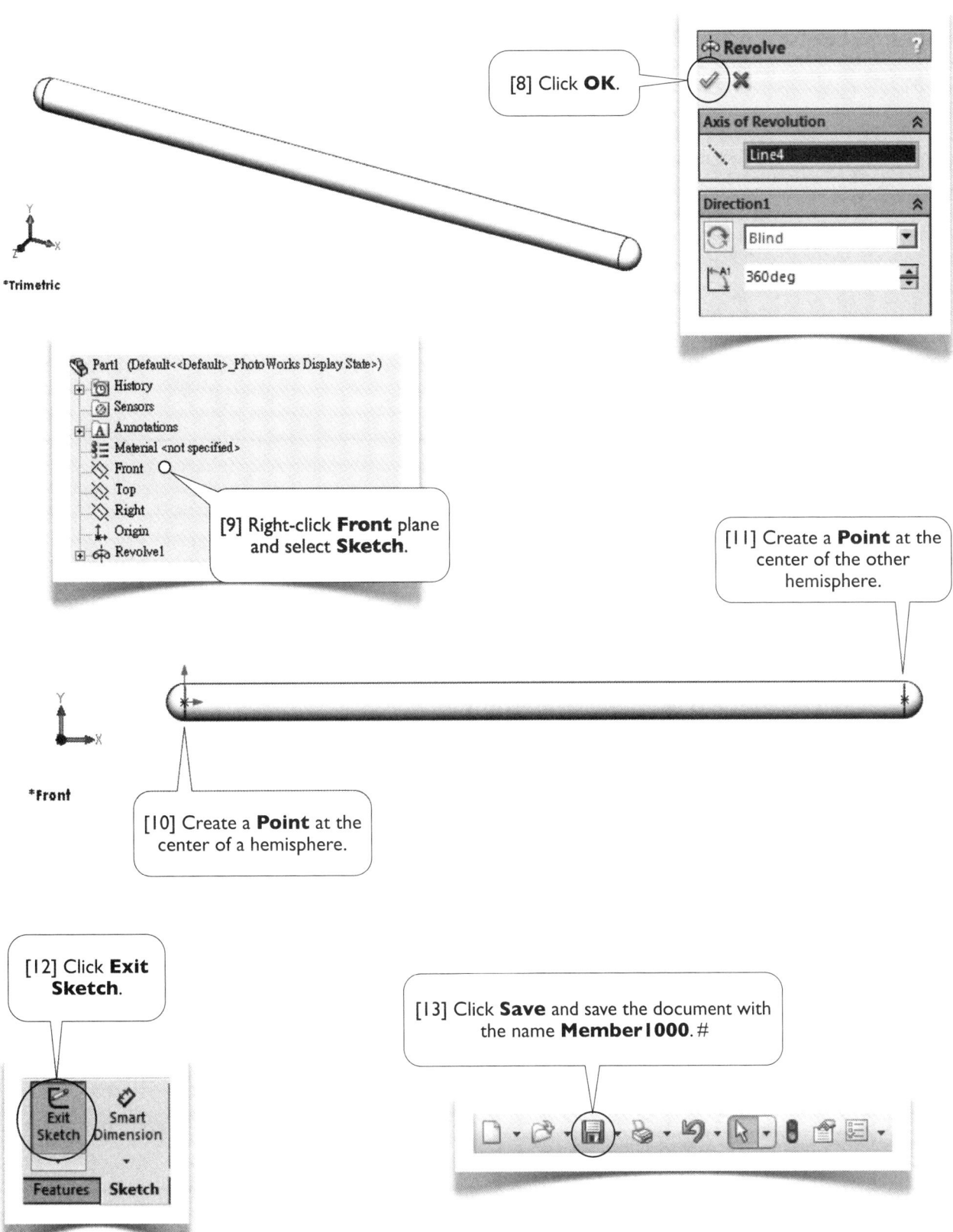

[8] Click **OK**.

Revolve

Axis of Revolution

Line4

Direction1

Blind

360deg

*Trimetric

Part1 (Default<<Default>_PhotoWorks Display State>)
- History
- Sensors
- Annotations
- Material <not specified>
- Front
- Top
- Right
- Origin
- Revolve1

[9] Right-click **Front** plane and select **Sketch**.

[11] Create a **Point** at the center of the other hemisphere.

*Front

[10] Create a **Point** at the center of a hemisphere.

[12] Click **Exit Sketch**.

[13] Click **Save** and save the document with the name **Member1000**. #

Exit Sketch   Smart Dimension

Features   Sketch

2.2-3 Create a Part: **Member1118**

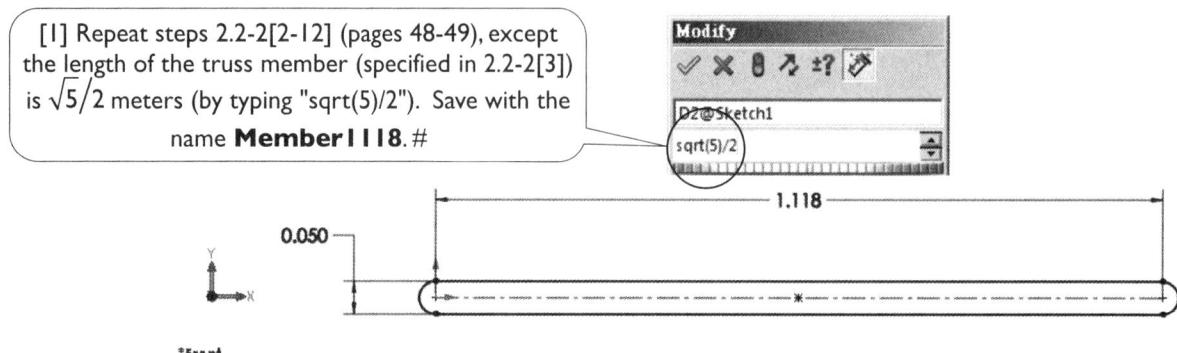

[1] Repeat steps 2.2-2[2-12] (pages 48-49), except the length of the truss member (specified in 2.2-2[3]) is $\sqrt{5}/2$ meters (by typing "sqrt(5)/2"). Save with the name **Member1118**. #

2.2-4 Create a Part: **Member1414**

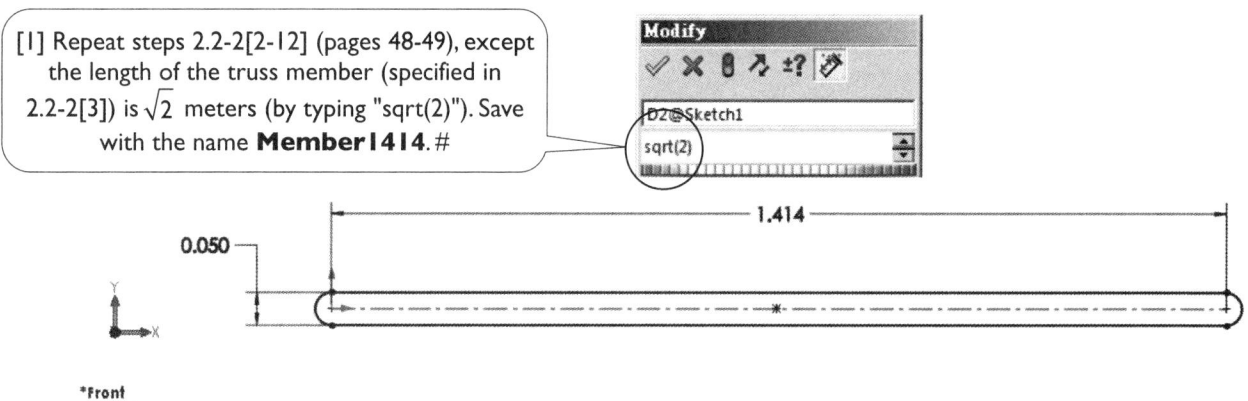

[1] Repeat steps 2.2-2[2-12] (pages 48-49), except the length of the truss member (specified in 2.2-2[3]) is $\sqrt{2}$ meters (by typing "sqrt(2)"). Save with the name **Member1414**. #

2.2-5 Create a Part: **Member1500**

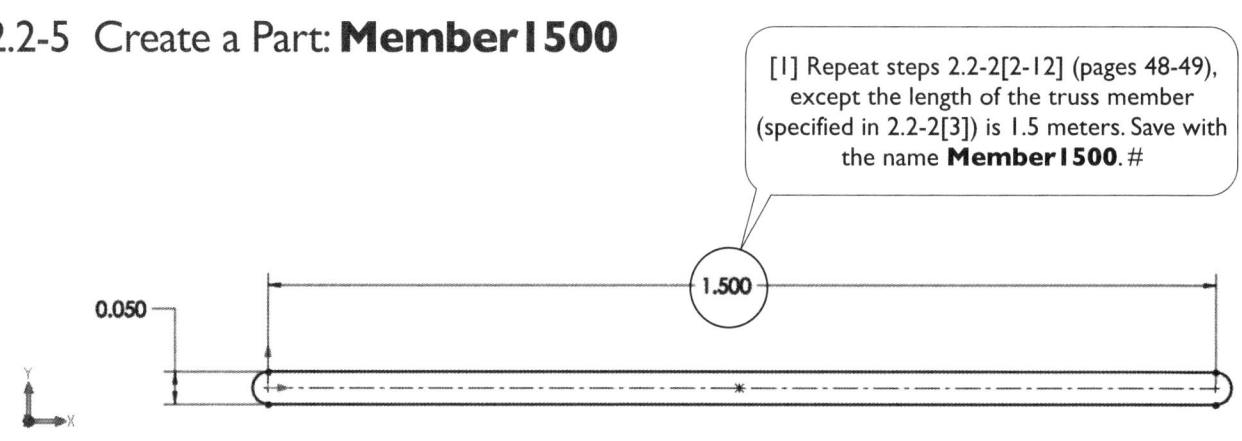

[1] Repeat steps 2.2-2[2-12] (pages 48-49), except the length of the truss member (specified in 2.2-2[3]) is 1.5 meters. Save with the name **Member1500**. #

2.2-6 Create an Assembly: **SpaceTruss**

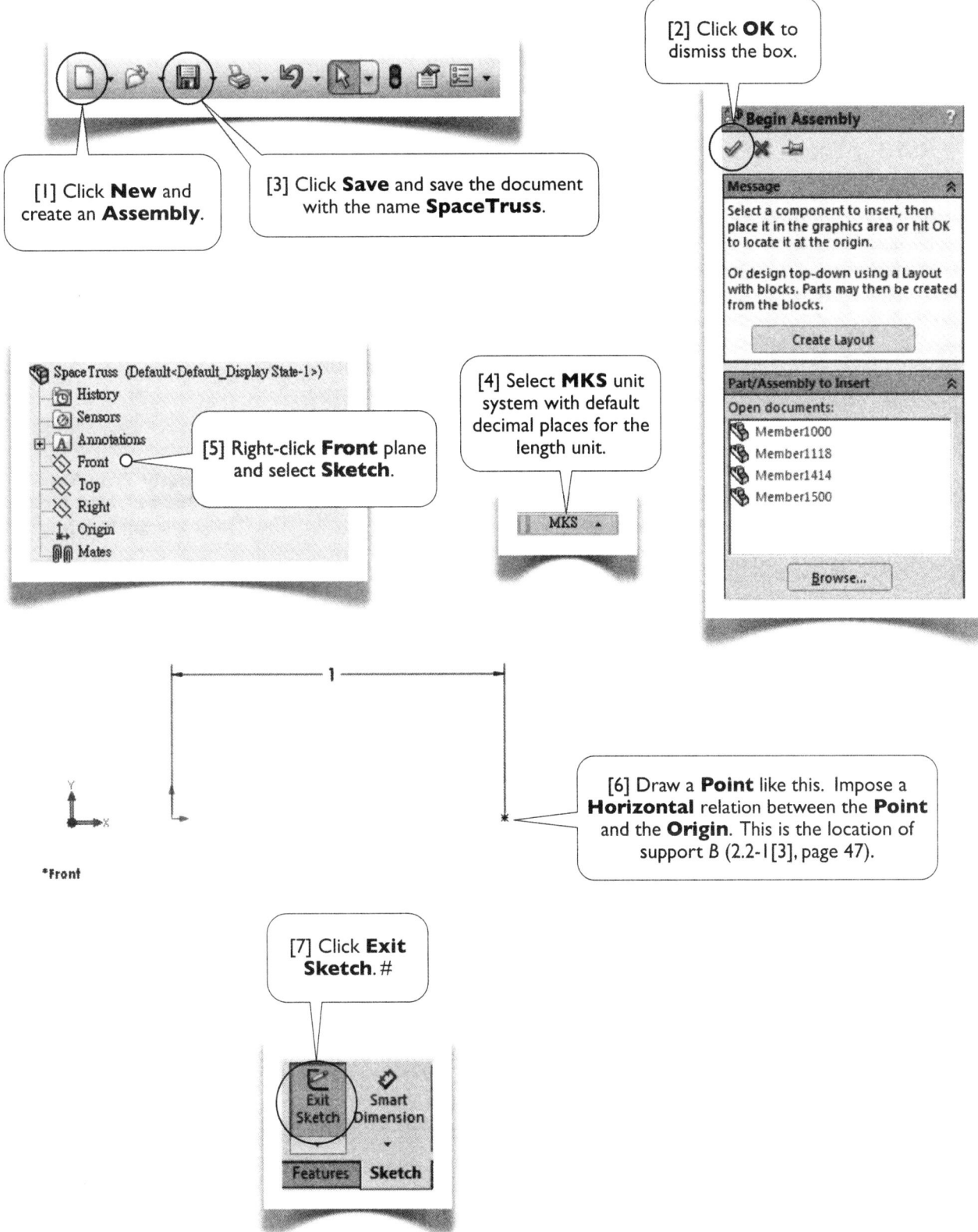

[2] Click **OK** to dismiss the box.

[1] Click **New** and create an **Assembly**.

[3] Click **Save** and save the document with the name **SpaceTruss**.

[4] Select **MKS** unit system with default decimal places for the length unit.

[5] Right-click **Front** plane and select **Sketch**.

[6] Draw a **Point** like this. Impose a **Horizontal** relation between the **Point** and the **Origin**. This is the location of support B (2.2-1[3], page 47).

[7] Click **Exit Sketch**. #

2.2-7 Insert and Duplicate **Parts**

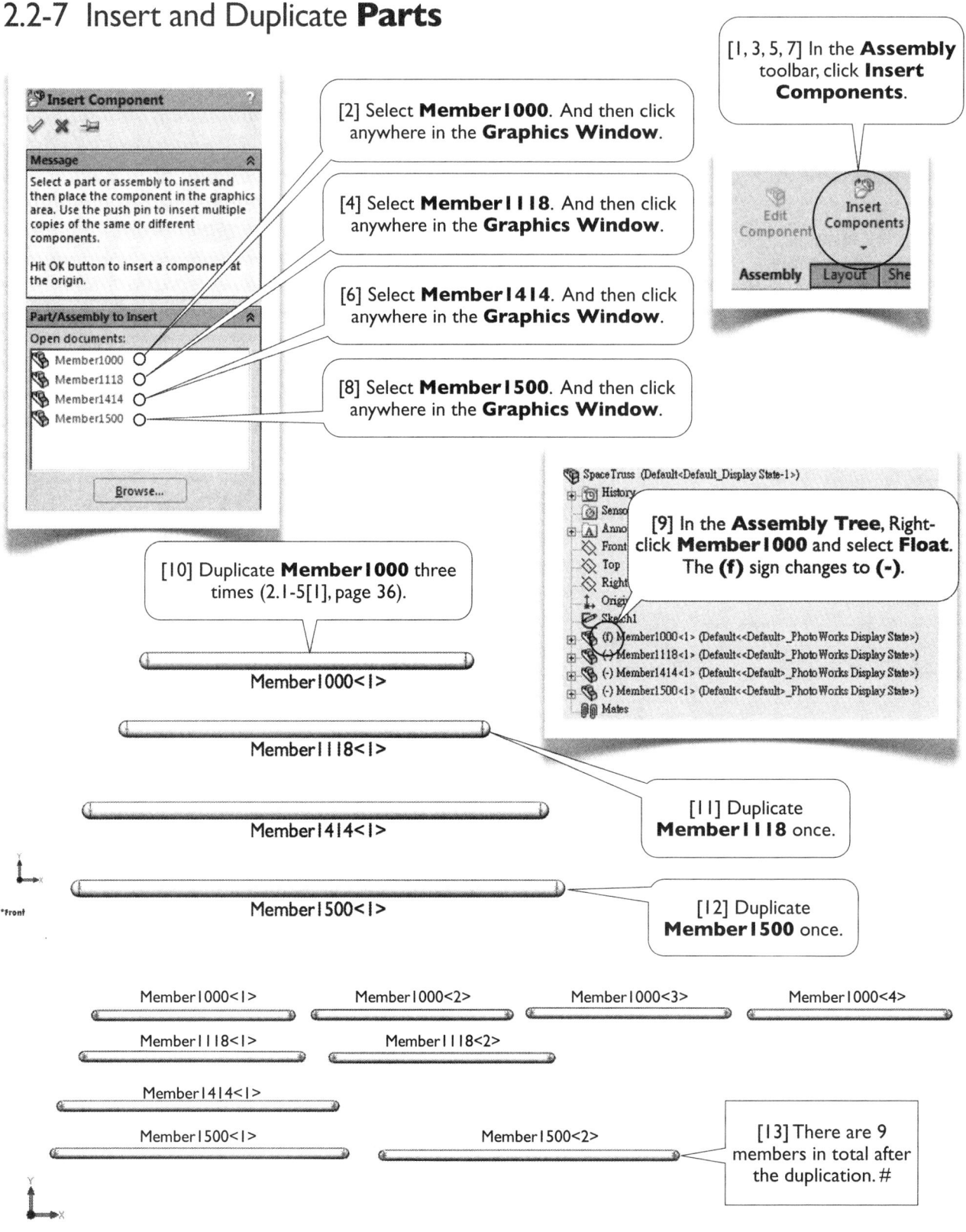

[1, 3, 5, 7] In the **Assembly** toolbar, click **Insert Components**.

Insert Component

Message

Select a part or assembly to insert and then place the component in the graphics area. Use the push pin to insert multiple copies of the same or different components.

Hit OK button to insert a component at the origin.

Part/Assembly to Insert

Open documents:

- Member1000
- Member1118
- Member1414
- Member1500

Browse...

[2] Select **Member1000**. And then click anywhere in the **Graphics Window**.

[4] Select **Member1118**. And then click anywhere in the **Graphics Window**.

[6] Select **Member1414**. And then click anywhere in the **Graphics Window**.

[8] Select **Member1500**. And then click anywhere in the **Graphics Window**.

SpaceTruss (Default<Default_Display State-1>)

- History
- Sensors
- Annotations
- Front
- Top
- Right
- Origin
- Sketch1
- (f) Member1000<1> (Default<<Default>_PhotoWorks Display State>)
- (-) Member1118<1> (Default<<Default>_PhotoWorks Display State>)
- (-) Member1414<1> (Default<<Default>_PhotoWorks Display State>)
- (-) Member1500<1> (Default<<Default>_PhotoWorks Display State>)
- Mates

[9] In the **Assembly Tree**, Right-click **Member1000** and select **Float**. The **(f)** sign changes to **(-)**.

[10] Duplicate **Member1000** three times (2.1-5[1], page 36).

Member1000<1>

Member1118<1>

Member1414<1>

Member1500<1>

*Front

[11] Duplicate **Member1118** once.

[12] Duplicate **Member1500** once.

Member1000<1>   Member1000<2>   Member1000<3>   Member1000<4>

Member1118<1>   Member1118<2>

Member1414<1>

Member1500<1>   Member1500<2>

[13] There are 9 members in total after the duplication. #

*Front

2.2-8 Set Up **Member1000**s and Supports

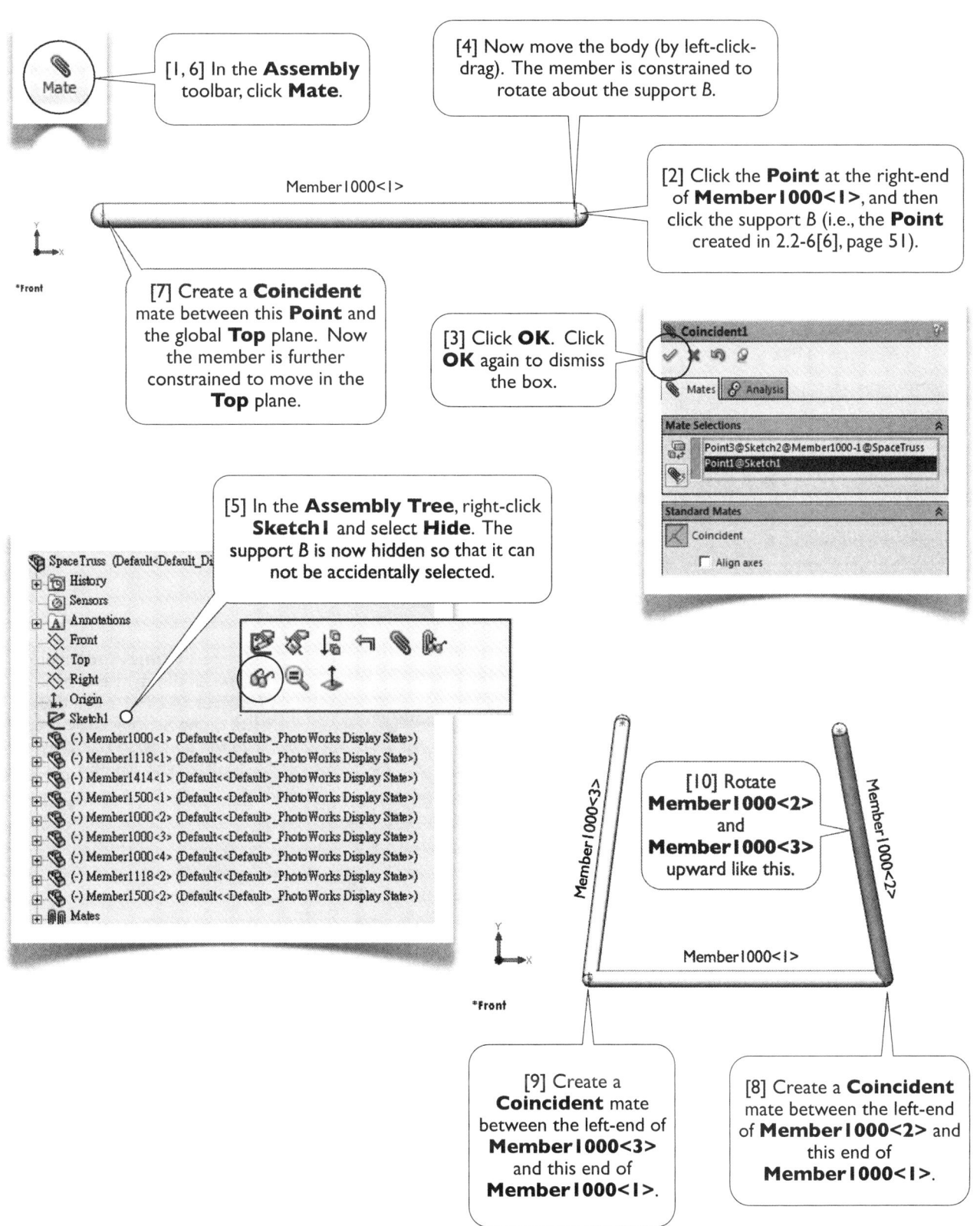

[1, 6] In the **Assembly** toolbar, click **Mate**.

[4] Now move the body (by left-click-drag). The member is constrained to rotate about the support *B*.

Member1000<1>

*Front

[2] Click the **Point** at the right-end of **Member1000<1>**, and then click the support *B* (i.e., the **Point** created in 2.2-6[6], page 51).

[7] Create a **Coincident** mate between this **Point** and the global **Top** plane. Now the member is further constrained to move in the **Top** plane.

[3] Click **OK**. Click **OK** again to dismiss the box.

Coincident1

Mates | Analysis

Mate Selections

Point3@Sketch2@Member1000-1@SpaceTruss
Point1@Sketch1

Standard Mates

Coincident

☐ Align axes

[5] In the **Assembly Tree**, right-click **Sketch1** and select **Hide**. The support *B* is now hidden so that it can not be accidentally selected.

SpaceTruss (Default<Default_Di
⊞ History
Sensors
⊞ A Annotations
◇ Front
◇ Top
◇ Right
Origin
Sketch1
⊞ (-) Member1000<1> (Default<<Default>_PhotoWorks Display State>)
⊞ (-) Member1118<1> (Default<<Default>_PhotoWorks Display State>)
⊞ (-) Member1414<1> (Default<<Default>_PhotoWorks Display State>)
⊞ (-) Member1500<1> (Default<<Default>_PhotoWorks Display State>)
⊞ (-) Member1000<2> (Default<<Default>_PhotoWorks Display State>)
⊞ (-) Member1000<3> (Default<<Default>_PhotoWorks Display State>)
⊞ (-) Member1000<4> (Default<<Default>_PhotoWorks Display State>)
⊞ (-) Member1118<2> (Default<<Default>_PhotoWorks Display State>)
⊞ (-) Member1500<2> (Default<<Default>_PhotoWorks Display State>)
⊞ Mates

Member1000<3>

[10] Rotate **Member1000<2>** and **Member1000<3>** upward like this.

Member1000<2>

Member1000<1>

*Front

[9] Create a **Coincident** mate between the left-end of **Member1000<3>** and this end of **Member1000<1>**.

[8] Create a **Coincident** mate between the left-end of **Member1000<2>** and this end of **Member1000<1>**.

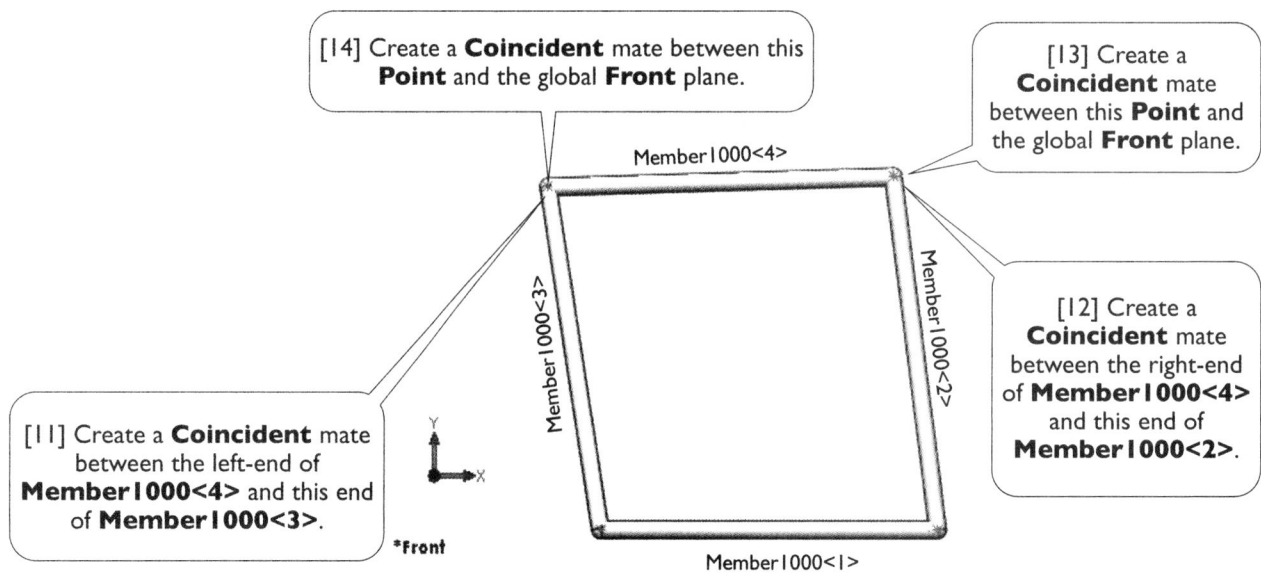

[14] Create a **Coincident** mate between this **Point** and the global **Front** plane.

[13] Create a **Coincident** mate between this **Point** and the global **Front** plane.

[11] Create a **Coincident** mate between the left-end of **Member1000<4>** and this end of **Member1000<3>**.

[12] Create a **Coincident** mate between the right-end of **Member1000<4>** and this end of **Member1000<2>**.

Member1000<4>

Member1000<3>

Member1000<2>

*Front

Member1000<1>

[15] Remember to save your document frequently. In cases you run into some difficulties when creating **Mates**, you always can go back to the point where you save the document. #

2.2-9 Set Up Other Members

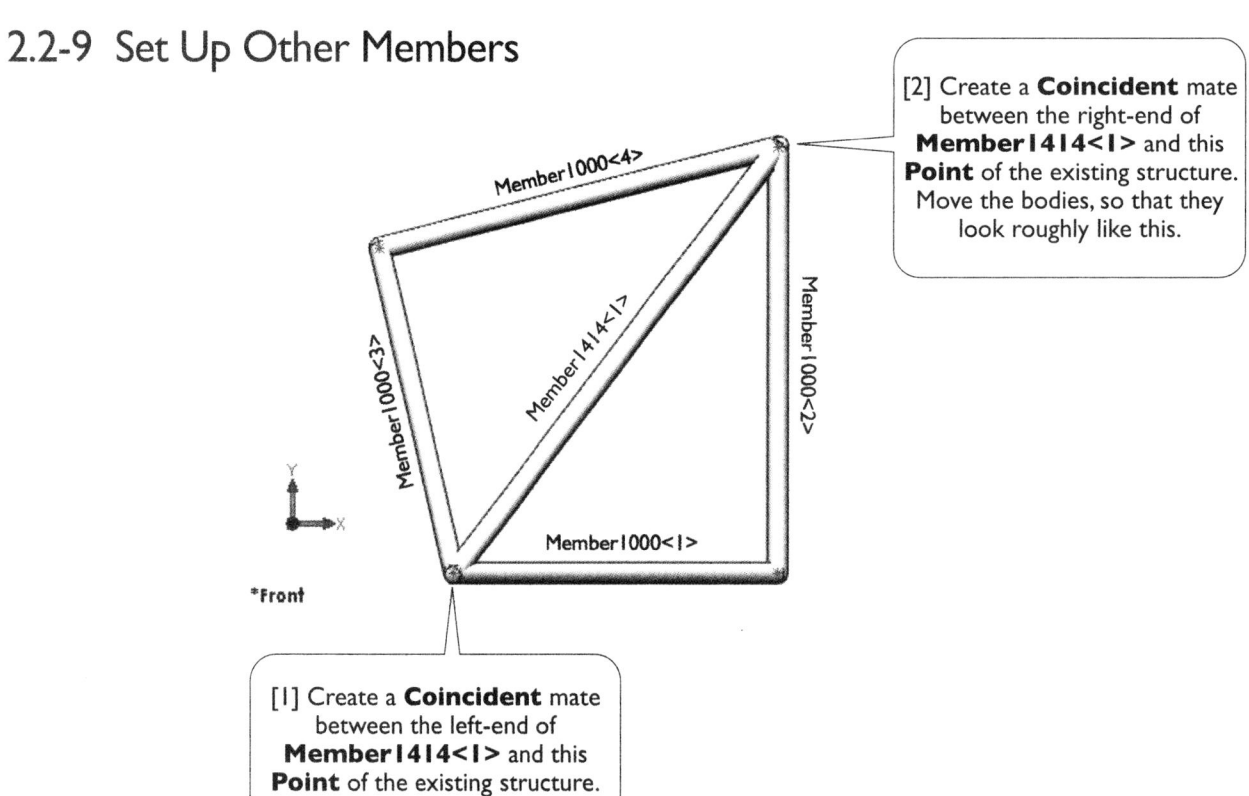

[2] Create a **Coincident** mate between the right-end of **Member1414<1>** and this **Point** of the existing structure. Move the bodies, so that they look roughly like this.

Member1000<4>

Member1000<3>

Member1414<1>

Member1000<2>

Member1000<1>

*Front

[1] Create a **Coincident** mate between the left-end of **Member1414<1>** and this **Point** of the existing structure.

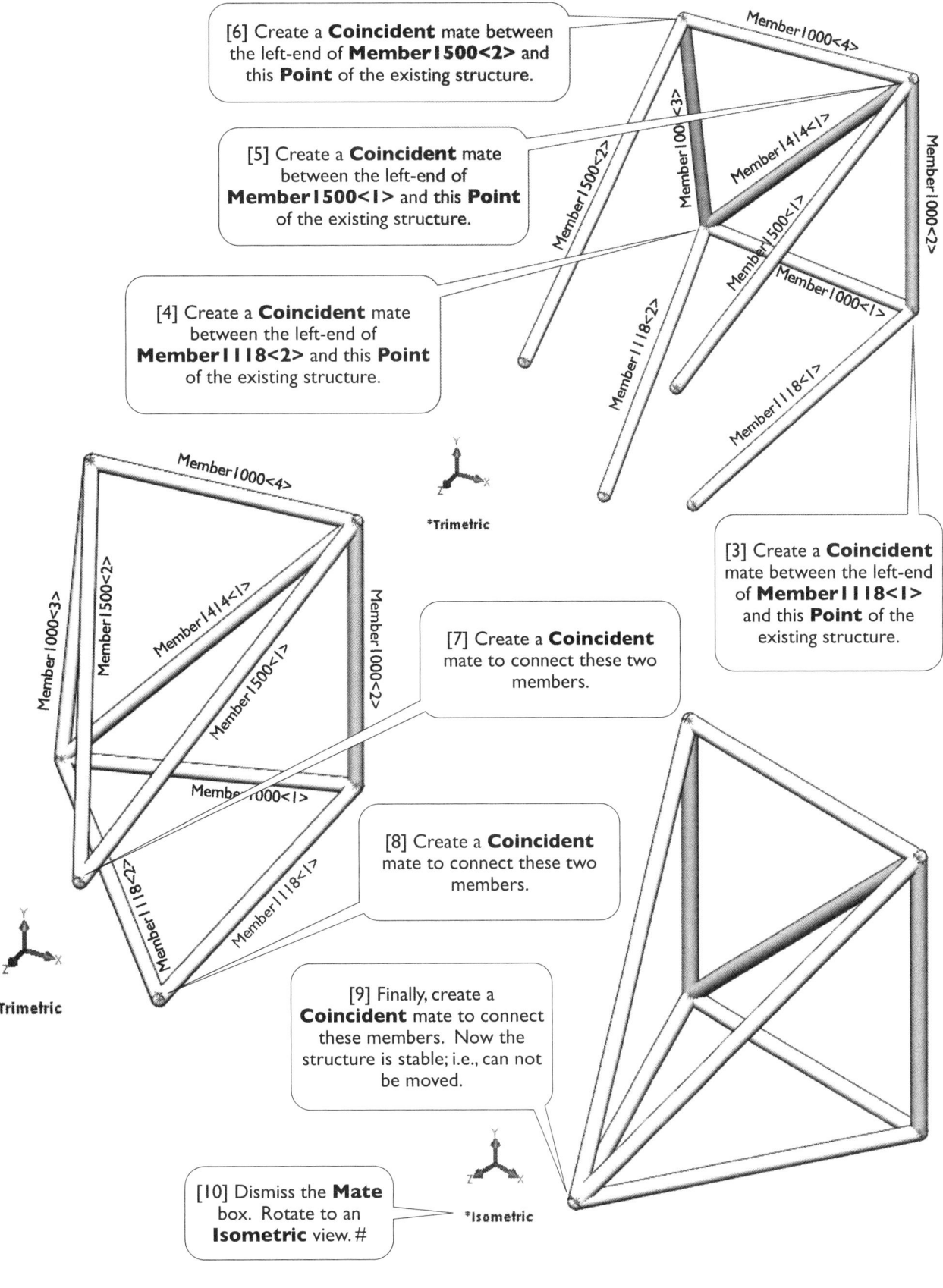

[6] Create a **Coincident** mate between the left-end of **Member1500<2>** and this **Point** of the existing structure.

[5] Create a **Coincident** mate between the left-end of **Member1500<1>** and this **Point** of the existing structure.

[4] Create a **Coincident** mate between the left-end of **Member1118<2>** and this **Point** of the existing structure.

[3] Create a **Coincident** mate between the left-end of **Member1118<1>** and this **Point** of the existing structure.

[7] Create a **Coincident** mate to connect these two members.

[8] Create a **Coincident** mate to connect these two members.

[9] Finally, create a **Coincident** mate to connect these members. Now the structure is stable; i.e., can not be moved.

[10] Dismiss the **Mate** box. Rotate to an **Isometric** view. #

*Trimetric

*Trimetric

*Isometric

Member1000<4>

Member1000<3>

Member1500<2>

Member1414<1>

Member1500<1>

Member1000<2>

Member1000<1>

Member1118<2>

Member1118<1>

2.2-10 Create a **Study**

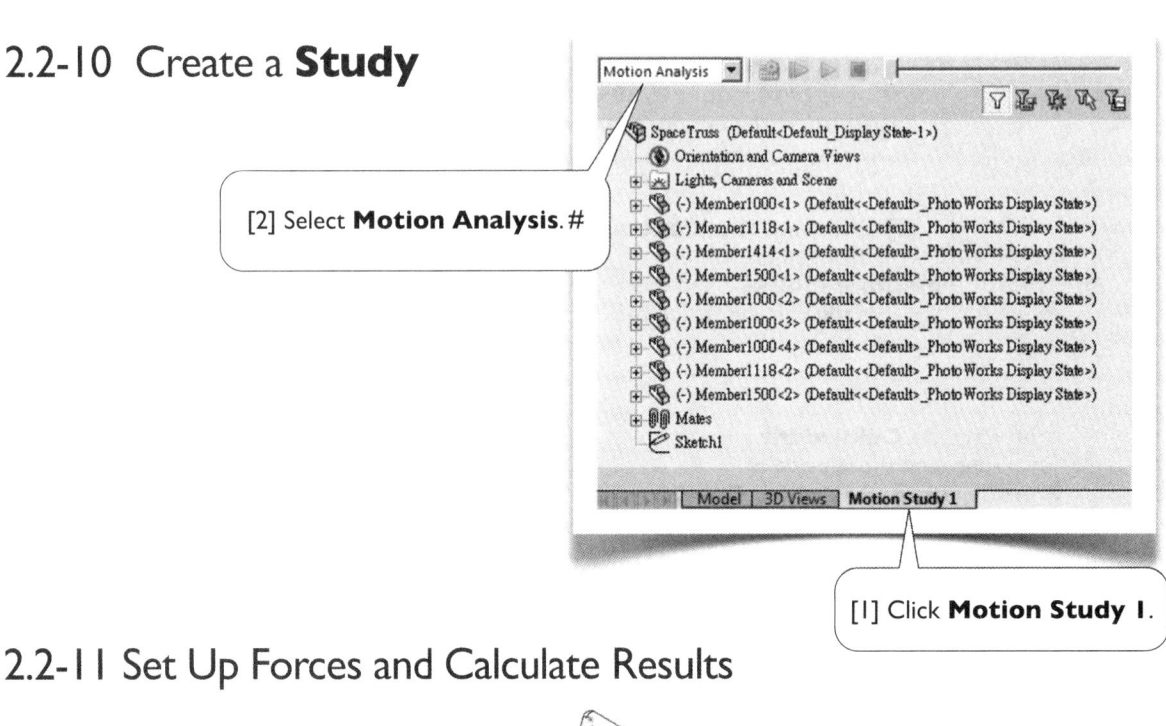

[2] Select **Motion Analysis**. #

[1] Click **Motion Study 1**.

2.2-11 Set Up Forces and Calculate Results

[1, 3] In **Motion** toolbar, click **Force**.

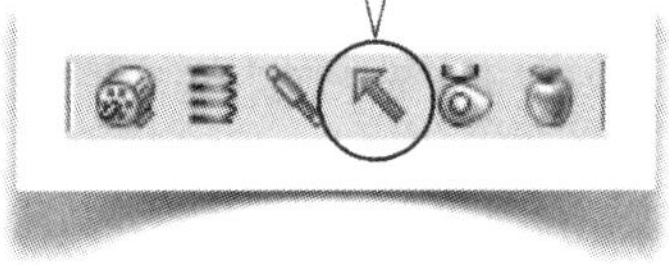

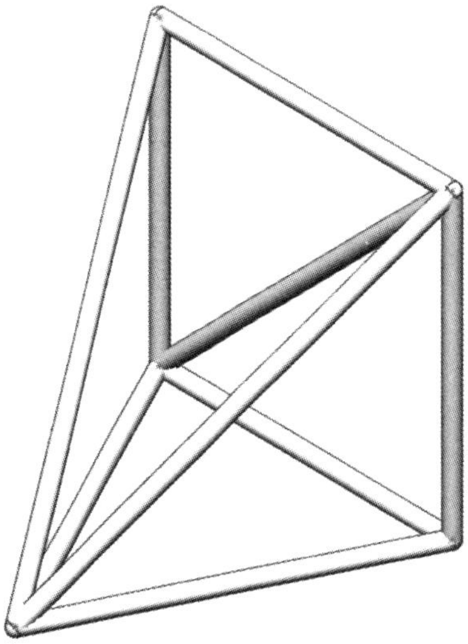

[2] Click this **Point** to define the location of the force. Click global **Top** plane to define the Force Direction; click **Reverse Direction**. Type 1200 (N) for the force value. Click **OK** to close the **Force/Torque** box.

[5] Click **Save**.

[4] Click this **Point** again to define the location of the force. Click global **Right** plane to define the **Force Direction**. Type 200 (N) for the force value. Click **OK** to close the **Force/Torque** box.

[6] Click **Calculate**. #

2.2-12 Retrieve Reaction Forces

⊟ 🔗 Mates
    ∠ Coincident1 (Member1000<1>,Sketch1)
    ∠ Coincident2 (Member1000<1>,Top)
    ∠ Coincident3 (Member1000<1>,Member1000<2>)
    ∠ Coincident4 (Member1000<1>,Member1000<3>)
    ∠ Coincident5 (Member1000<3>,Member1000<4>)
    ∠ Coincident6 (Member1000<2>,Member1000<4>)
    ∠ Coincident7 (Member1000<2>,Front)
    ∠ Coincident8 (Member1000<3>,Front)
    ∠ Coincident9 (Member1000<1>,Member1414<1>)
    ∠ Coincident10 (Member1414<1>,Member1000<4>)
    ∠ Coincident11 (Member1000<1>,Member1118<1>)
    ∠ Coincident12 (Member1414<1>,Member1118<2>)
    ∠ Coincident13 (Member1414<1>,Member1500<1>)
    ∠ Coincident14 (Member1000<4>,Member1500<2>)
    ∠ Coincident15 (Member1500<1>,Member1500<2>)
    ∠ Coincident16 (Member1118<1>,Member1118<2>)
    ∠ Coincident17 (Member1118<1>,Member1500<2>)

[1, 3, 5, 7] In **Motion** toolbar, click **Results and Plots**.

[2] Retrieve the three components of the **Reaction Force** at support B.

[4] Retrieve the **Y Component** of the **Reaction Force** at support A.

[6] Retrieve the **Z Component** of the **Reaction Force** at support C.

[8] Retrieve the **Z Component** of the **Reaction Force** at support D.

Do It Yourself

[9] The retrieved reaction forces are summarized here. The validity of these results may be verified by checking force and moment equilibria, regarding the truss structure as an integral body. (See 1.2-11[2], page 30.)

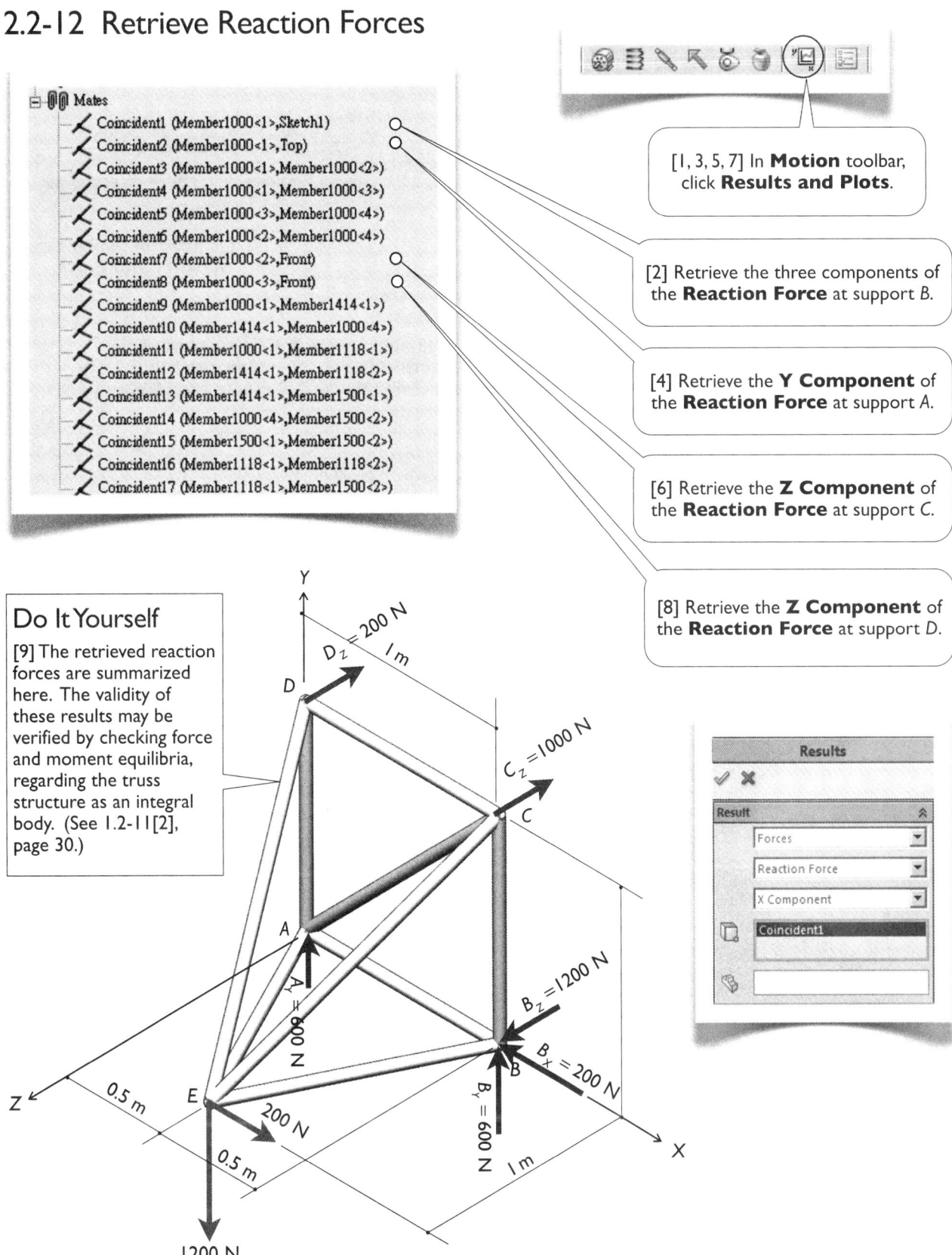

Results

Result

Forces

Reaction Force

X Component

Coincident1

2.2-13 Retrieve the Member Force of **Member1000<2>**

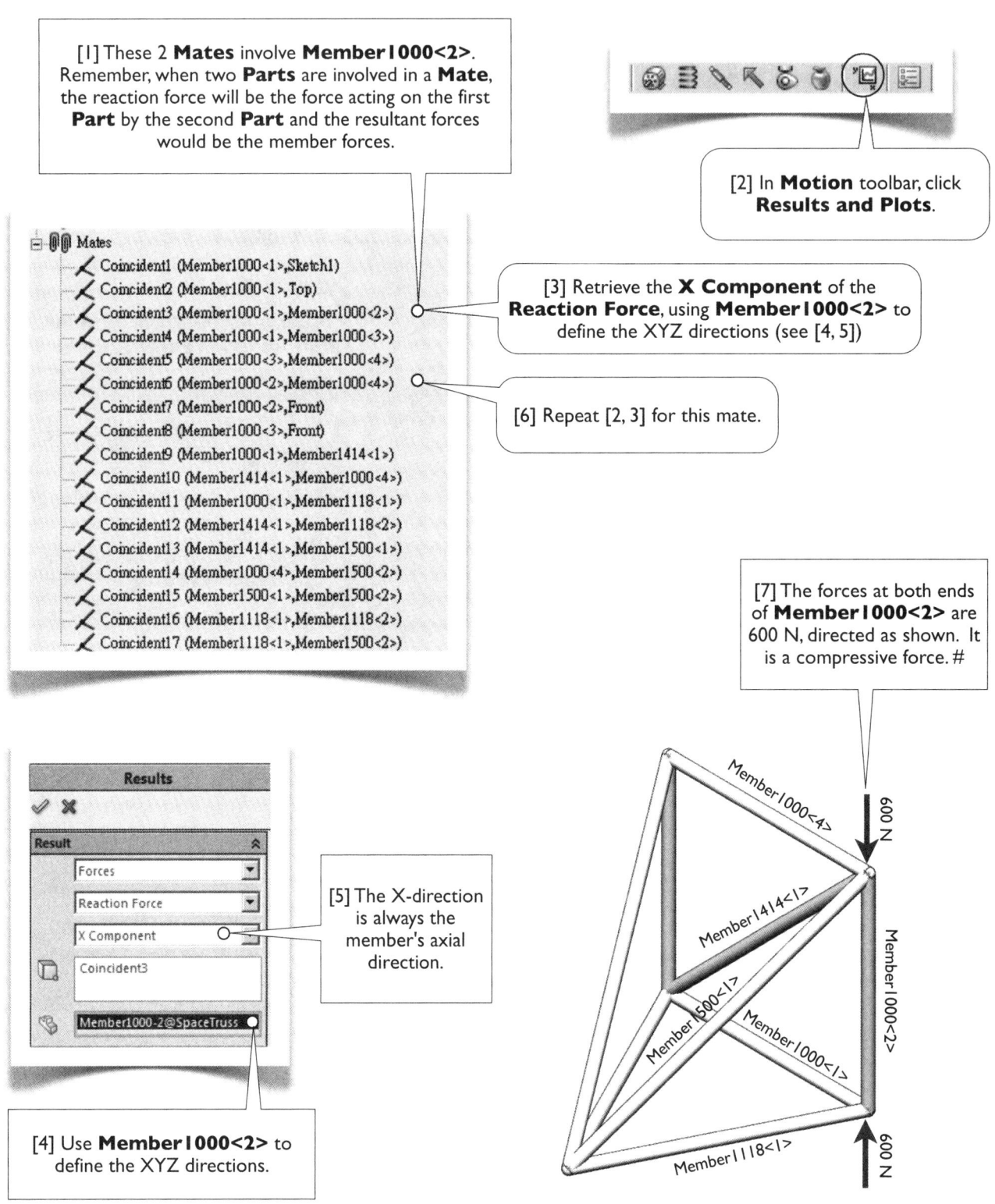

[1] These 2 **Mates** involve **Member1000<2>**. Remember, when two **Parts** are involved in a **Mate**, the reaction force will be the force acting on the first **Part** by the second **Part** and the resultant forces would be the member forces.

[2] In **Motion** toolbar, click **Results and Plots**.

[3] Retrieve the **X Component** of the **Reaction Force**, using **Member1000<2>** to define the XYZ directions (see [4, 5])

[6] Repeat [2, 3] for this mate.

Mates
- Coincident1 (Member1000<1>,Sketch1)
- Coincident2 (Member1000<1>,Top)
- Coincident3 (Member1000<1>,Member1000<2>)
- Coincident4 (Member1000<1>,Member1000<3>)
- Coincident5 (Member1000<3>,Member1000<4>)
- Coincident6 (Member1000<2>,Member1000<4>)
- Coincident7 (Member1000<2>,Front)
- Coincident8 (Member1000<3>,Front)
- Coincident9 (Member1000<1>,Member1414<1>)
- Coincident10 (Member1414<1>,Member1000<4>)
- Coincident11 (Member1000<1>,Member1118<1>)
- Coincident12 (Member1414<1>,Member1118<2>)
- Coincident13 (Member1414<1>,Member1500<1>)
- Coincident14 (Member1000<4>,Member1500<2>)
- Coincident15 (Member1500<1>,Member1500<2>)
- Coincident16 (Member1118<1>,Member1118<2>)
- Coincident17 (Member1118<1>,Member1500<2>)

[7] The forces at both ends of **Member1000<2>** are 600 N, directed as shown. It is a compressive force. #

Results

Result

Forces

Reaction Force

X Component

Coincident3

Member1000-2@SpaceTruss

[5] The X-direction is always the member's axial direction.

[4] Use **Member1000<2>** to define the XYZ directions.

Member1000<4>

Member1414<1>

Member1500<1>

Member1000<1>

Member1000<2>

Member1118<1>

600 N

600 N

2.2-14 Do It Yourself: Other Member Forces

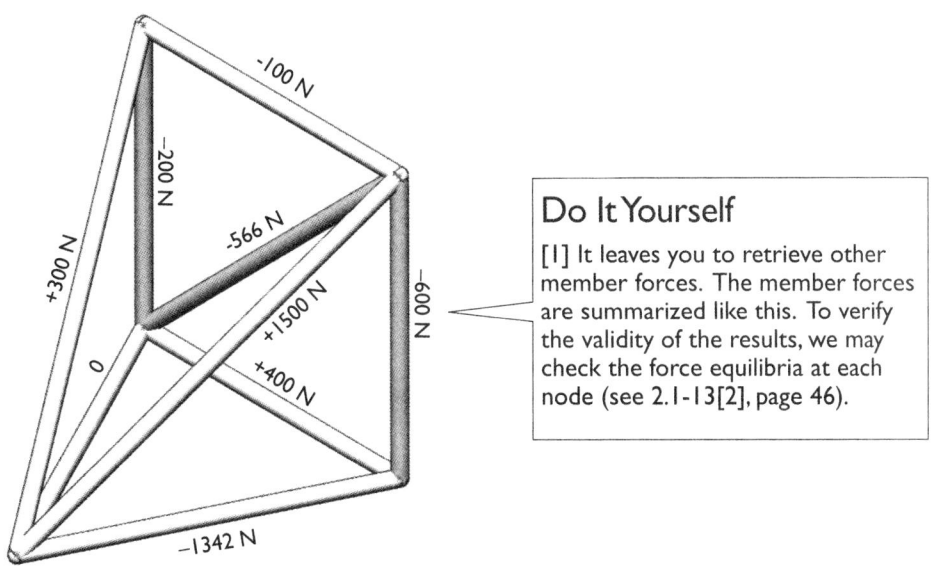

Do It Yourself

[1] It leaves you to retrieve other member forces. The member forces are summarized like this. To verify the validity of the results, we may check the force equilibria at each node (see 2.1-13[2], page 46).

[2] From the **Pull-Down Menus**, Select **File>Exit** to quit **SOLIDWORKS**. Click **Save all.** Click **Rebuild and save the document**. #

Chapter 3

Frames

As mentioned in the opening of Chapter 2, a truss is defined as a structure in which (a) the members are pin-jointed or ball-and-socket-jointed, and (b) the external forces are applied at joints. If any of the above conditions are not met, the members are no longer two-force members and the structure is called a **frame**.

In Section 3.1, we'll show you a way to solve a plane frame with **SOLIDWORKS** and, in Section 3.2, solve a space frame. Remember, in this book, we discuss only statically determinate structures.

Section 3.1

Plane Frame

3.1-1 Introduction

[1] In this section, we consider a plane frame supported by a hinge at the left and a roller at the right and subject to a downward force of 2400 N as shown. We want to find the reaction forces and the forces acting on each member.

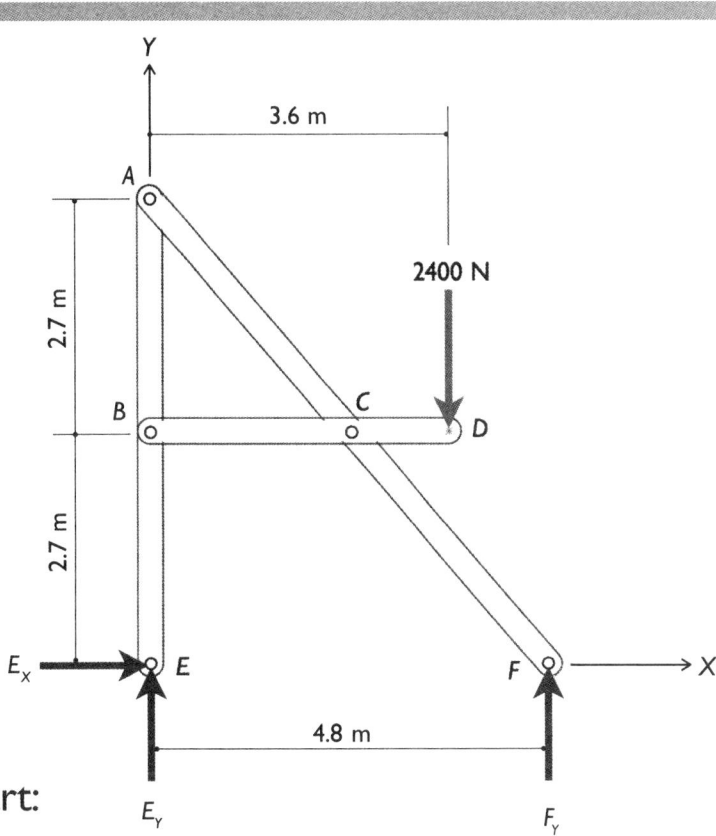

3.1-2 Start Up and Create a Part: **Horizontal**

[1] Launch **SOLIDWORKS**.

[2] Create a new **Part**. Select **MKS** unit system with three decimal places for the length unit.

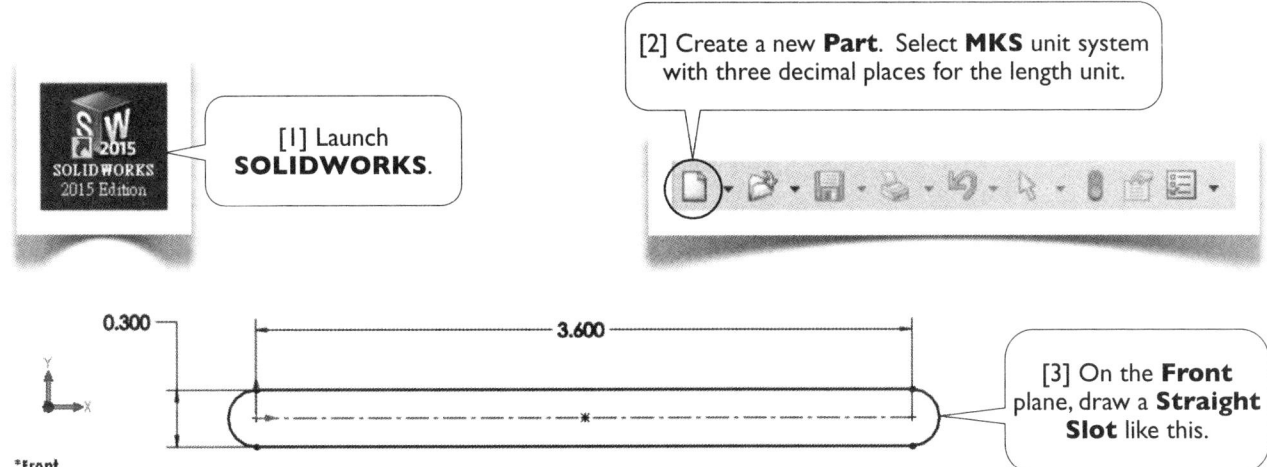

[3] On the **Front** plane, draw a **Straight Slot** like this.

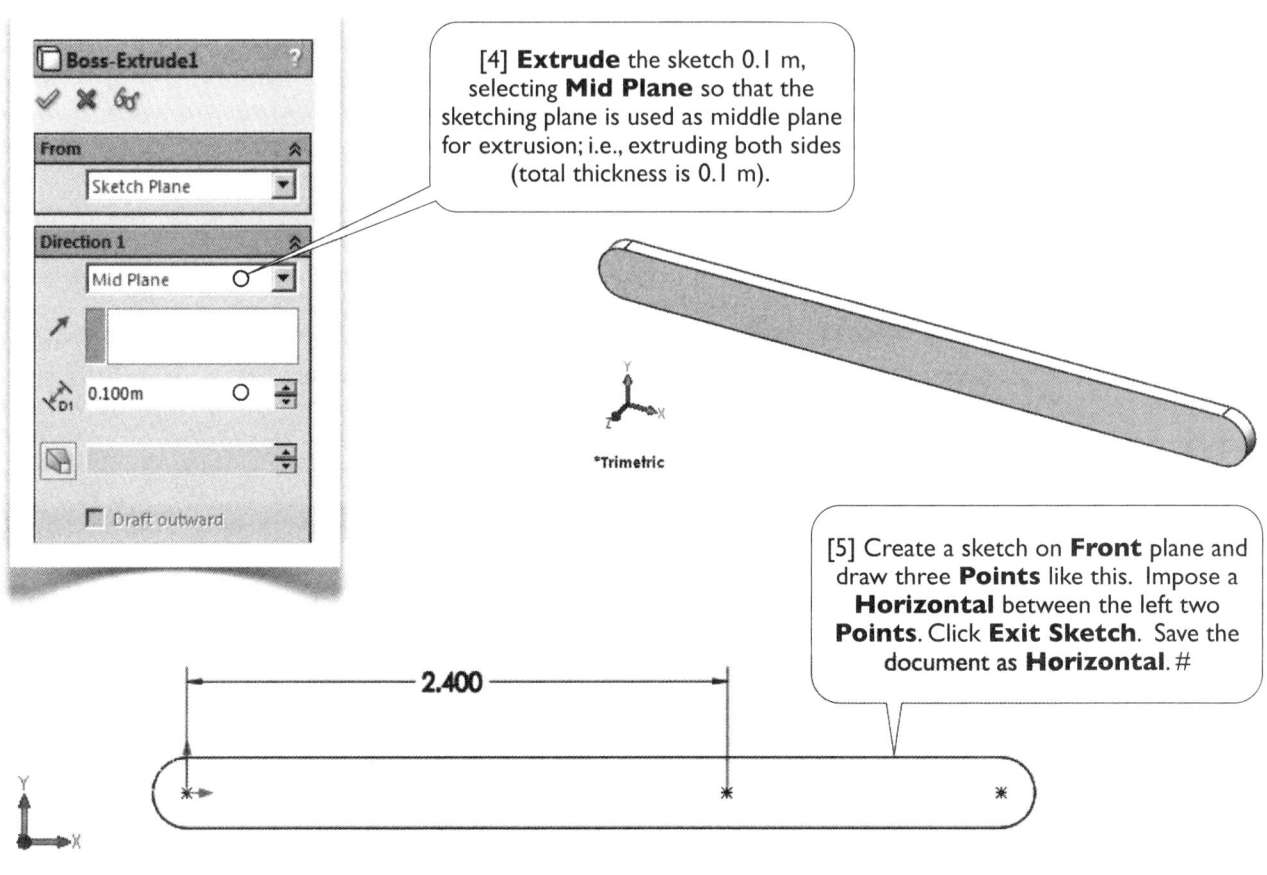

[4] **Extrude** the sketch 0.1 m, selecting **Mid Plane** so that the sketching plane is used as middle plane for extrusion; i.e., extruding both sides (total thickness is 0.1 m).

[5] Create a sketch on **Front** plane and draw three **Points** like this. Impose a **Horizontal** between the left two **Points**. Click **Exit Sketch**. Save the document as **Horizontal**. #

3.1-3 Create a Part: **Vertical**

[1] Repeat steps 3.1-2[2-4], except the length of the member is 5.4 meters.

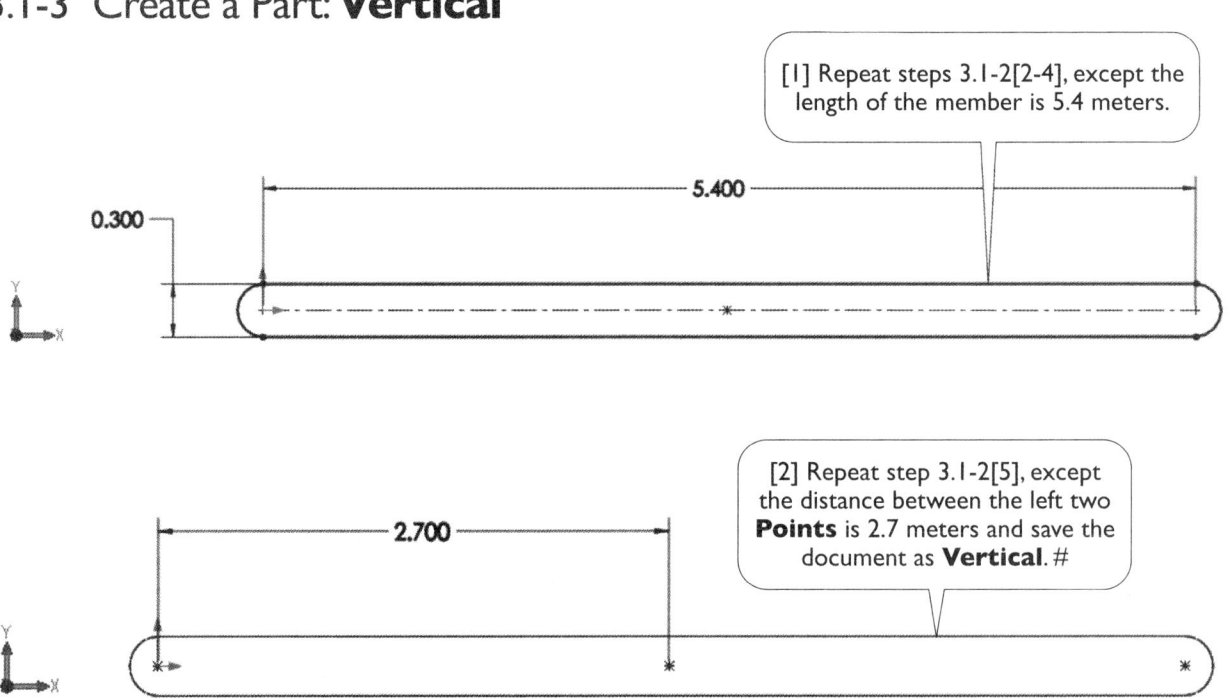

[2] Repeat step 3.1-2[5], except the distance between the left two **Points** is 2.7 meters and save the document as **Vertical**. #

3.1-4 Create a Part: **Diagonal**

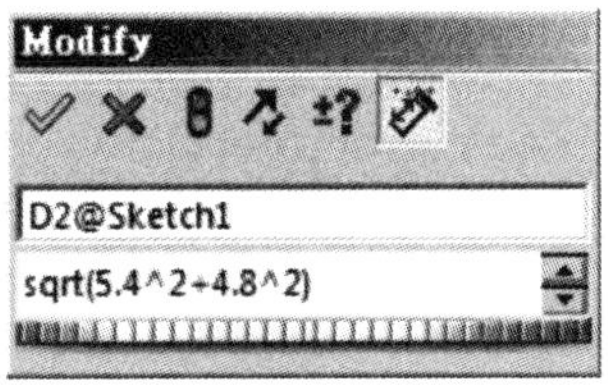

Modify

D2@Sketch1

sqrt(5.4^2+4.8^2)

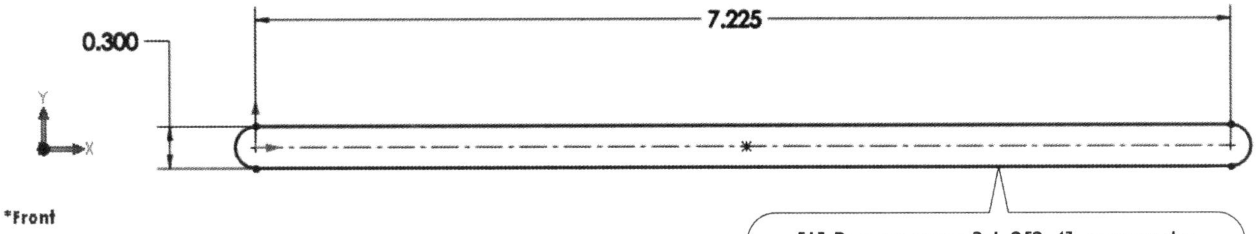

7.225

0.300

*Front

[1] Repeat steps 3.1-2[2-4], except the length of the member is $\sqrt{5.4^2 + 4.8^2}$ meters (by typing "sqrt(5.4^2+4.8^2)").

Modify

D1@Sketch2

sqrt(5.4^2+4.8^2)/2

3.612

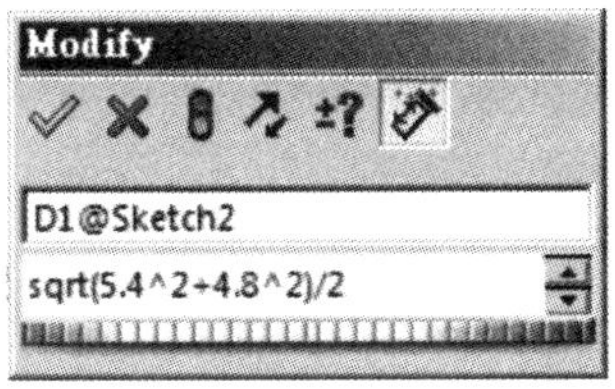

[2] Repeat step 3.1-2[5], except the distance between the left two **Points** is $\sqrt{5.4^2 + 4.8^2}/2$ meters (by typing "sqrt(5.4^2+4.8^2)/2") and save the document as **Diagonal**. #

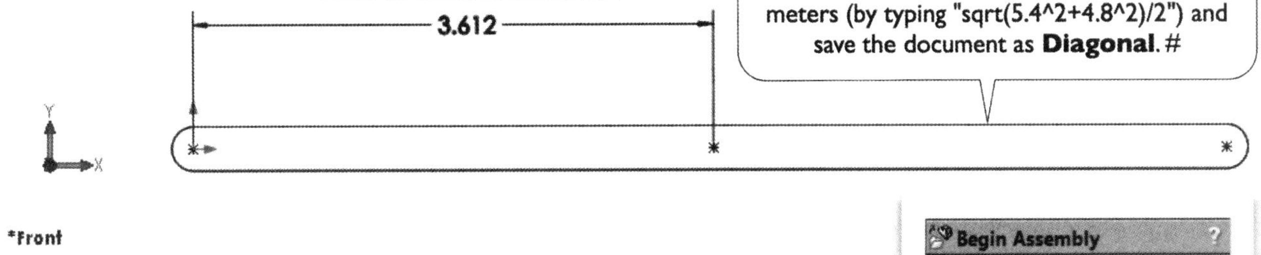

*Front

3.1-5 Create an Assembly: **PlaneFrame**

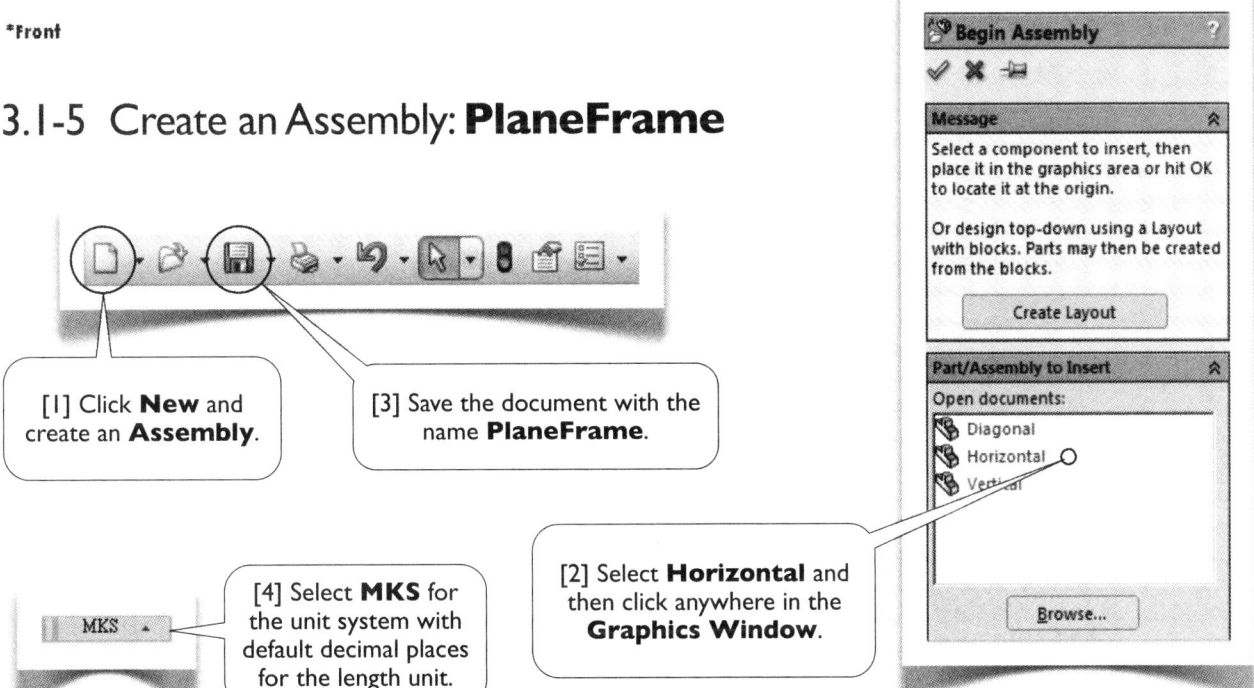

Begin Assembly

Message

Select a component to insert, then place it in the graphics area or hit OK to locate it at the origin.

Or design top-down using a Layout with blocks. Parts may then be created from the blocks.

Create Layout

Part/Assembly to Insert

Open documents:

Diagonal
Horizontal
Vertical

Browse...

[1] Click **New** and create an **Assembly**.

[3] Save the document with the name **PlaneFrame**.

[4] Select **MKS** for the unit system with default decimal places for the length unit.

MKS

[2] Select **Horizontal** and then click anywhere in the **Graphics Window**.

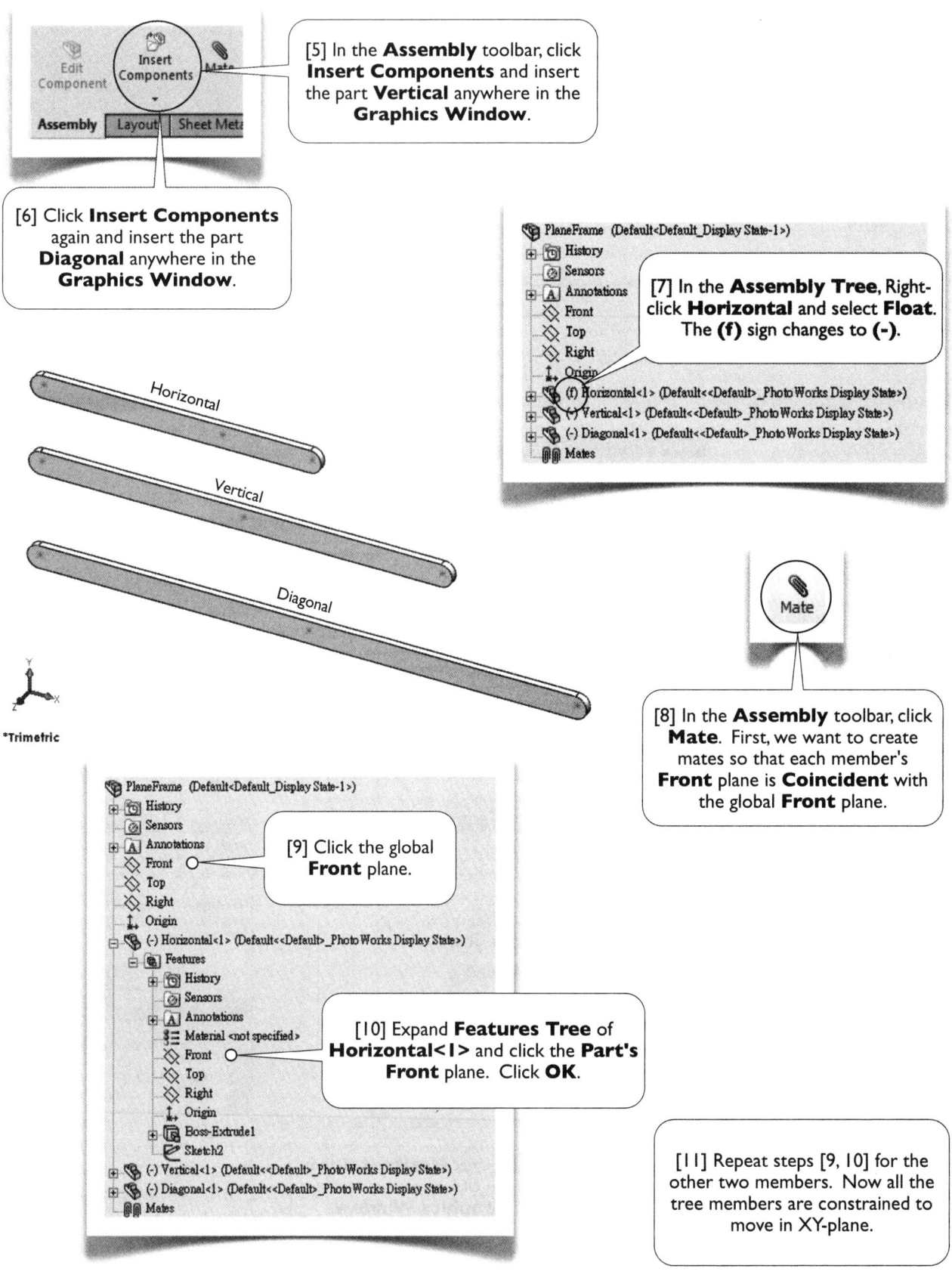

[5] In the **Assembly** toolbar, click **Insert Components** and insert the part **Vertical** anywhere in the **Graphics Window**.

[6] Click **Insert Components** again and insert the part **Diagonal** anywhere in the **Graphics Window**.

[7] In the **Assembly Tree**, Right-click **Horizontal** and select **Float**. The **(f)** sign changes to **(-)**.

Horizontal

Vertical

Diagonal

*Trimetric

[8] In the **Assembly** toolbar, click **Mate**. First, we want to create mates so that each member's **Front** plane is **Coincident** with the global **Front** plane.

[9] Click the global **Front** plane.

[10] Expand **Features Tree** of **Horizontal<1>** and click the **Part's Front** plane. Click **OK**.

[11] Repeat steps [9, 10] for the other two members. Now all the tree members are constrained to move in XY-plane.

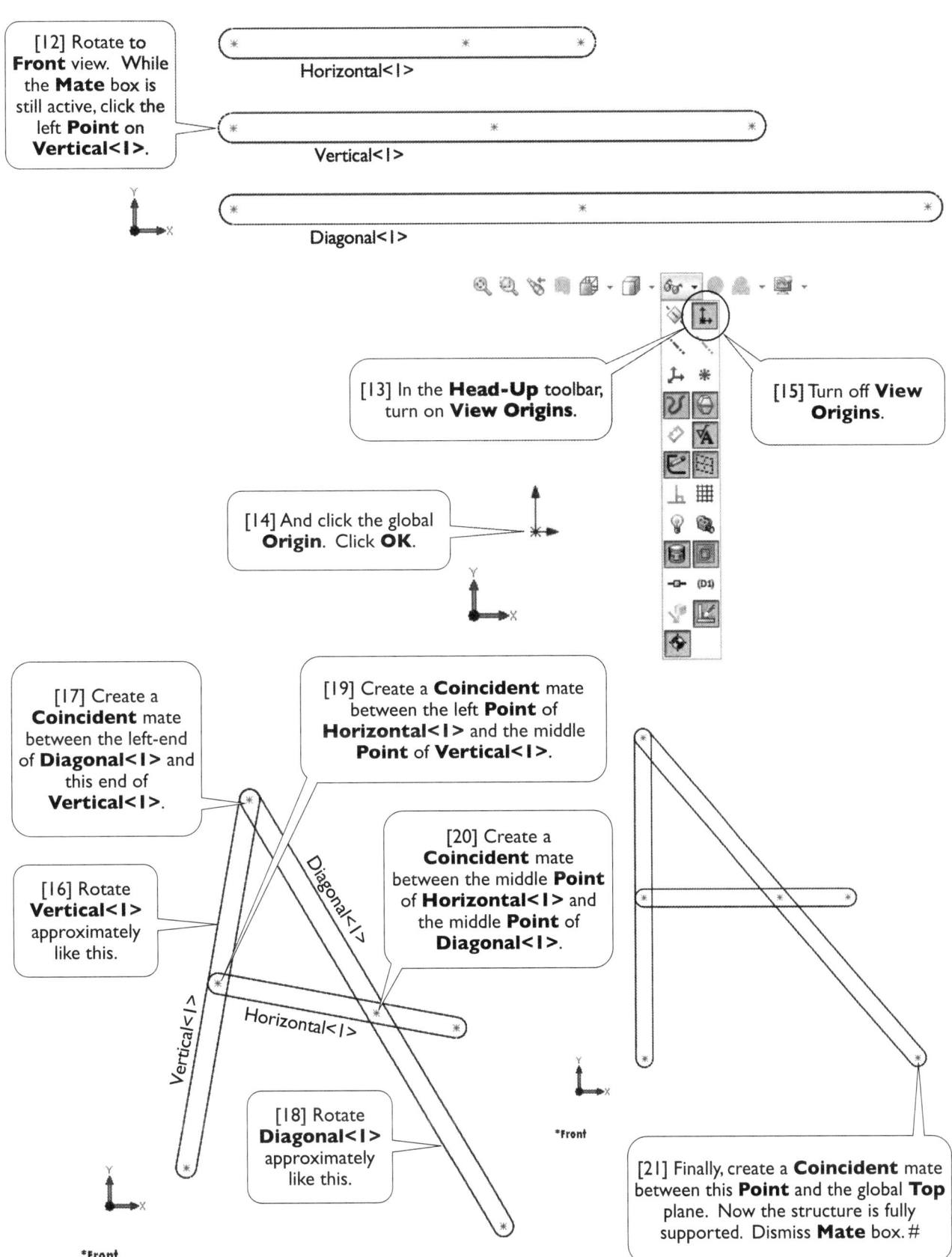

[12] Rotate to **Front** view. While the **Mate** box is still active, click the left **Point** on **Vertical<1>**.

Horizontal<1>

Vertical<1>

Diagonal<1>

[13] In the **Head-Up** toolbar, turn on **View Origins**.

[15] Turn off **View Origins**.

[14] And click the global **Origin**. Click **OK**.

[17] Create a **Coincident** mate between the left-end of **Diagonal<1>** and this end of **Vertical<1>**.

[19] Create a **Coincident** mate between the left **Point** of **Horizontal<1>** and the middle **Point** of **Vertical<1>**.

[20] Create a **Coincident** mate between the middle **Point** of **Horizontal<1>** and the middle **Point** of **Diagonal<1>**.

[16] Rotate **Vertical<1>** approximately like this.

Diagonal<1>

Vertical<1>

Horizontal<1>

[18] Rotate **Diagonal<1>** approximately like this.

*Front

[21] Finally, create a **Coincident** mate between this **Point** and the global **Top** plane. Now the structure is fully supported. Dismiss **Mate** box. #

*Front

3.1-6 Create a **Study**

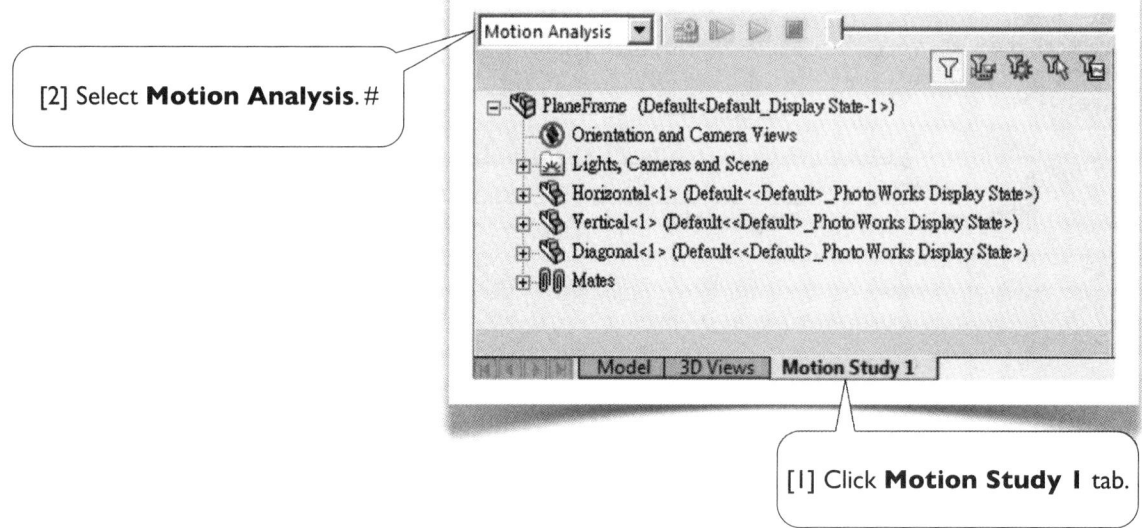

[2] Select **Motion Analysis**. #

[1] Click **Motion Study 1** tab.

3.1-7 Set Up Forces and Calculate Results

[1] In **Motion** toolbar, click **Force**.

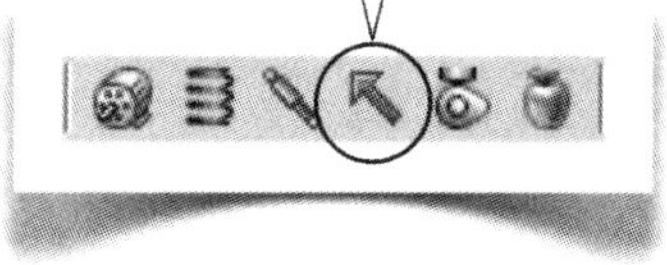

[2] Click this **Point** to define the location of the force. Click global **Top** plane to define the **Force Direction**; click **Reverse Direction**. Type 2400 (N) for the force value. Click **OK** to close the **Force/Torque** box.

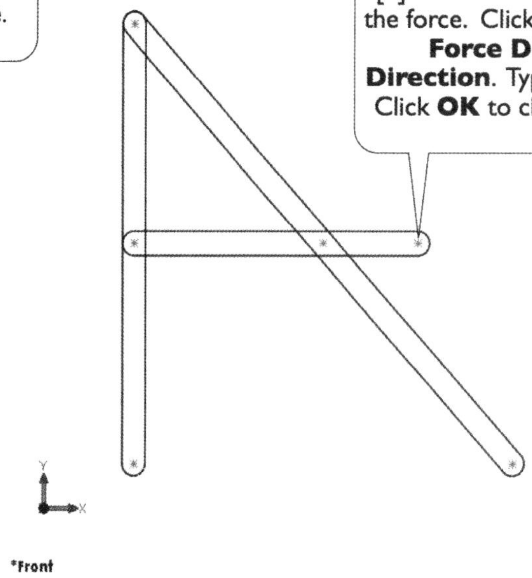

*Front

[3] Click **Save**.

[4] Click **Calculate**. #

3.1-8 Retrieve Reaction Forces

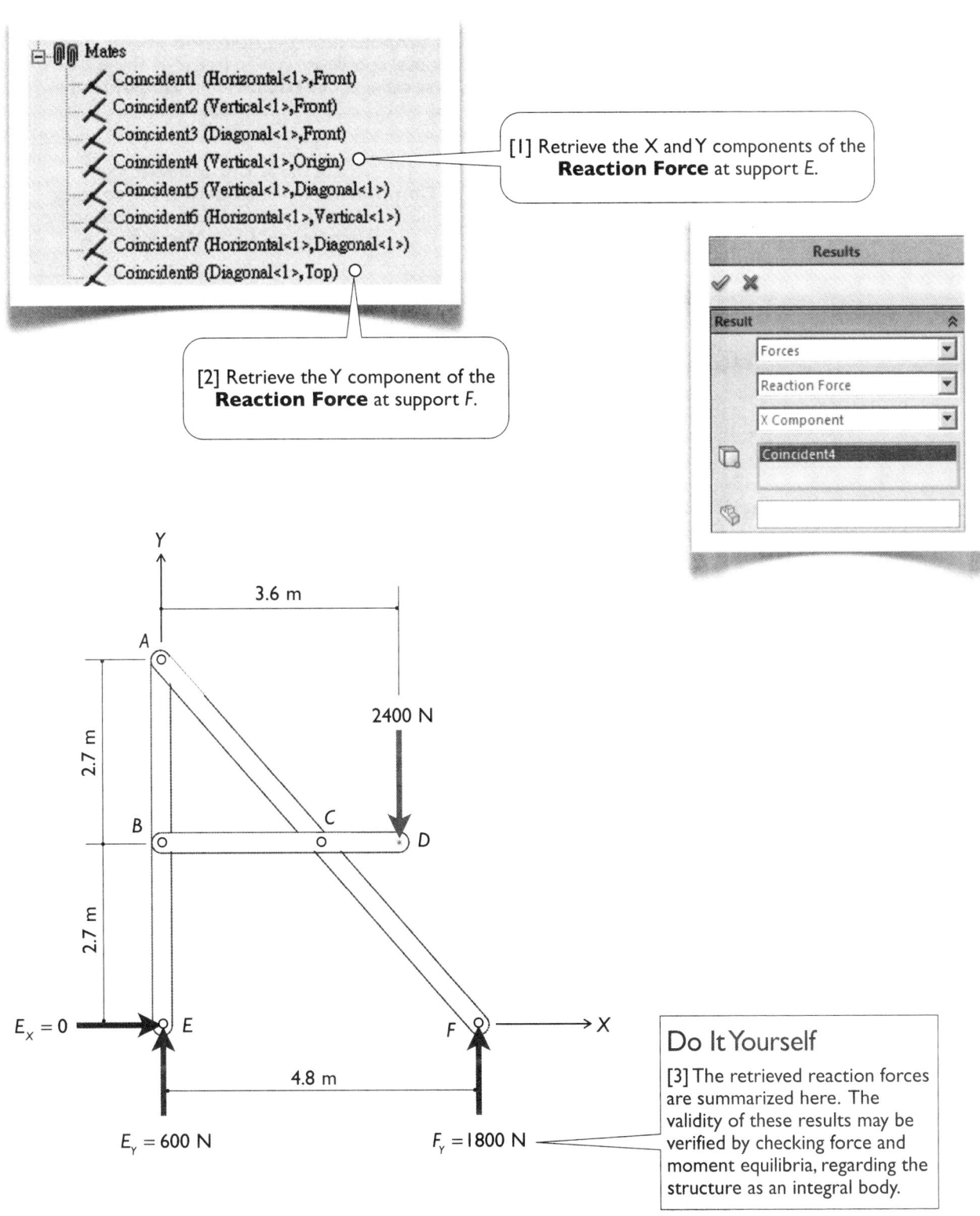

Mates
Coincident1 (Horizontal<1>,Front)
Coincident2 (Vertical<1>,Front)
Coincident3 (Diagonal<1>,Front)
Coincident4 (Vertical<1>,Origin)
Coincident5 (Vertical<1>,Diagonal<1>)
Coincident6 (Horizontal<1>,Vertical<1>)
Coincident7 (Horizontal<1>,Diagonal<1>)
Coincident8 (Diagonal<1>,Top)

[1] Retrieve the X and Y components of the **Reaction Force** at support E.

[2] Retrieve the Y component of the **Reaction Force** at support F.

Results

Result

Forces

Reaction Force

X Component

Coincident4

$E_x = 0$

$E_Y = 600$ N

$F_Y = 1800$ N

3.6 m

2.7 m

2.7 m

2400 N

4.8 m

Do It Yourself

[3] The retrieved reaction forces are summarized here. The validity of these results may be verified by checking force and moment equilibria, regarding the structure as an integral body.

3.1-9 Retrieve Other Forces on Each Members

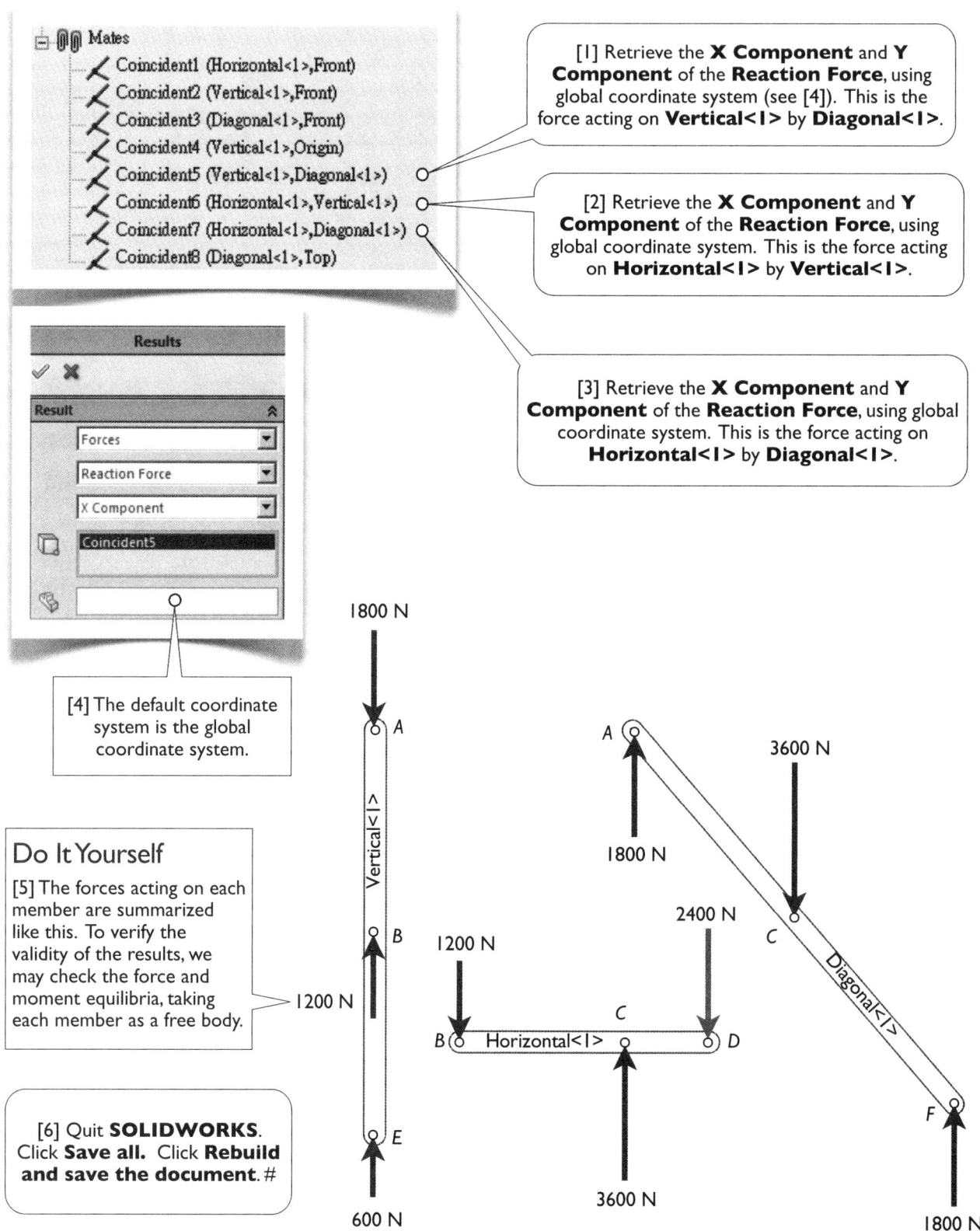

[1] Retrieve the **X Component** and **Y Component** of the **Reaction Force**, using global coordinate system (see [4]). This is the force acting on **Vertical<1>** by **Diagonal<1>**.

[2] Retrieve the **X Component** and **Y Component** of the **Reaction Force**, using global coordinate system. This is the force acting on **Horizontal<1>** by **Vertical<1>**.

[3] Retrieve the **X Component** and **Y Component** of the **Reaction Force**, using global coordinate system. This is the force acting on **Horizontal<1>** by **Diagonal<1>**.

[4] The default coordinate system is the global coordinate system.

Do It Yourself

[5] The forces acting on each member are summarized like this. To verify the validity of the results, we may check the force and moment equilibria, taking each member as a free body.

[6] Quit **SOLIDWORKS**. Click **Save all**. Click **Rebuild and save the document**. #

Section 3.2

Space Frame

3.2-1 Introduction

[1] In this section, we consider four beams, each of length 4 m, connected together at their mid-points and supported vertically at D, E, and H to form a space frame system as shown. We want to find the reaction forces at the supports.

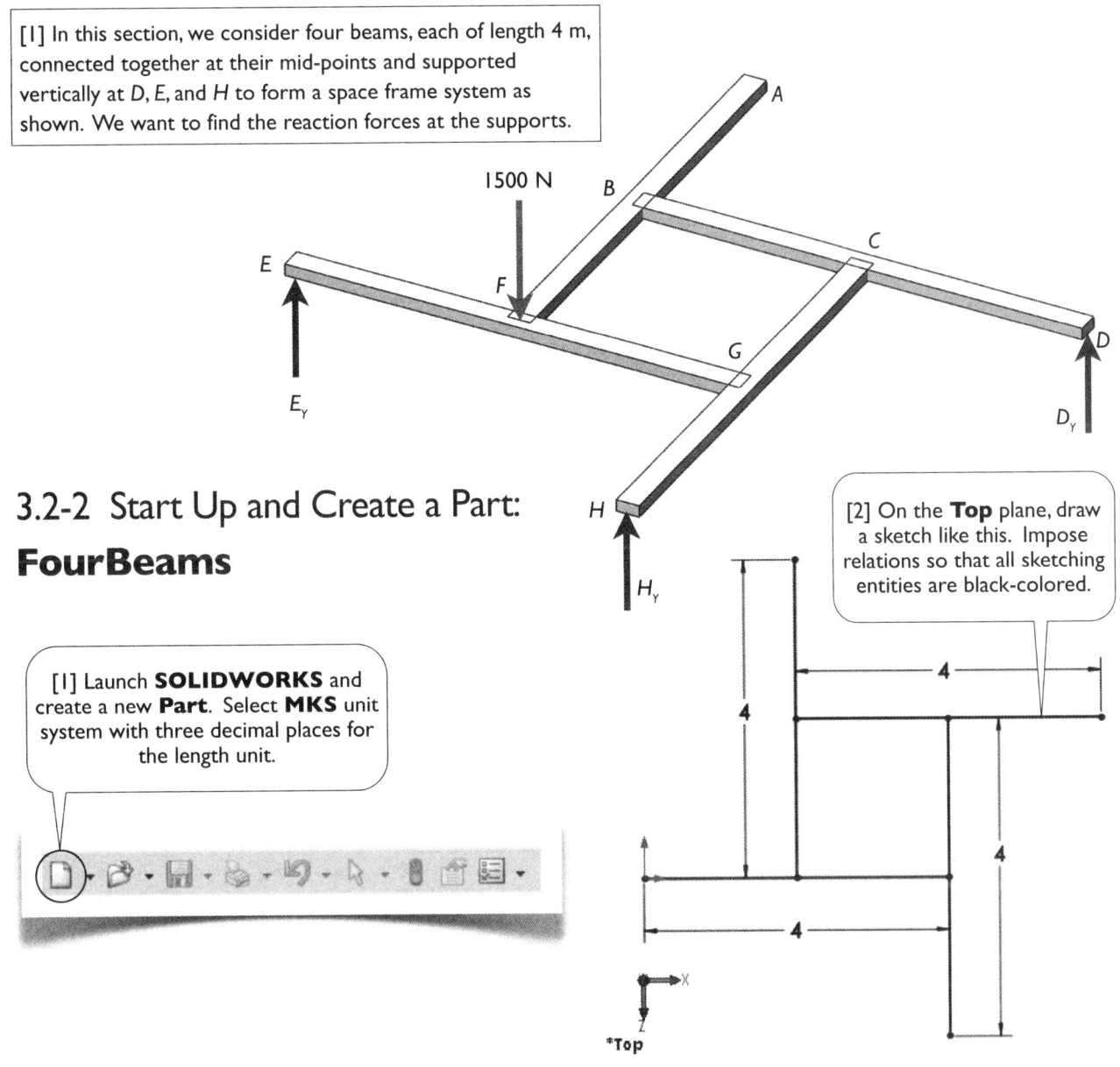

1500 N

3.2-2 Start Up and Create a Part: **FourBeams**

[1] Launch **SOLIDWORKS** and create a new **Part**. Select **MKS** unit system with three decimal places for the length unit.

[2] On the **Top** plane, draw a sketch like this. Impose relations so that all sketching entities are black-colored.

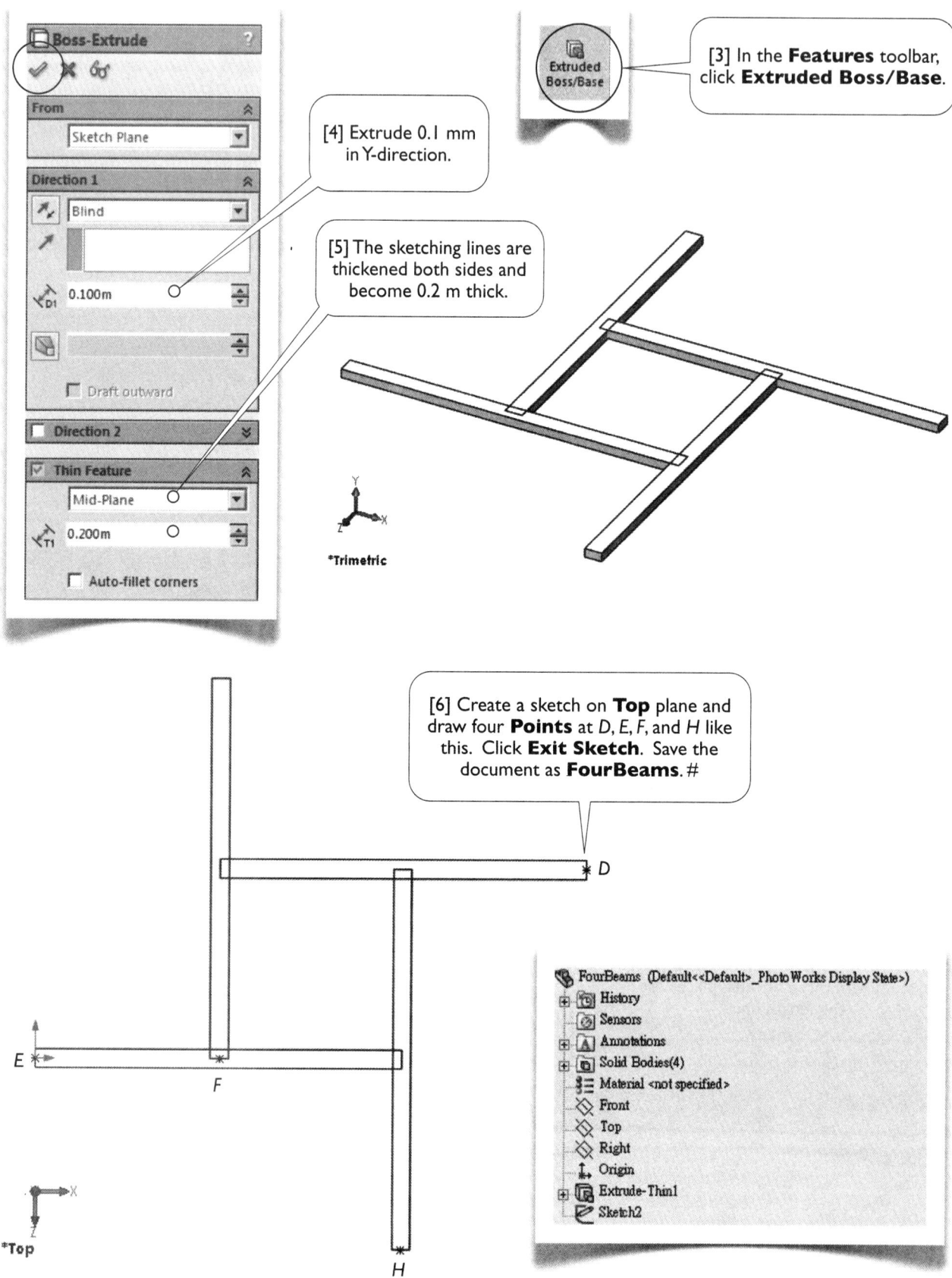

Boss-Extrude

From
Sketch Plane

Direction 1
Blind

0.100m

☐ Draft outward

☐ **Direction 2**

☑ **Thin Feature**
Mid-Plane
0.200m

☐ Auto-fillet corners

[3] In the **Features** toolbar, click **Extruded Boss/Base**.

Extruded
Boss/Base

[4] Extrude 0.1 mm in Y-direction.

[5] The sketching lines are thickened both sides and become 0.2 m thick.

*Trimetric

[6] Create a sketch on **Top** plane and draw four **Points** at D, E, F, and H like this. Click **Exit Sketch**. Save the document as **FourBeams**. #

E

F

H

*Top

D

FourBeams (Default<<Default>_PhotoWorks Display State>)
History
Sensors
Annotations
Solid Bodies(4)
Material <not specified>
Front
Top
Right
Origin
Extrude-Thin1
Sketch2

3.2-3 Create an Assembly: **SpaceFrame**

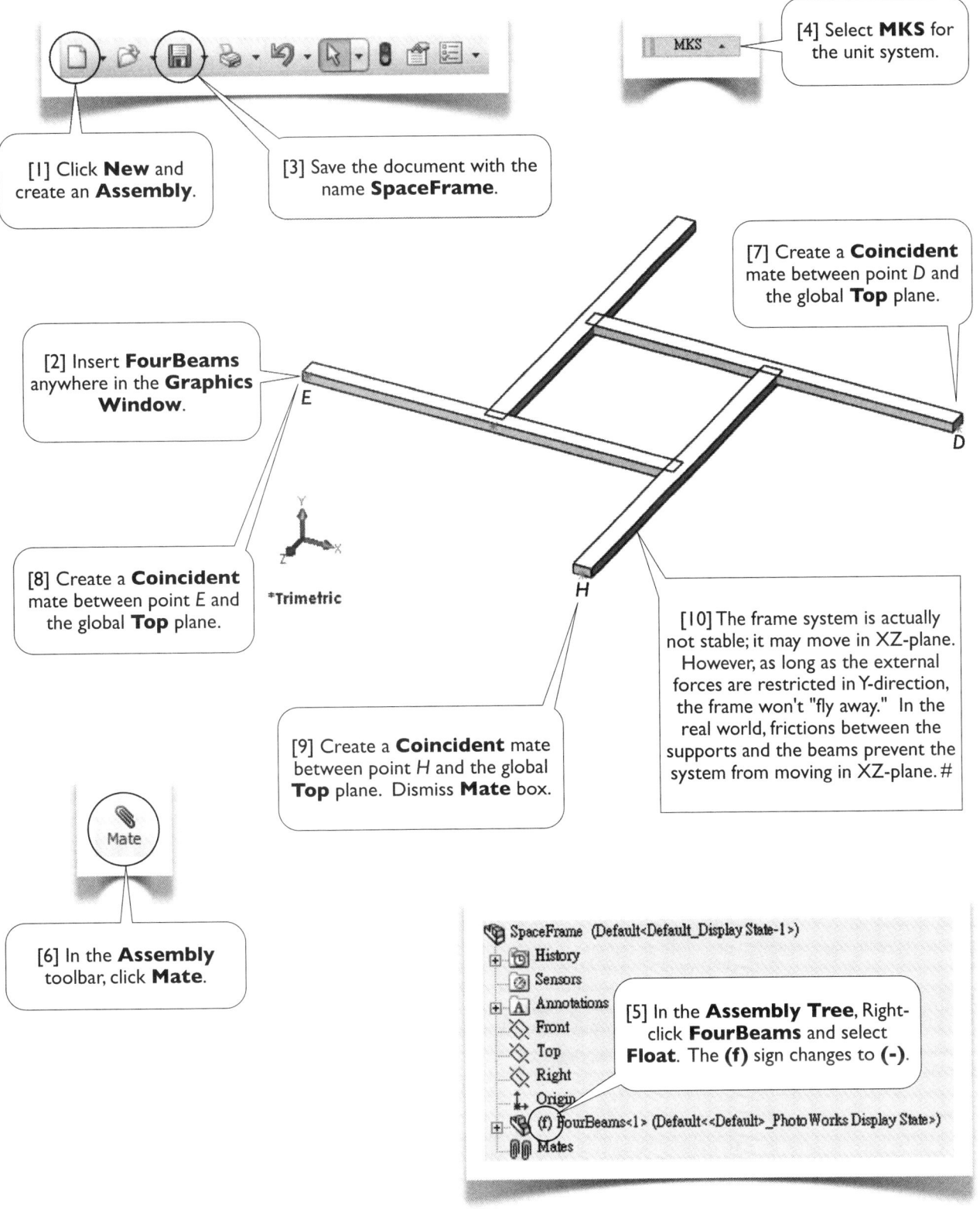

[1] Click **New** and create an **Assembly**.

[3] Save the document with the name **SpaceFrame**.

[4] Select **MKS** for the unit system.

[7] Create a **Coincident** mate between point *D* and the global **Top** plane.

[2] Insert **FourBeams** anywhere in the **Graphics Window**.

*Trimetric

[8] Create a **Coincident** mate between point *E* and the global **Top** plane.

[9] Create a **Coincident** mate between point *H* and the global **Top** plane. Dismiss **Mate** box.

[10] The frame system is actually not stable; it may move in XZ-plane. However, as long as the external forces are restricted in Y-direction, the frame won't "fly away." In the real world, frictions between the supports and the beams prevent the system from moving in XZ-plane. #

Mate

[6] In the **Assembly** toolbar, click **Mate**.

SpaceFrame (Default<Default_Display State-1>)
History
Sensors
Annotations
Front
Top
Right
Origin
(f) FourBeams<1> (Default<<Default>_PhotoWorks Display State>)
Mates

[5] In the **Assembly Tree**, Right-click **FourBeams** and select **Float**. The **(f)** sign changes to **(-)**.

3.2-4 Create a **Study**

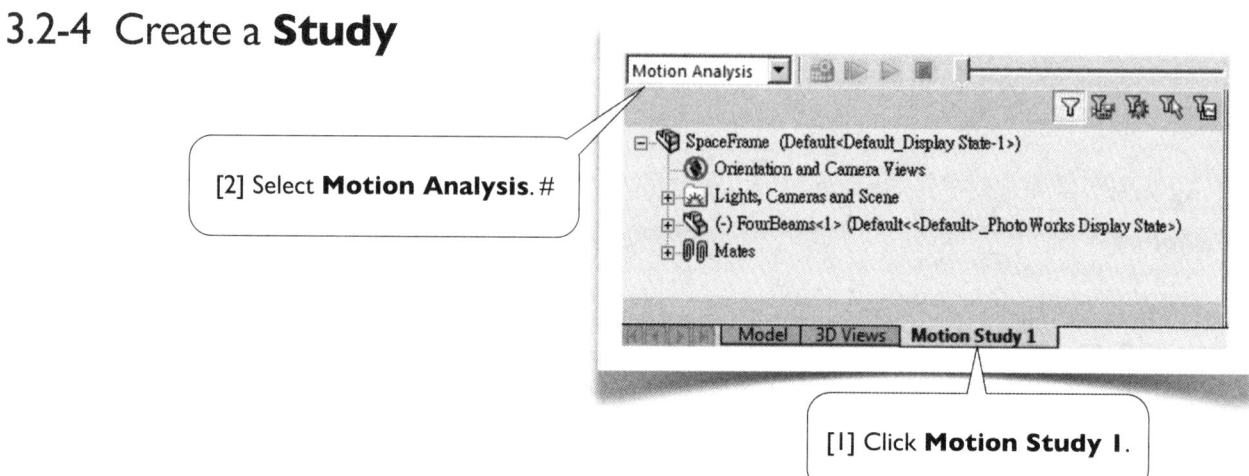

[2] Select **Motion Analysis**. #

[1] Click **Motion Study 1**.

3.2-5 Set Up Forces and Calculate Results

[1] In **Motion** toolbar, click **Force**.

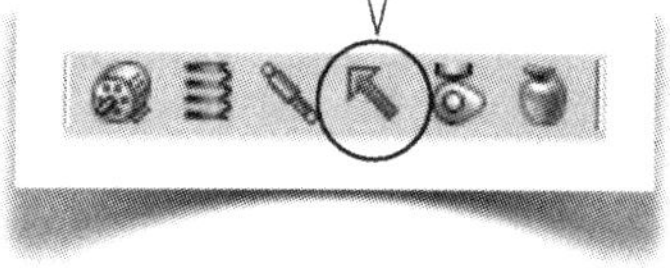

[2] Click **Point** *F* to define the location of the force. Click global **Top** plane to define the **Force Direction**; click **Reverse Direction**. Type 1500 (N) for the force value. Click **OK** to close the **Force/ Torque** box.

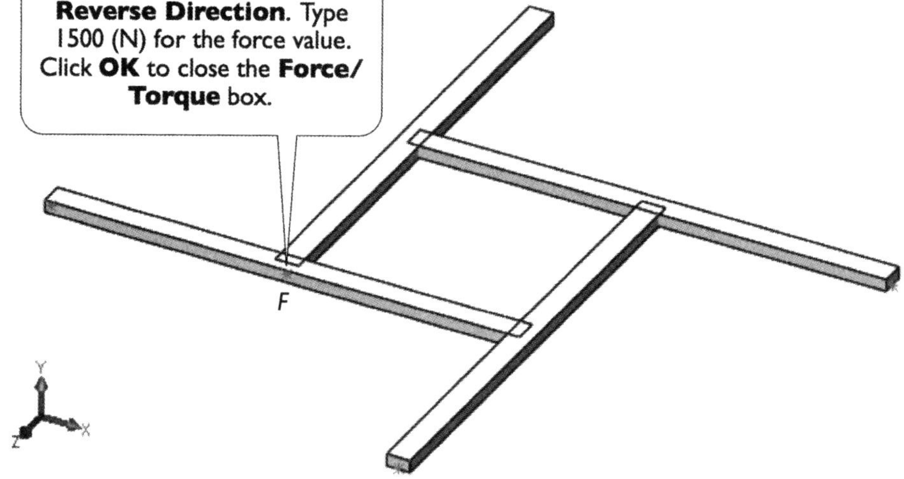

F

*Trimetric

[3] Click **Save**.

[4] Click **Calculate**. #

Motion Analysis

3.2-6 Retrieve Reaction Forces

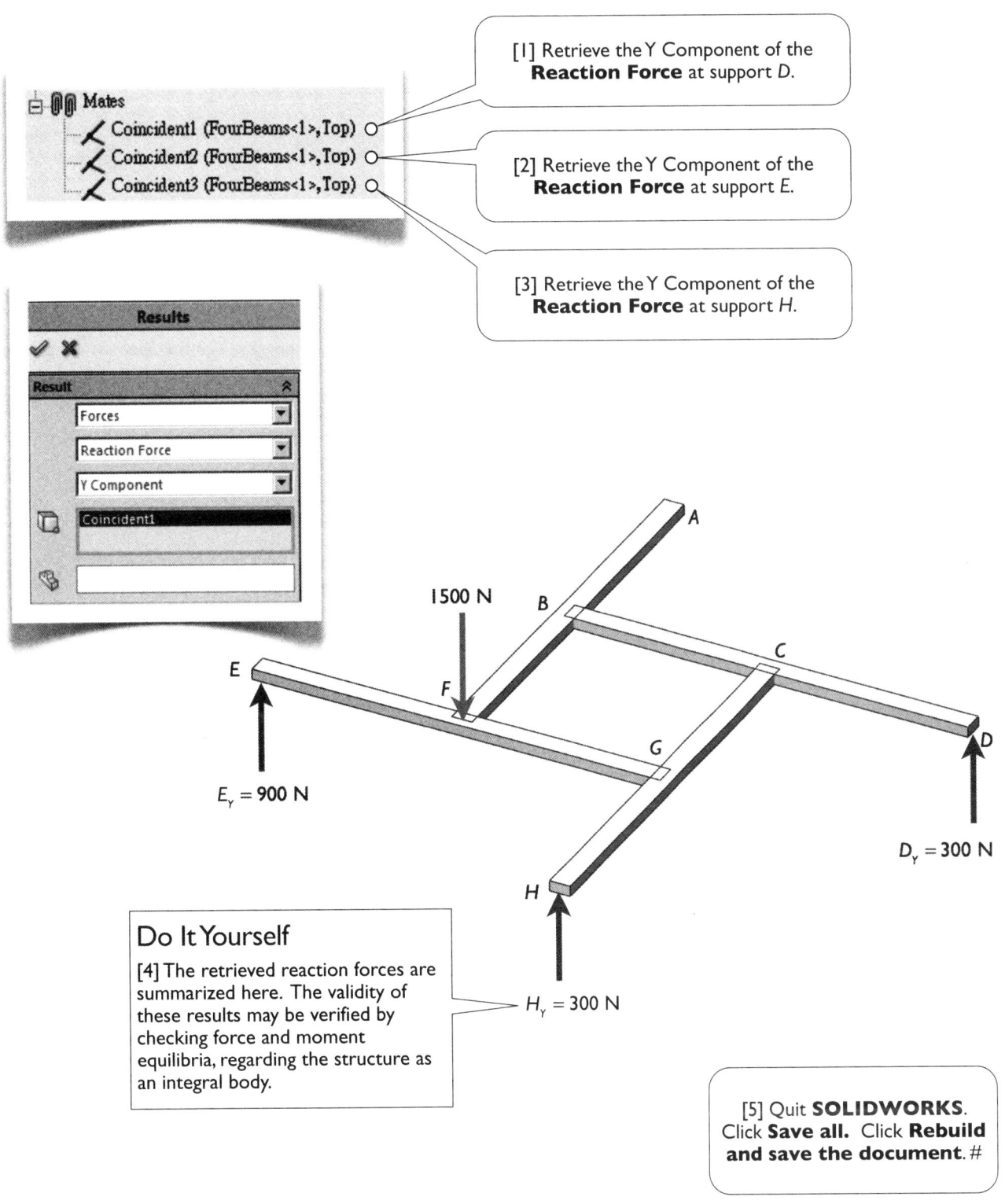

[1] Retrieve the Y Component of the **Reaction Force** at support *D*.

[2] Retrieve the Y Component of the **Reaction Force** at support *E*.

[3] Retrieve the Y Component of the **Reaction Force** at support *H*.

1500 N

$E_Y = 900$ N

$D_Y = 300$ N

$H_Y = 300$ N

Do It Yourself

[4] The retrieved reaction forces are summarized here. The validity of these results may be verified by checking force and moment equilibria, regarding the structure as an integral body.

[5] Quit **SOLIDWORKS**. Click **Save all.** Click **Rebuild and save the document.** #

Chapter 4
Machines

Chapters 2 and 3 discuss structures. The function of a structure is to withstand external loads and maintain its stability; i.e., it doesn't move. It does deform but the deformation is too small to affect the function of the structure.

On the other hand, a machine is designed to transform forces, displacements, or energy. The members of a machine may move substantially.

In this chapter, we consider machines in static equilibrium states, so that we may solve the forces with Newton's equations of static equilibrium.

Section 4.1

Press

4.1-1 Introduction

[1] In this section, we consider a press machine [2-9]. The press is used to emboss a seal [5]. Knowing that $P = 250$ N, we want to find the vertical component of the force exerted on the seal and the reaction forces at A.

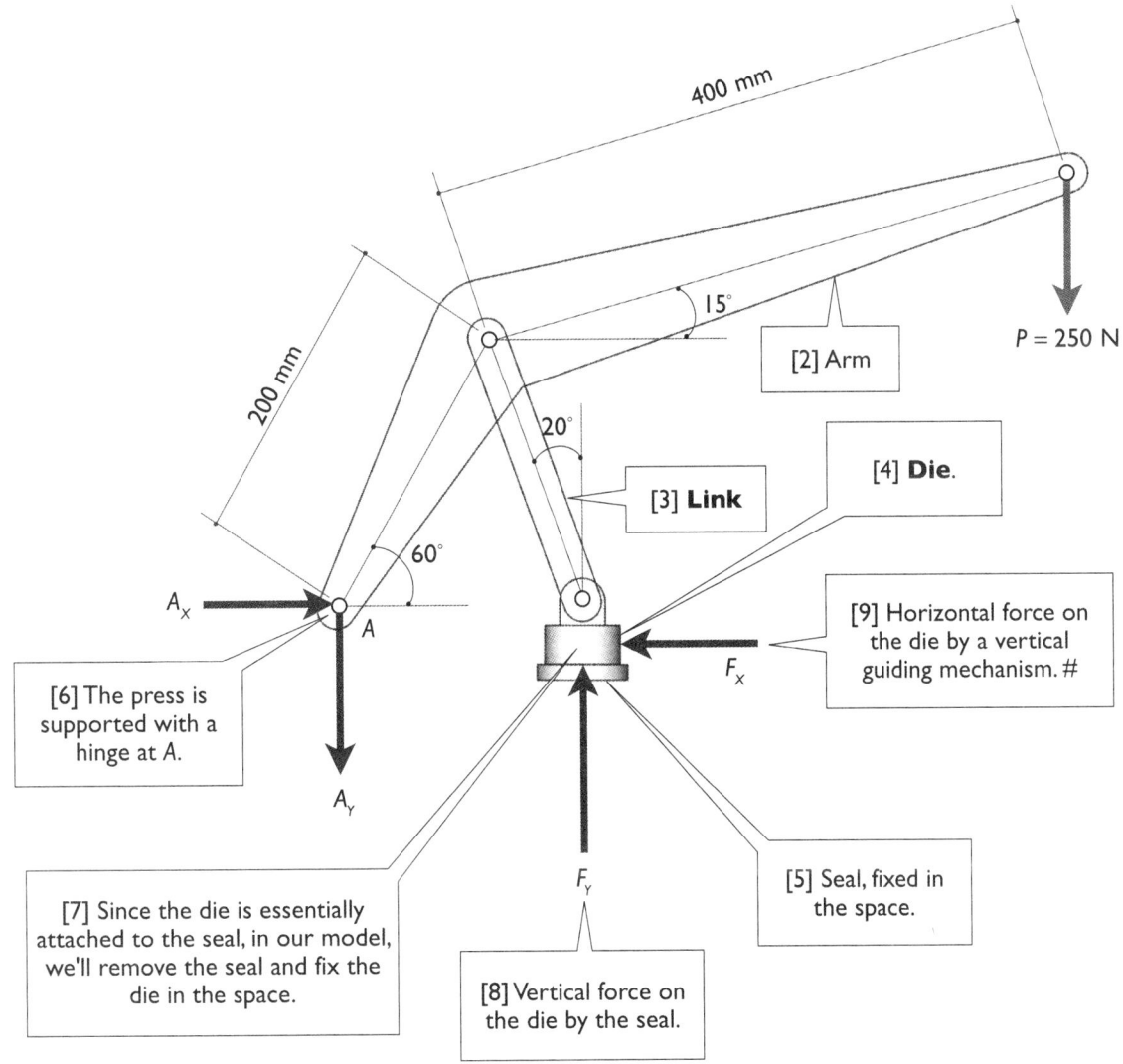

400 mm

200 mm

15°

[2] Arm

$P = 250$ N

20°

[4] **Die**.

[3] **Link**

[9] Horizontal force on the die by a vertical guiding mechanism. #

60°

A_X

A

F_X

[6] The press is supported with a hinge at A.

A_Y

[7] Since the die is essentially attached to the seal, in our model, we'll remove the seal and fix the die in the space.

F_Y

[5] Seal, fixed in the space.

[8] Vertical force on the die by the seal.

4.1-2 Start Up and Create a Part: **Arm**

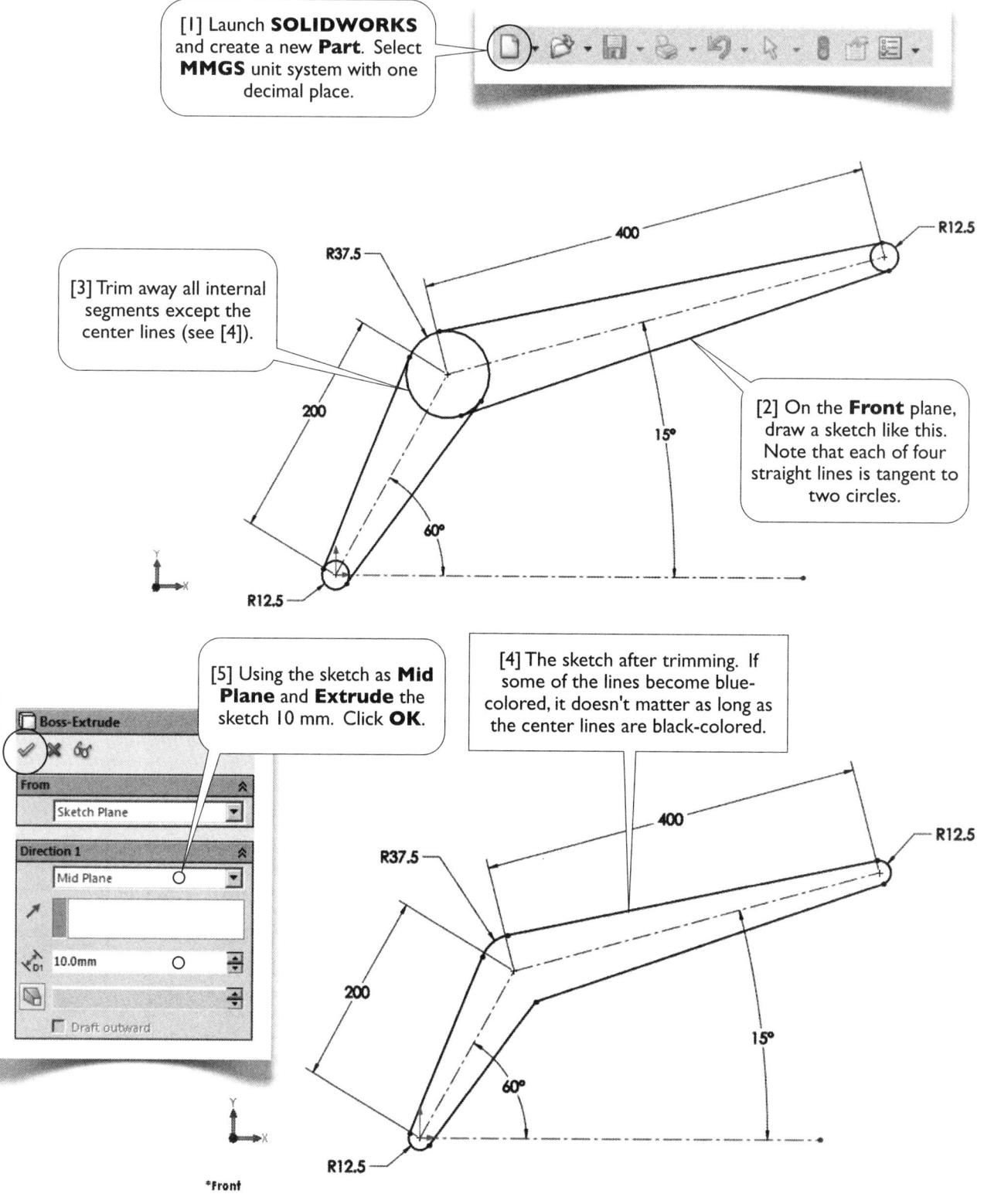

[1] Launch **SOLIDWORKS** and create a new **Part**. Select **MMGS** unit system with one decimal place.

[3] Trim away all internal segments except the center lines (see [4]).

[2] On the **Front** plane, draw a sketch like this. Note that each of four straight lines is tangent to two circles.

R37.5

R12.5

400

200

15°

60°

R12.5

[5] Using the sketch as **Mid Plane** and **Extrude** the sketch 10 mm. Click **OK**.

[4] The sketch after trimming. If some of the lines become blue-colored, it doesn't matter as long as the center lines are black-colored.

Boss-Extrude

From
Sketch Plane

Direction 1
Mid Plane

10.0mm

Draft outward

R37.5

R12.5

400

200

15°

60°

R12.5

*Front

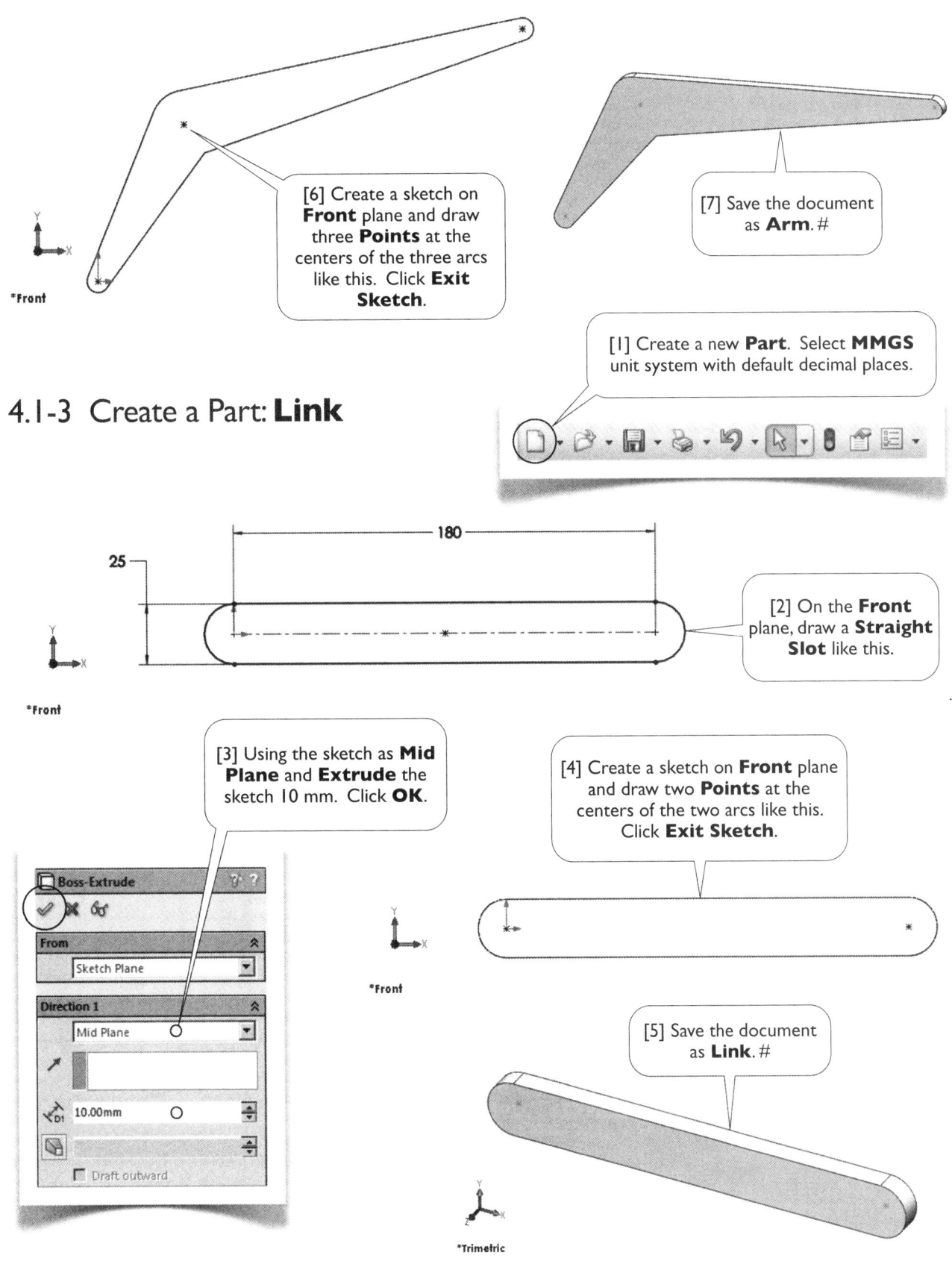

[6] Create a sketch on **Front** plane and draw three **Points** at the centers of the three arcs like this. Click **Exit Sketch**.

*Front

[7] Save the document as **Arm**. #

[1] Create a new **Part**. Select **MMGS** unit system with default decimal places.

4.1-3 Create a Part: **Link**

180

25

*Front

[2] On the **Front** plane, draw a **Straight Slot** like this.

[3] Using the sketch as **Mid Plane** and **Extrude** the sketch 10 mm. Click **OK**.

[4] Create a sketch on **Front** plane and draw two **Points** at the centers of the two arcs like this. Click **Exit Sketch**.

*Front

Boss-Extrude

From

Sketch Plane

Direction 1

Mid Plane

10.00mm

Draft outward

[5] Save the document as **Link**. #

*Trimetric

4.1-4 Create a Part: **Die**

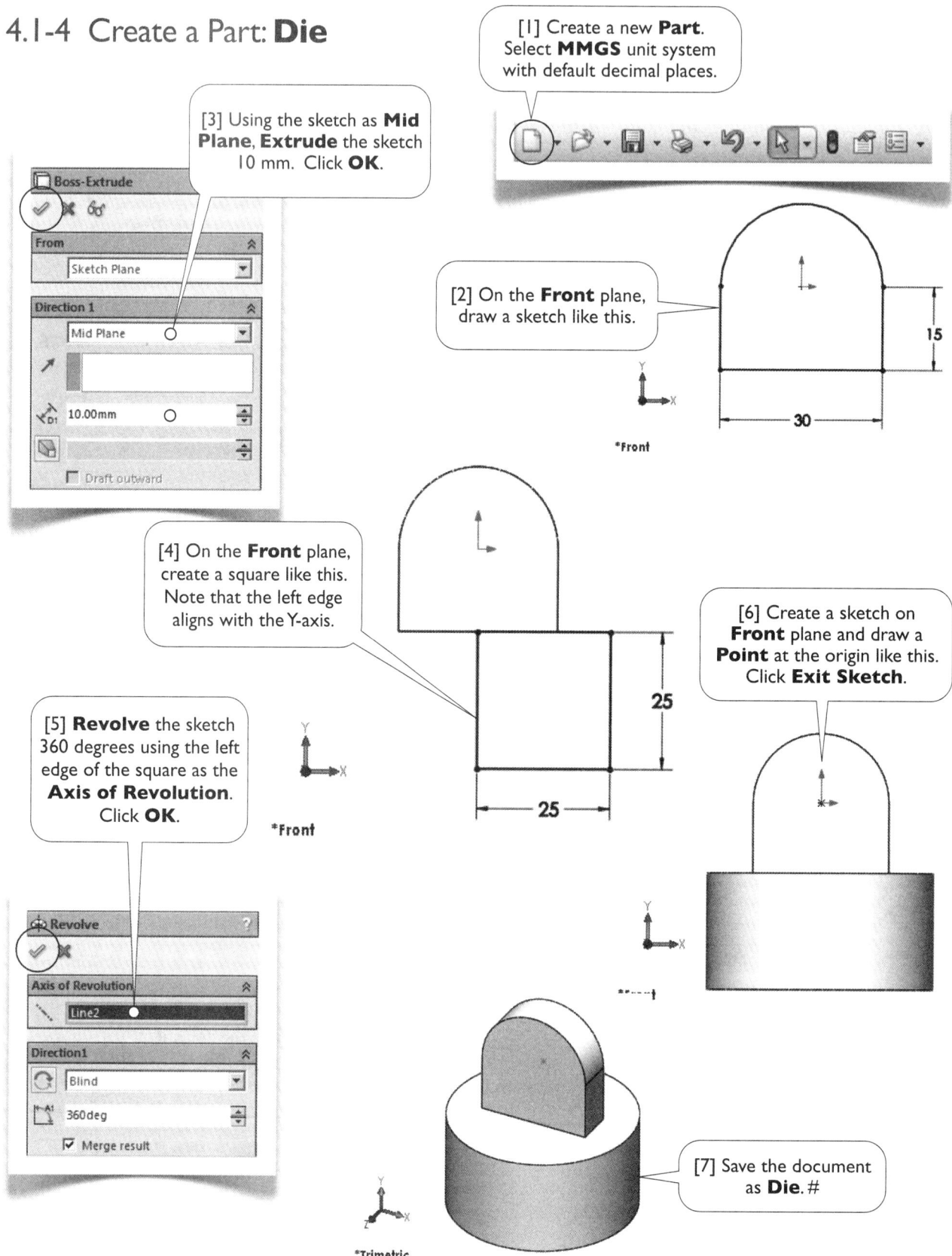

[1] Create a new **Part**. Select **MMGS** unit system with default decimal places.

[2] On the **Front** plane, draw a sketch like this.

*Front

[3] Using the sketch as **Mid Plane**, **Extrude** the sketch 10 mm. Click **OK**.

Boss-Extrude

From
Sketch Plane

Direction 1
Mid Plane

10.00mm

Draft outward

[4] On the **Front** plane, create a square like this. Note that the left edge aligns with the Y-axis.

*Front

[5] **Revolve** the sketch 360 degrees using the left edge of the square as the **Axis of Revolution**. Click **OK**.

Revolve

Axis of Revolution
Line2

Direction1
Blind

360deg

Merge result

[6] Create a sketch on **Front** plane and draw a **Point** at the origin like this. Click **Exit Sketch**.

[7] Save the document as **Die**. #

*Trimetric

4.1-5 Create an Assembly: **Press**

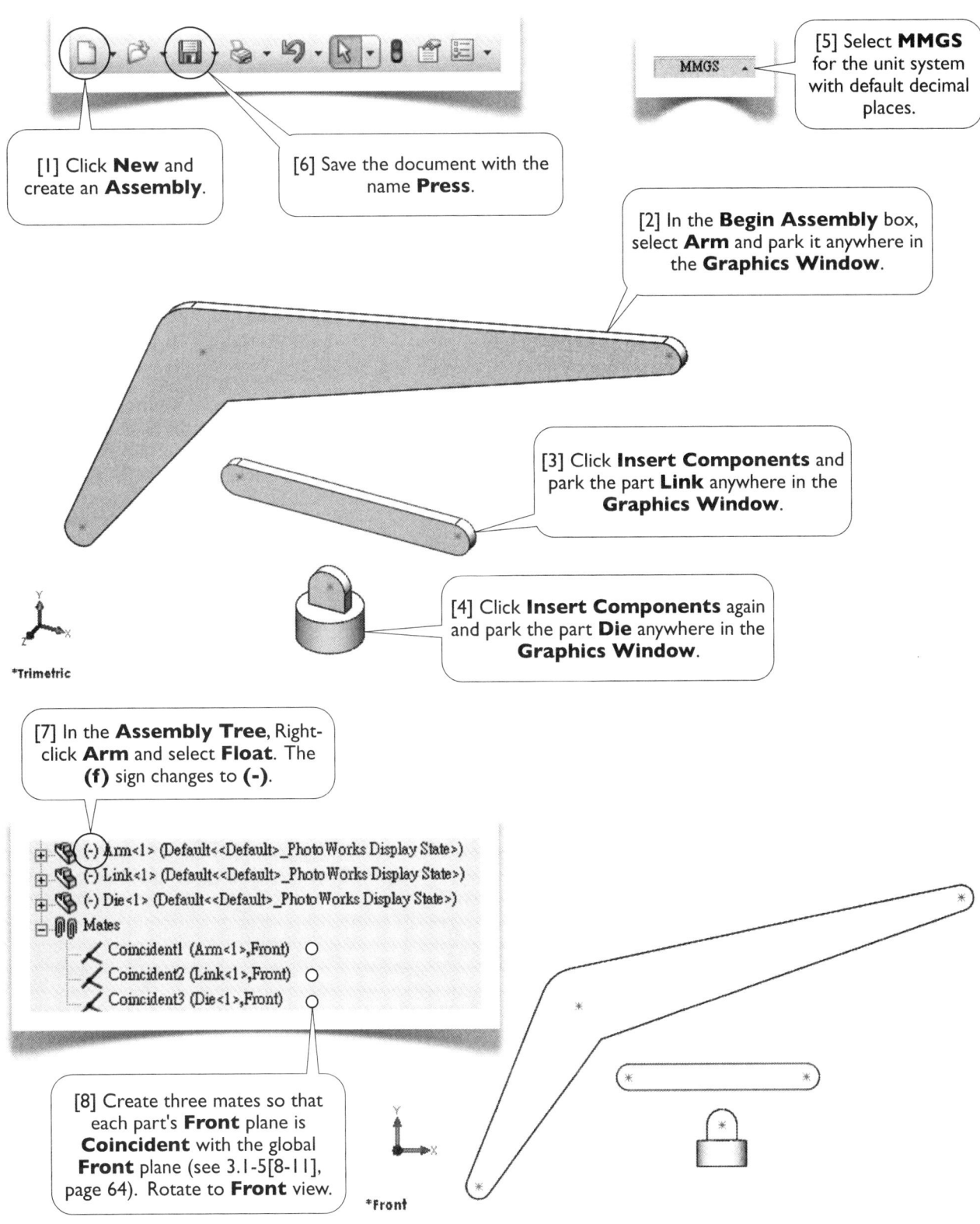

[5] Select **MMGS** for the unit system with default decimal places.

[1] Click **New** and create an **Assembly**.

[6] Save the document with the name **Press**.

[2] In the **Begin Assembly** box, select **Arm** and park it anywhere in the **Graphics Window**.

[3] Click **Insert Components** and park the part **Link** anywhere in the **Graphics Window**.

[4] Click **Insert Components** again and park the part **Die** anywhere in the **Graphics Window**.

*Trimetric

[7] In the **Assembly Tree**, Right-click **Arm** and select **Float**. The **(f)** sign changes to **(-)**.

(-) Arm<1> (Default<<Default>_PhotoWorks Display State>)
(-) Link<1> (Default<<Default>_PhotoWorks Display State>)
(-) Die<1> (Default<<Default>_PhotoWorks Display State>)
Mates
 Coincident1 (Arm<1>,Front) ○
 Coincident2 (Link<1>,Front) ○
 Coincident3 (Die<1>,Front) ○

[8] Create three mates so that each part's **Front** plane is **Coincident** with the global **Front** plane (see 3.1-5[8-11], page 64). Rotate to **Front** view.

*Front

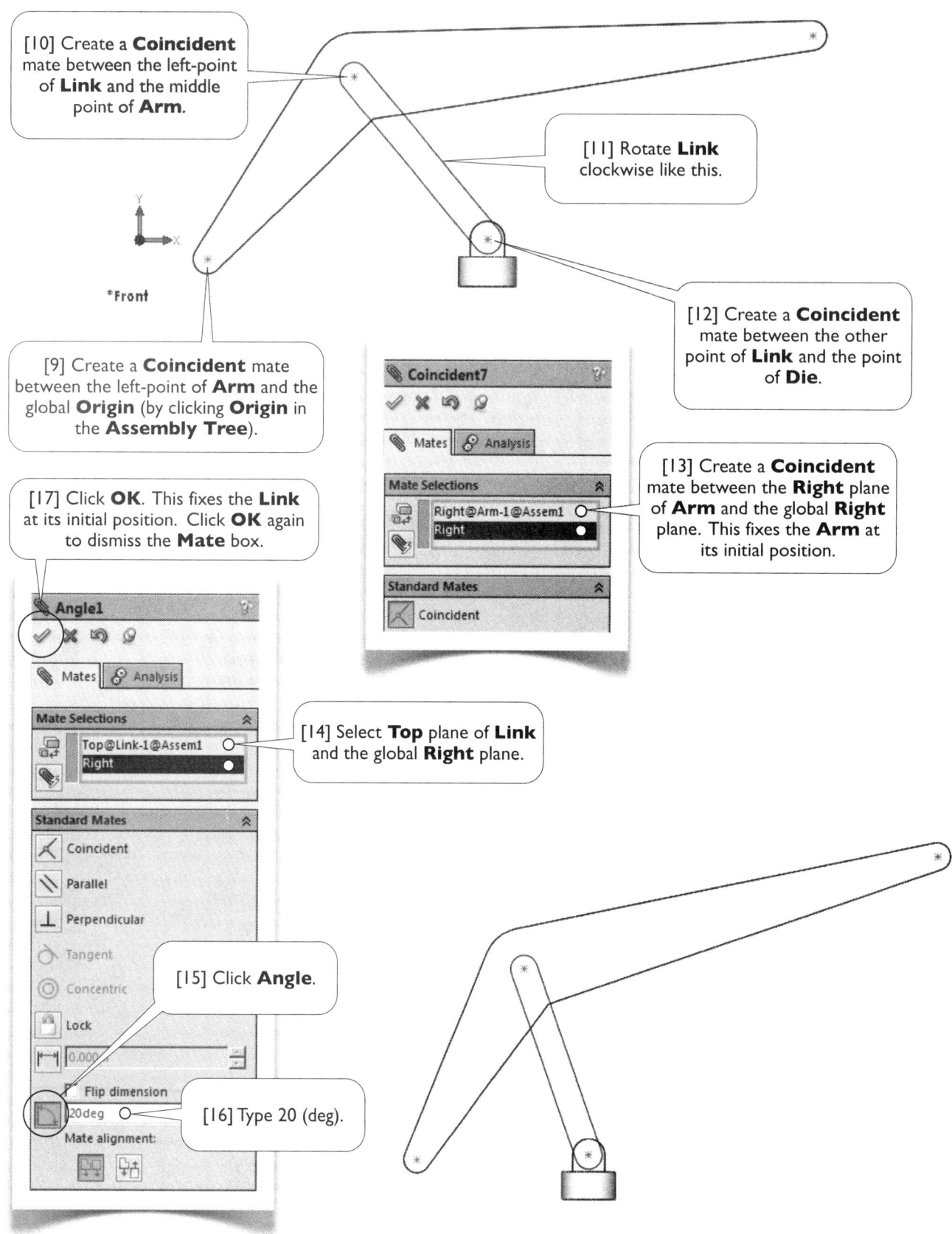

[10] Create a **Coincident** mate between the left-point of **Link** and the middle point of **Arm**.

[11] Rotate **Link** clockwise like this.

*Front

[12] Create a **Coincident** mate between the other point of **Link** and the point of **Die**.

[9] Create a **Coincident** mate between the left-point of **Arm** and the global **Origin** (by clicking **Origin** in the **Assembly Tree**).

[17] Click **OK**. This fixes the **Link** at its initial position. Click **OK** again to dismiss the **Mate** box.

Coincident7

Mates | Analysis

Mate Selections

Right@Arm-1@Assem1
Right

Standard Mates

Coincident

[13] Create a **Coincident** mate between the **Right** plane of **Arm** and the global **Right** plane. This fixes the **Arm** at its initial position.

Angle1

Mates | Analysis

Mate Selections

Top@Link-1@Assem1
Right

[14] Select **Top** plane of **Link** and the global **Right** plane.

Standard Mates

Coincident
Parallel
Perpendicular
Tangent
Concentric
Lock

0.000

Flip dimension

20deg

Mate alignment:

[15] Click **Angle**.

[16] Type 20 (deg).

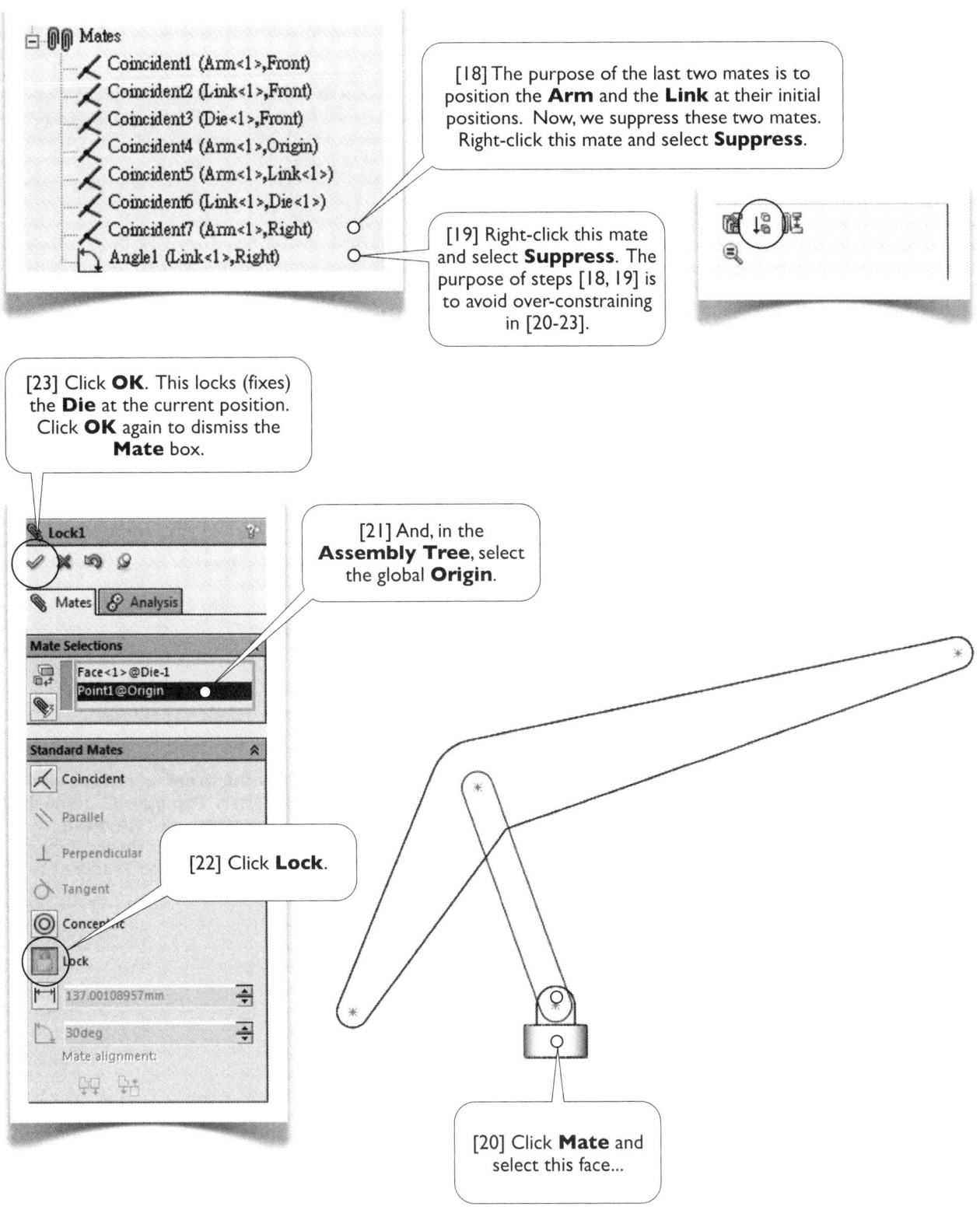

Mates
Coincident1 (Arm<1>,Front)
Coincident2 (Link<1>,Front)
Coincident3 (Die<1>,Front)
Coincident4 (Arm<1>,Origin)
Coincident5 (Arm<1>,Link<1>)
Coincident6 (Link<1>,Die<1>)
Coincident7 (Arm<1>,Right)
Angle1 (Link<1>,Right)

[18] The purpose of the last two mates is to position the **Arm** and the **Link** at their initial positions. Now, we suppress these two mates. Right-click this mate and select **Suppress**.

[19] Right-click this mate and select **Suppress**. The purpose of steps [18, 19] is to avoid over-constraining in [20-23].

[23] Click **OK**. This locks (fixes) the **Die** at the current position. Click **OK** again to dismiss the **Mate** box.

Lock1

Mates | Analysis

Mate Selections
Face<1>@Die-1
Point1@Origin

[21] And, in the **Assembly Tree**, select the global **Origin**.

Standard Mates
Coincident
Parallel
Perpendicular
Tangent
Concentric
Lock
137.00108957mm
30deg
Mate alignment:

[22] Click **Lock**.

[20] Click **Mate** and select this face...

4.1-6 Create a **Study**

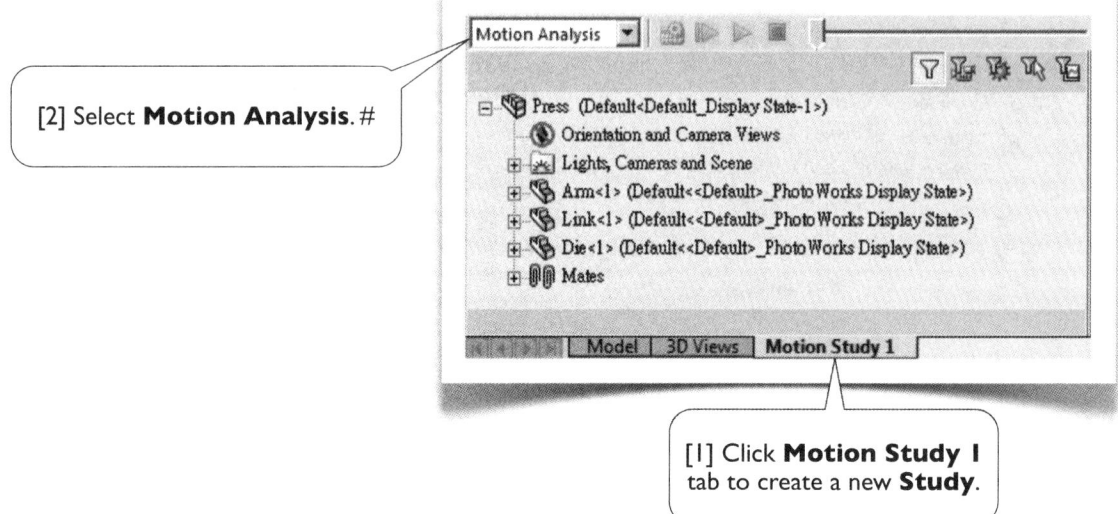

[2] Select **Motion Analysis**. #

[1] Click **Motion Study 1** tab to create a new **Study**.

4.1-7 Set Up Forces and Calculate Results

[1] In **Motion** toolbar, click **Force**.

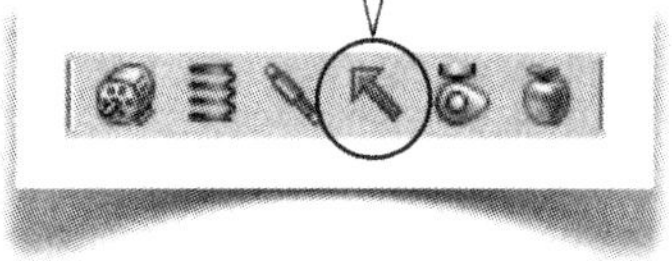

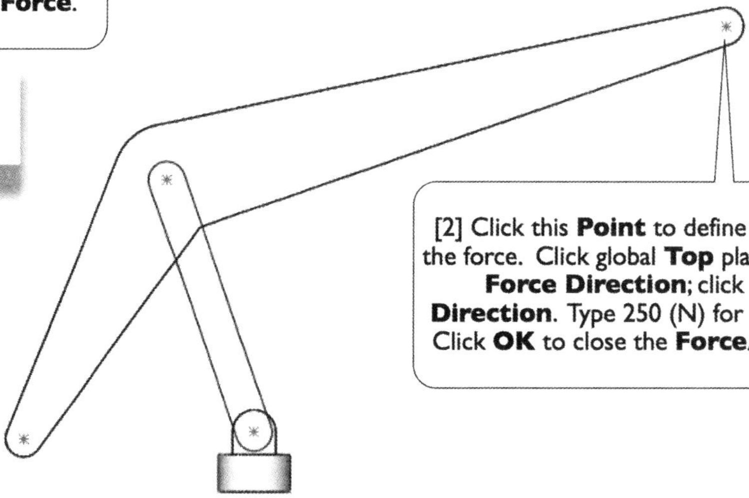

[2] Click this **Point** to define the location of the force. Click global **Top** plane to define the **Force Direction**; click **Reverse Direction**. Type 250 (N) for the force value. Click **OK** to close the **Force/Torque** box.

[3] Click **Save**.

[4] Click **Calculate**. #

4.1-8 Retrieve the Results

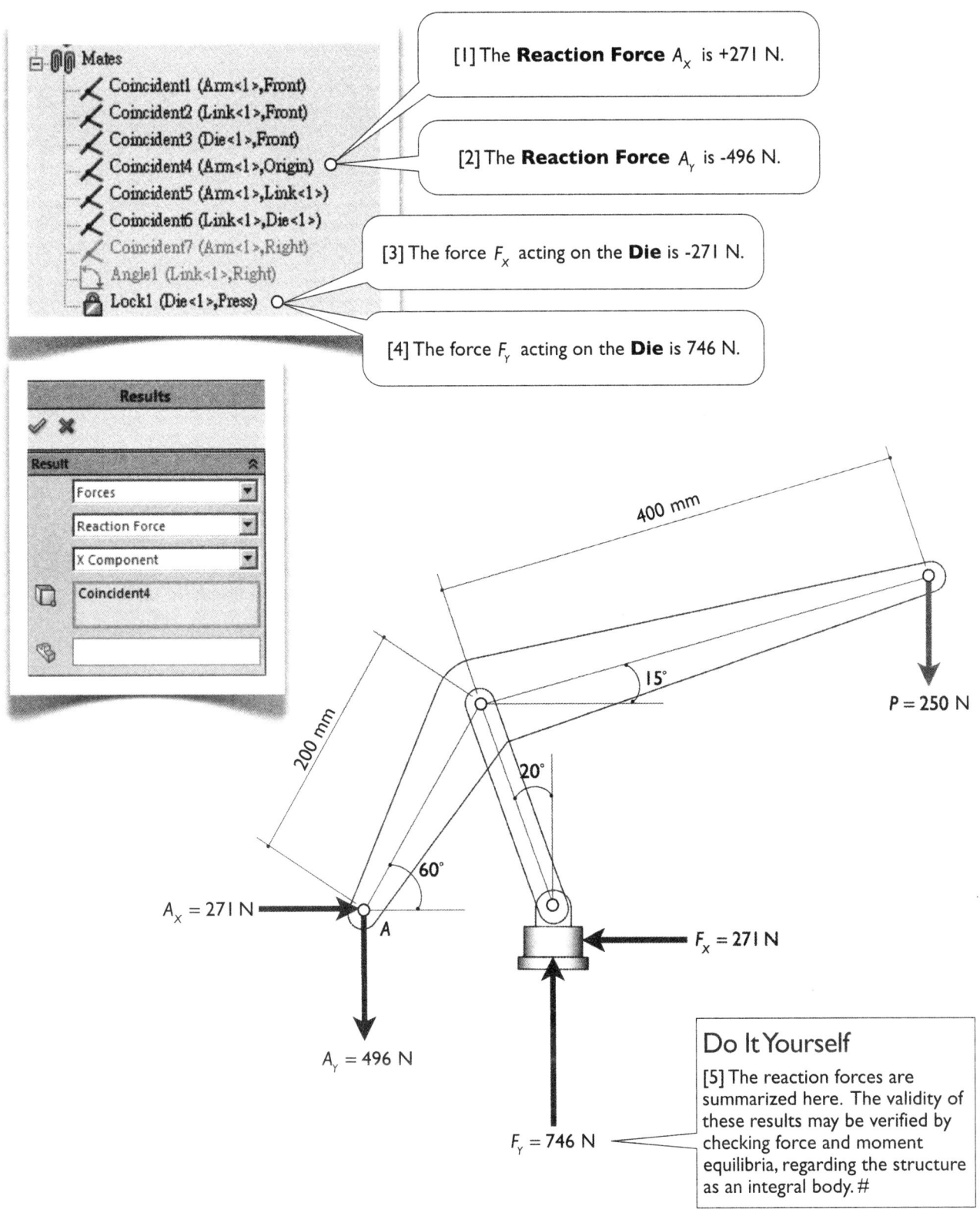

[1] The **Reaction Force** A_x is +271 N.

[2] The **Reaction Force** A_y is -496 N.

[3] The force F_x acting on the **Die** is -271 N.

[4] The force F_y acting on the **Die** is 746 N.

Mates
- Coincident1 (Arm<1>,Front)
- Coincident2 (Link<1>,Front)
- Coincident3 (Die<1>,Front)
- Coincident4 (Arm<1>,Origin)
- Coincident5 (Arm<1>,Link<1>)
- Coincident6 (Link<1>,Die<1>)
- Coincident7 (Arm<1>,Right)
- Angle1 (Link<1>,Right)
- Lock1 (Die<1>,Press)

Results

Result
- Forces
- Reaction Force
- X Component
- Coincident4

400 mm

15°

$P = 250$ N

200 mm

20°

60°

$A_x = 271$ N

A

$F_x = 271$ N

$A_y = 496$ N

$F_y = 746$ N

Do It Yourself

[5] The reaction forces are summarized here. The validity of these results may be verified by checking force and moment equilibria, regarding the structure as an integral body. #

4.1-9 Do It Yourself: Forces on Each Members

Do It Yourself

[1] Retrieve forces acting on each member. Draw free body diagrams for each member. Verify the results by checking the force and moment equilibria, taking each member as a free body.

[2] Quit **SOLIDWORKS**. Click **Save all.** Click **Rebuild and save the document**. #

Section 4.2

Cutter

4.2-1 Introduction

[1] In this section, we consider a bolt cutter [2-6]. In using the bolt cutter, a worker applies two 300-N forces to the handles [7]. We want to find the magnitude of the forces exerted by the cutter on the bolt [8].

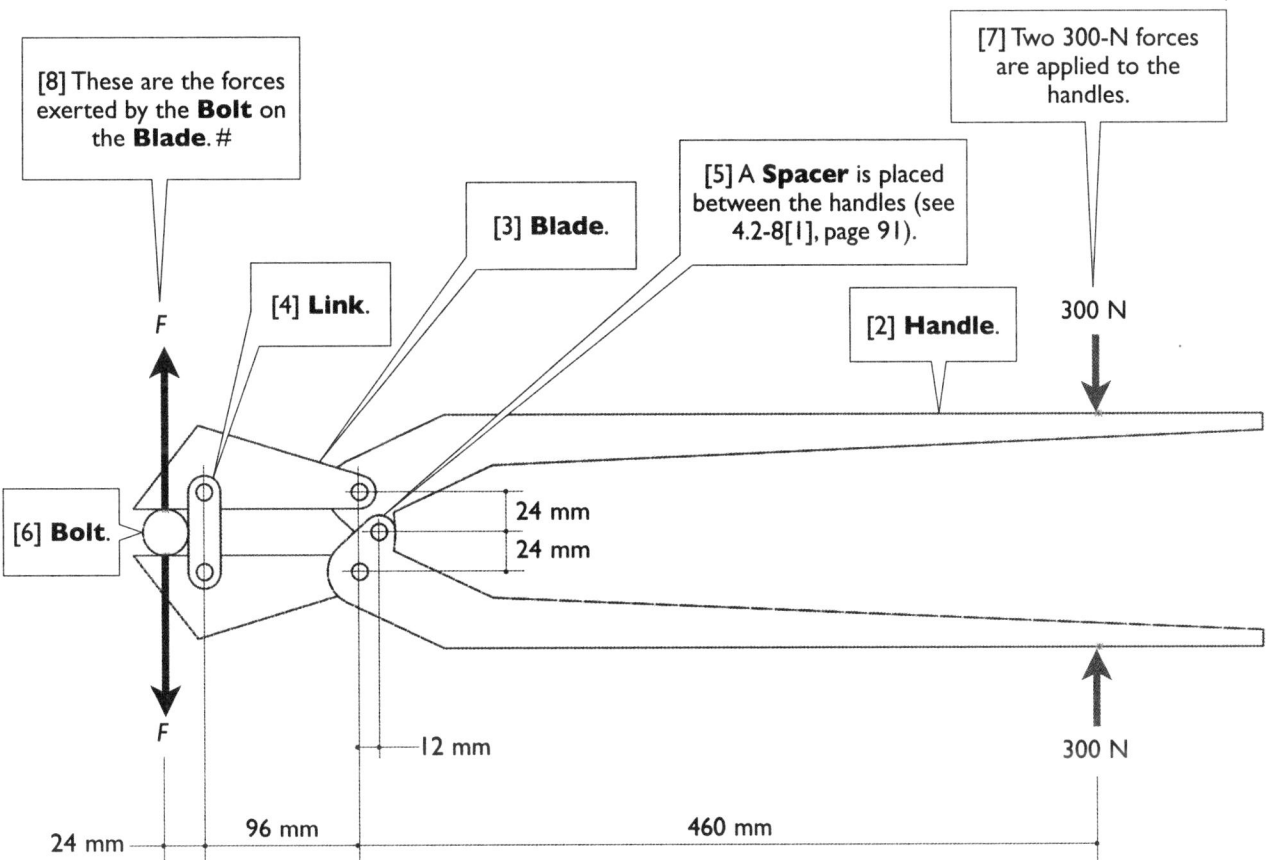

[7] Two 300-N forces are applied to the handles.

[8] These are the forces exerted by the **Bolt** on the **Blade**. #

[5] A **Spacer** is placed between the handles (see 4.2-8[1], page 91).

[3] **Blade**.

[4] **Link**.

[2] **Handle**.

[6] **Bolt**.

300 N

300 N

F

F

24 mm

24 mm

12 mm

24 mm

96 mm

460 mm

4.2-2 Start Up and Create a Part: **Handle**

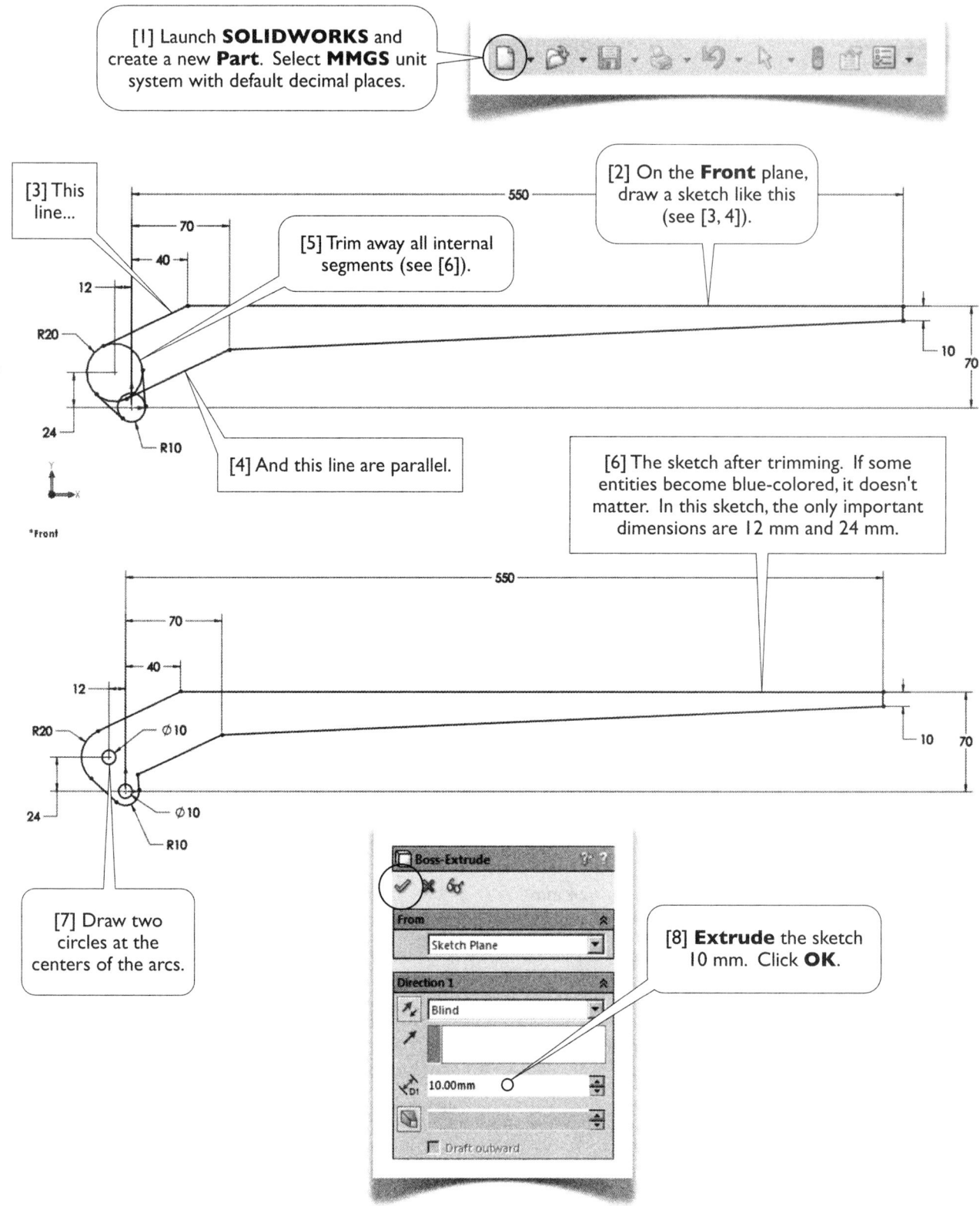

[1] Launch **SOLIDWORKS** and create a new **Part**. Select **MMGS** unit system with default decimal places.

[2] On the **Front** plane, draw a sketch like this (see [3, 4]).

[3] This line...

[5] Trim away all internal segments (see [6]).

R20

12

24   R10

[4] And this line are parallel.

*Front

[6] The sketch after trimming. If some entities become blue-colored, it doesn't matter. In this sketch, the only important dimensions are 12 mm and 24 mm.

550

70

40

12

R20   Ø10

24   Ø10

R10

[7] Draw two circles at the centers of the arcs.

Boss-Extrude

From
Sketch Plane

Direction 1
Blind

10.00mm

Draft outward

[8] **Extrude** the sketch 10 mm. Click **OK**.

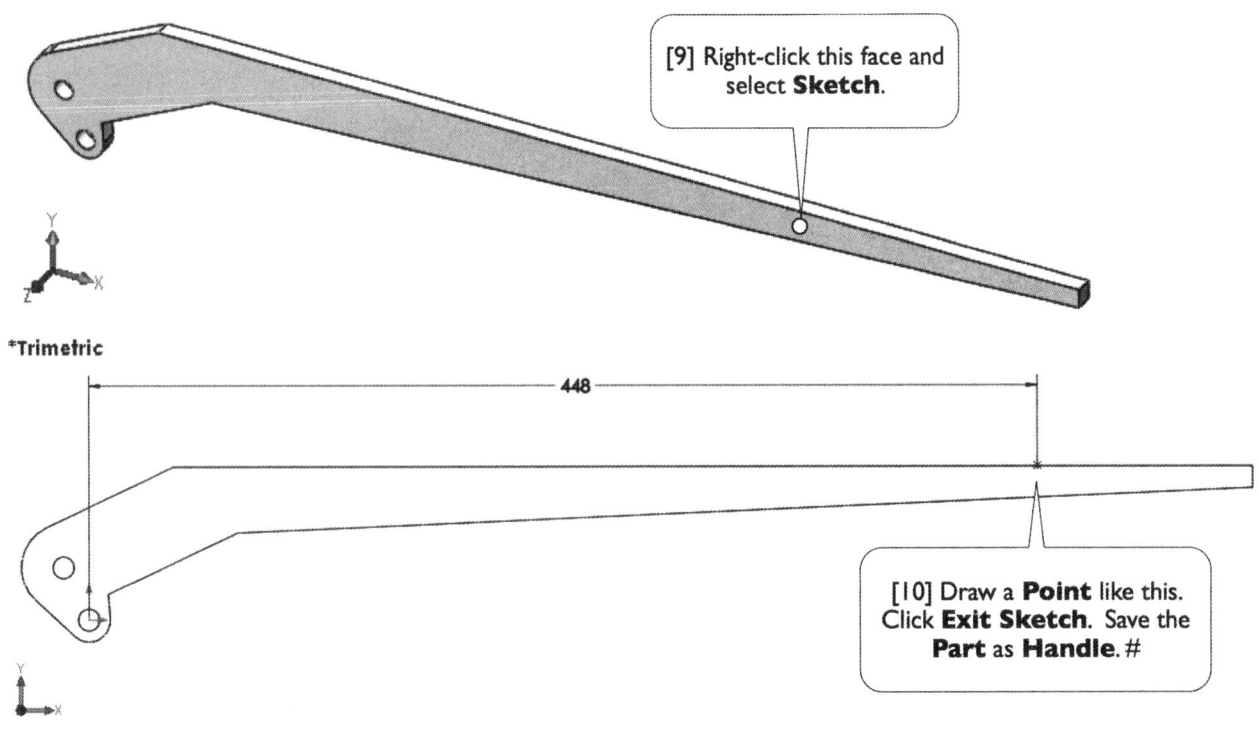

*Trimetric

[9] Right-click this face and select **Sketch**.

[10] Draw a **Point** like this. Click **Exit Sketch**. Save the **Part** as **Handle**. #

4.2-3 Create a Part: **Blade**

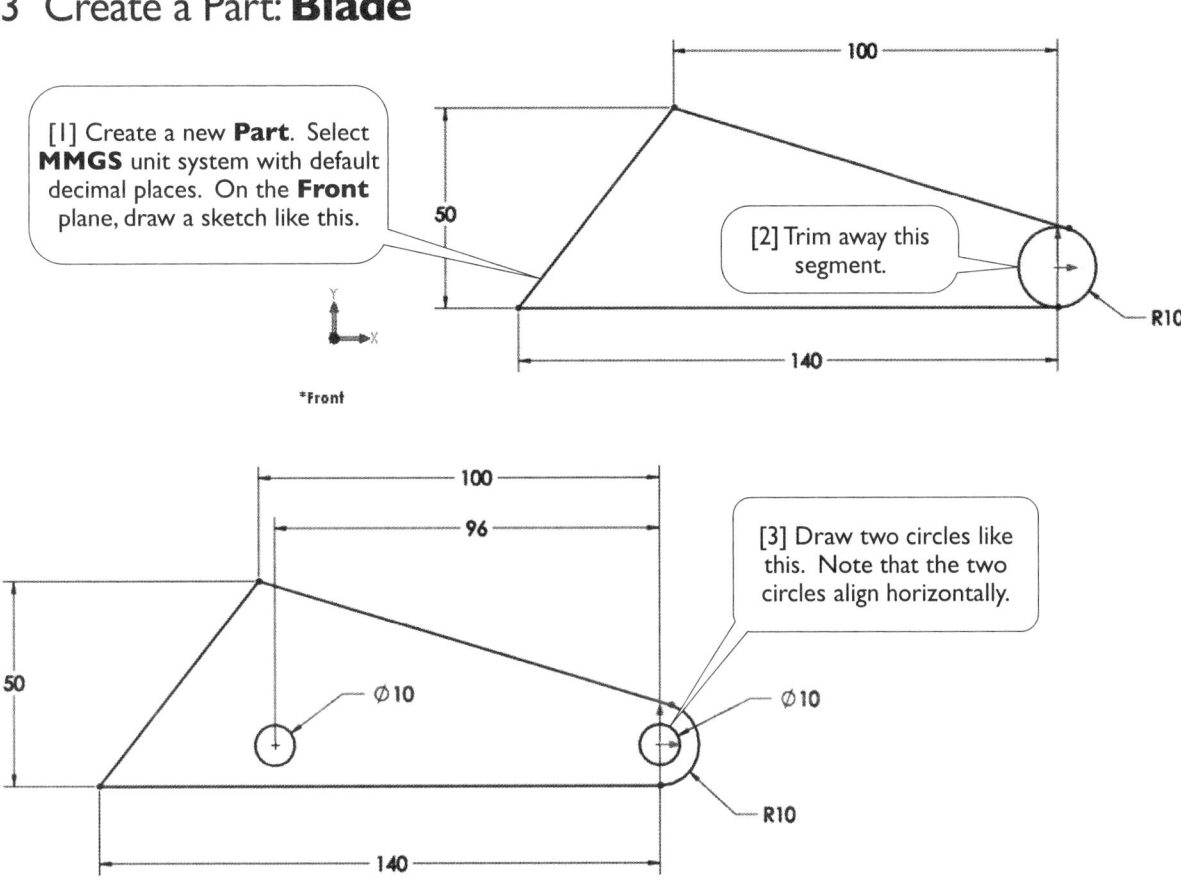

[1] Create a new **Part**. Select **MMGS** unit system with default decimal places. On the **Front** plane, draw a sketch like this.

*Front

[2] Trim away this segment.

[3] Draw two circles like this. Note that the two circles align horizontally.

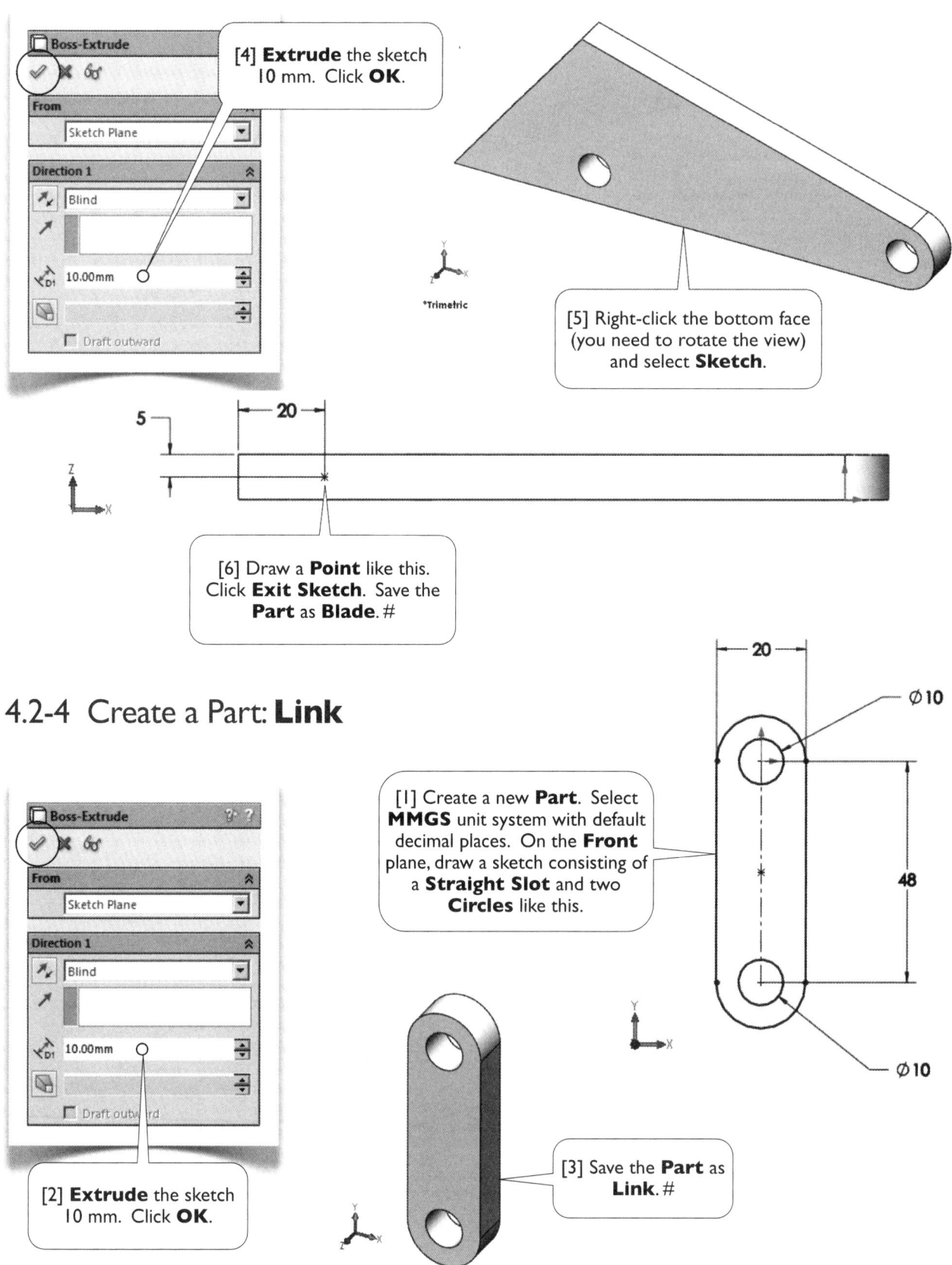

[4] **Extrude** the sketch 10 mm. Click **OK**.

*Trimetric

[5] Right-click the bottom face (you need to rotate the view) and select **Sketch**.

5

20

Z

[6] Draw a **Point** like this. Click **Exit Sketch**. Save the **Part** as **Blade**. #

4.2-4 Create a Part: **Link**

[1] Create a new **Part**. Select **MMGS** unit system with default decimal places. On the **Front** plane, draw a sketch consisting of a **Straight Slot** and two **Circles** like this.

20

Ø10

48

Ø10

[2] **Extrude** the sketch 10 mm. Click **OK**.

[3] Save the **Part** as **Link**. #

*Trimetric

4.2-5 Create a Part: **Spacer**

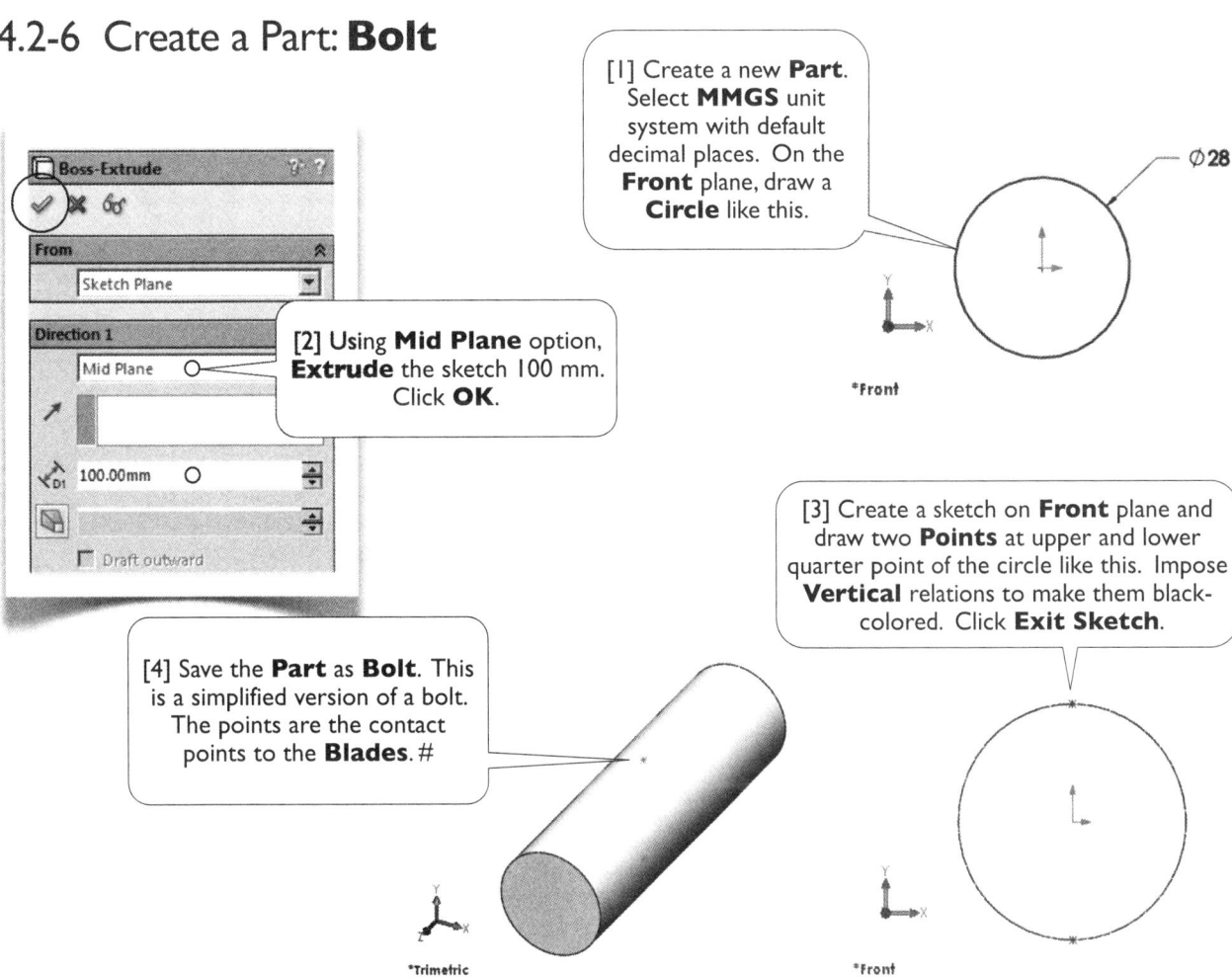

[1] Create a new **Part**. Select **MMGS** unit system with default decimal places. On the **Front** plane, draw a sketch consisting of two **Circles** like this.

Ø20

Ø10

*Front

[2] Using **Mid Plane** option, **Extrude** the sketch 10 mm. Click **OK**.

[3] Save the **Part** as **Spacer**. #

*Trimetric

4.2-6 Create a Part: **Bolt**

[1] Create a new **Part**. Select **MMGS** unit system with default decimal places. On the **Front** plane, draw a **Circle** like this.

Ø28

*Front

[2] Using **Mid Plane** option, **Extrude** the sketch 100 mm. Click **OK**.

[3] Create a sketch on **Front** plane and draw two **Points** at upper and lower quarter point of the circle like this. Impose **Vertical** relations to make them black-colored. Click **Exit Sketch**.

[4] Save the **Part** as **Bolt**. This is a simplified version of a bolt. The points are the contact points to the **Blades**. #

*Trimetric

*Front

4.2-7 Create an Assembly: **Cutter**

[2] In the **Head-Up** toolbar, turn on **View Origins**.

[4] Turn off **View Origins**.

[1] Click **New** and create an **Assembly**.

[13] Save the document with the name **Cutter**. #

[5] Click **Insert Components** and park the part **Handle** anywhere in the **Graphics Window**.

Handle<1>

[3] In the **Begin Assembly** box, select **Spacer** and click the global **Origin**. It is fixed at the **Origin**.

Handle<2>

Blade<1>

[7] Click **Insert Components** again and park the part **Blade** anywhere in the **Graphics Window**.

Blade<2>

Spacer<1>

[6] Duplicate the **Handle**.

[8] Duplicate the **Blade**.

Link<1>

[9] Click **Insert Components** again and park the part **Link** anywhere in the **Graphics Window**.

Link<2>

[12] Select **MMGS** for the unit system with default decimal places.

Bolt<1>

[10] Duplicate the **Link**.

*Trimetric

[11] Click **Insert Components** again and park the part **Bolt** anywhere in the **Graphics Window**.

MMGS

4.2-8 Assemble the **Parts**

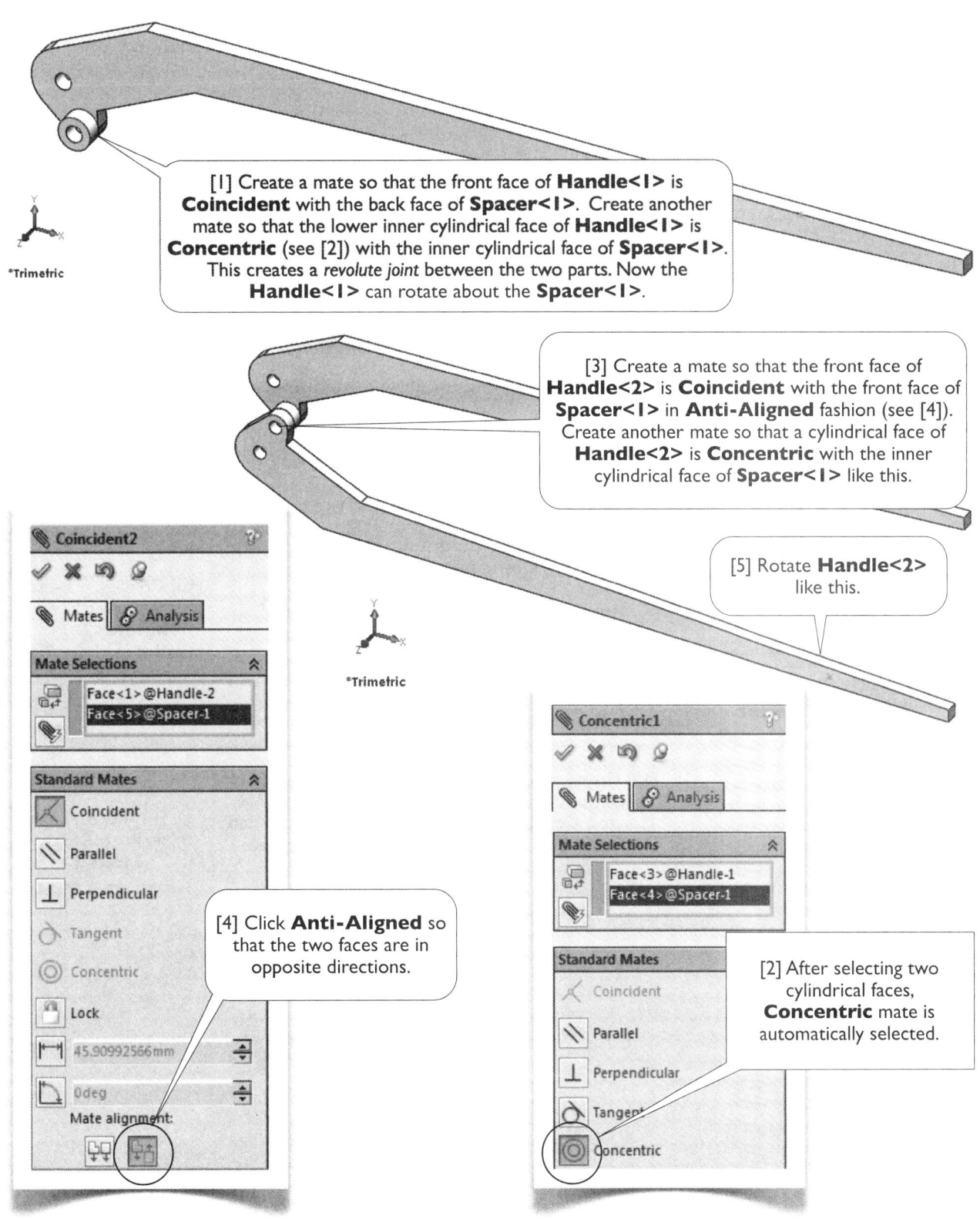

[1] Create a mate so that the front face of **Handle<1>** is **Coincident** with the back face of **Spacer<1>**. Create another mate so that the lower inner cylindrical face of **Handle<1>** is **Concentric** (see [2]) with the inner cylindrical face of **Spacer<1>**. This creates a *revolute joint* between the two parts. Now the **Handle<1>** can rotate about the **Spacer<1>**.

*Trimetric

[3] Create a mate so that the front face of **Handle<2>** is **Coincident** with the front face of **Spacer<1>** in **Anti-Aligned** fashion (see [4]). Create another mate so that a cylindrical face of **Handle<2>** is **Concentric** with the inner cylindrical face of **Spacer<1>** like this.

[5] Rotate **Handle<2>** like this.

*Trimetric

Coincident2

Mates   Analysis

Mate Selections
Face<1>@Handle-2
Face<5>@Spacer-1

Standard Mates
Coincident
Parallel
Perpendicular
Tangent
Concentric
Lock
45.90992566mm
0deg
Mate alignment:

[4] Click **Anti-Aligned** so that the two faces are in opposite directions.

Concentric1

Mates   Analysis

Mate Selections
Face<3>@Handle-1
Face<4>@Spacer-1

Standard Mates
Coincident
Parallel
Perpendicular
Tangent
Concentric

[2] After selecting two cylindrical faces, **Concentric** mate is automatically selected.

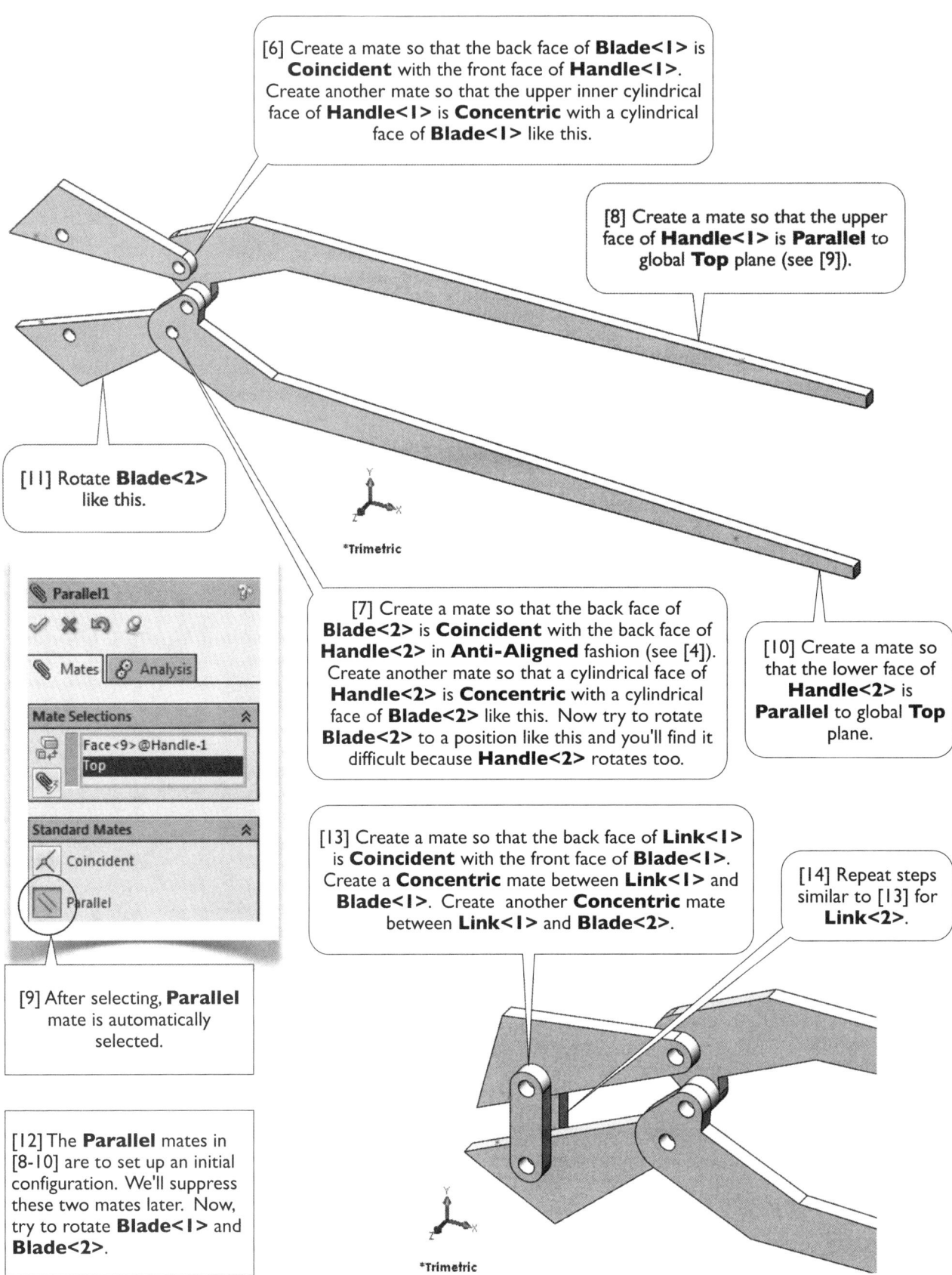

[6] Create a mate so that the back face of **Blade<1>** is **Coincident** with the front face of **Handle<1>**. Create another mate so that the upper inner cylindrical face of **Handle<1>** is **Concentric** with a cylindrical face of **Blade<1>** like this.

[8] Create a mate so that the upper face of **Handle<1>** is **Parallel** to global **Top** plane (see [9]).

[11] Rotate **Blade<2>** like this.

*Trimetric

Parallel1

Mates | Analysis

Mate Selections

Face<9>@Handle-1
Top

Standard Mates

Coincident

Parallel

[7] Create a mate so that the back face of **Blade<2>** is **Coincident** with the back face of **Handle<2>** in **Anti-Aligned** fashion (see [4]). Create another mate so that a cylindrical face of **Handle<2>** is **Concentric** with a cylindrical face of **Blade<2>** like this. Now try to rotate **Blade<2>** to a position like this and you'll find it difficult because **Handle<2>** rotates too.

[10] Create a mate so that the lower face of **Handle<2>** is **Parallel** to global **Top** plane.

[9] After selecting, **Parallel** mate is automatically selected.

[13] Create a mate so that the back face of **Link<1>** is **Coincident** with the front face of **Blade<1>**. Create a **Concentric** mate between **Link<1>** and **Blade<1>**. Create another **Concentric** mate between **Link<1>** and **Blade<2>**.

[14] Repeat steps similar to [13] for **Link<2>**.

[12] The **Parallel** mates in [8-10] are to set up an initial configuration. We'll suppress these two mates later. Now, try to rotate **Blade<1>** and **Blade<2>**.

*Trimetric

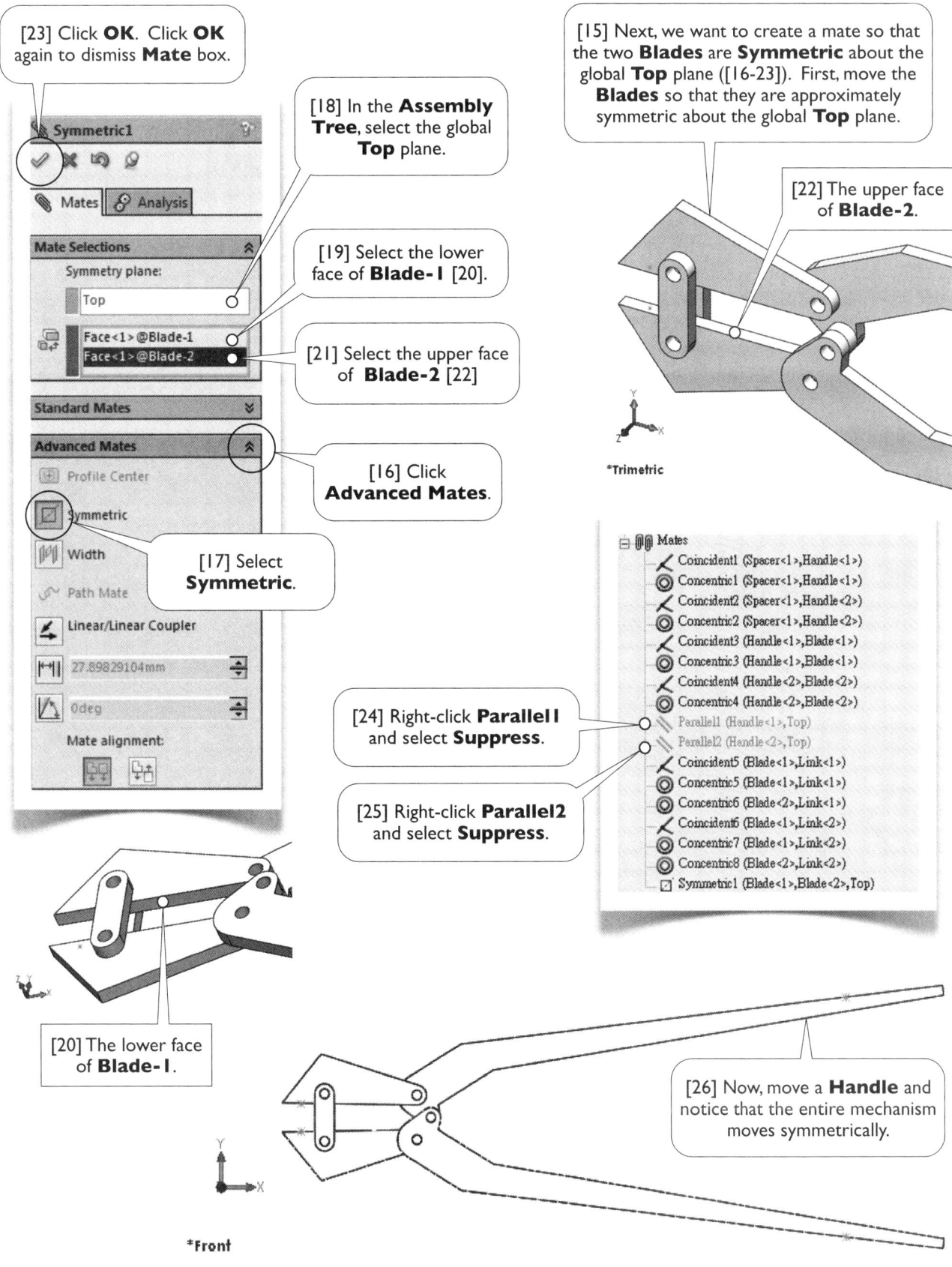

[23] Click **OK**. Click **OK** again to dismiss **Mate** box.

[15] Next, we want to create a mate so that the two **Blades** are **Symmetric** about the global **Top** plane ([16-23]). First, move the **Blades** so that they are approximately symmetric about the global **Top** plane.

[18] In the **Assembly Tree**, select the global **Top** plane.

[22] The upper face of **Blade-2**.

Symmetric1

Mates   Analysis

Mate Selections

Symmetry plane:

Top

[19] Select the lower face of **Blade-1** [20].

Face<1>@Blade-1
Face<1>@Blade-2

[21] Select the upper face of **Blade-2** [22]

Standard Mates

Advanced Mates

Profile Center

[16] Click **Advanced Mates**.

Symmetric

Width

[17] Select **Symmetric**.

Path Mate

Linear/Linear Coupler

27.89829104mm

0deg

Mate alignment:

[24] Right-click **Parallel1** and select **Suppress**.

[25] Right-click **Parallel2** and select **Suppress**.

Mates
Coincident1 (Spacer<1>,Handle<1>)
Concentric1 (Spacer<1>,Handle<1>)
Coincident2 (Spacer<1>,Handle<2>)
Concentric2 (Spacer<1>,Handle<2>)
Coincident3 (Handle<1>,Blade<1>)
Concentric3 (Handle<1>,Blade<1>)
Coincident4 (Handle<2>,Blade<2>)
Concentric4 (Handle<2>,Blade<2>)
Parallel1 (Handle<1>,Top)
Parallel2 (Handle<2>,Top)
Coincident5 (Blade<1>,Link<1>)
Concentric5 (Blade<1>,Link<1>)
Concentric6 (Blade<2>,Link<1>)
Coincident6 (Blade<1>,Link<2>)
Concentric7 (Blade<1>,Link<2>)
Concentric8 (Blade<2>,Link<2>)
Symmetric1 (Blade<1>,Blade<2>,Top)

[20] The lower face of **Blade-1**.

[26] Now, move a **Handle** and notice that the entire mechanism moves symmetrically.

*Trimetric

*Front

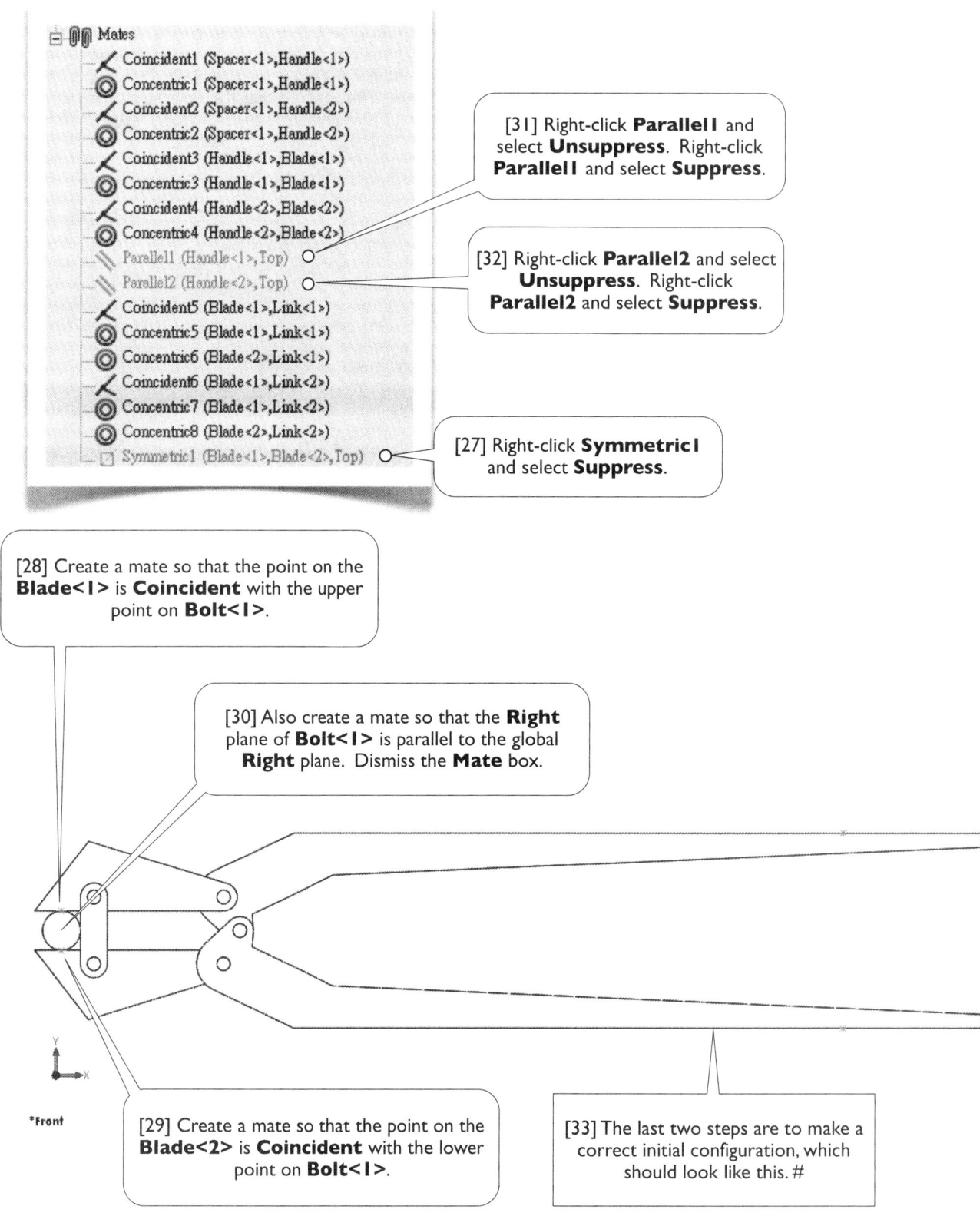

Mates
- Coincident1 (Spacer<1>,Handle<1>)
- Concentric1 (Spacer<1>,Handle<1>)
- Coincident2 (Spacer<1>,Handle<2>)
- Concentric2 (Spacer<1>,Handle<2>)
- Coincident3 (Handle<1>,Blade<1>)
- Concentric3 (Handle<1>,Blade<1>)
- Coincident4 (Handle<2>,Blade<2>)
- Concentric4 (Handle<2>,Blade<2>)
- Parallel1 (Handle<1>,Top)
- Parallel2 (Handle<2>,Top)
- Coincident5 (Blade<1>,Link<1>)
- Concentric5 (Blade<1>,Link<1>)
- Concentric6 (Blade<2>,Link<1>)
- Coincident6 (Blade<1>,Link<2>)
- Concentric7 (Blade<1>,Link<2>)
- Concentric8 (Blade<2>,Link<2>)
- Symmetric1 (Blade<1>,Blade<2>,Top)

[31] Right-click **Parallel1** and select **Unsuppress**. Right-click **Parallel1** and select **Suppress**.

[32] Right-click **Parallel2** and select **Unsuppress**. Right-click **Parallel2** and select **Suppress**.

[27] Right-click **Symmetric1** and select **Suppress**.

[28] Create a mate so that the point on the **Blade<1>** is **Coincident** with the upper point on **Bolt<1>**.

[30] Also create a mate so that the **Right** plane of **Bolt<1>** is parallel to the global **Right** plane. Dismiss the **Mate** box.

[29] Create a mate so that the point on the **Blade<2>** is **Coincident** with the lower point on **Bolt<1>**.

[33] The last two steps are to make a correct initial configuration, which should look like this. #

*Front

4.2-9 Create a **Study**

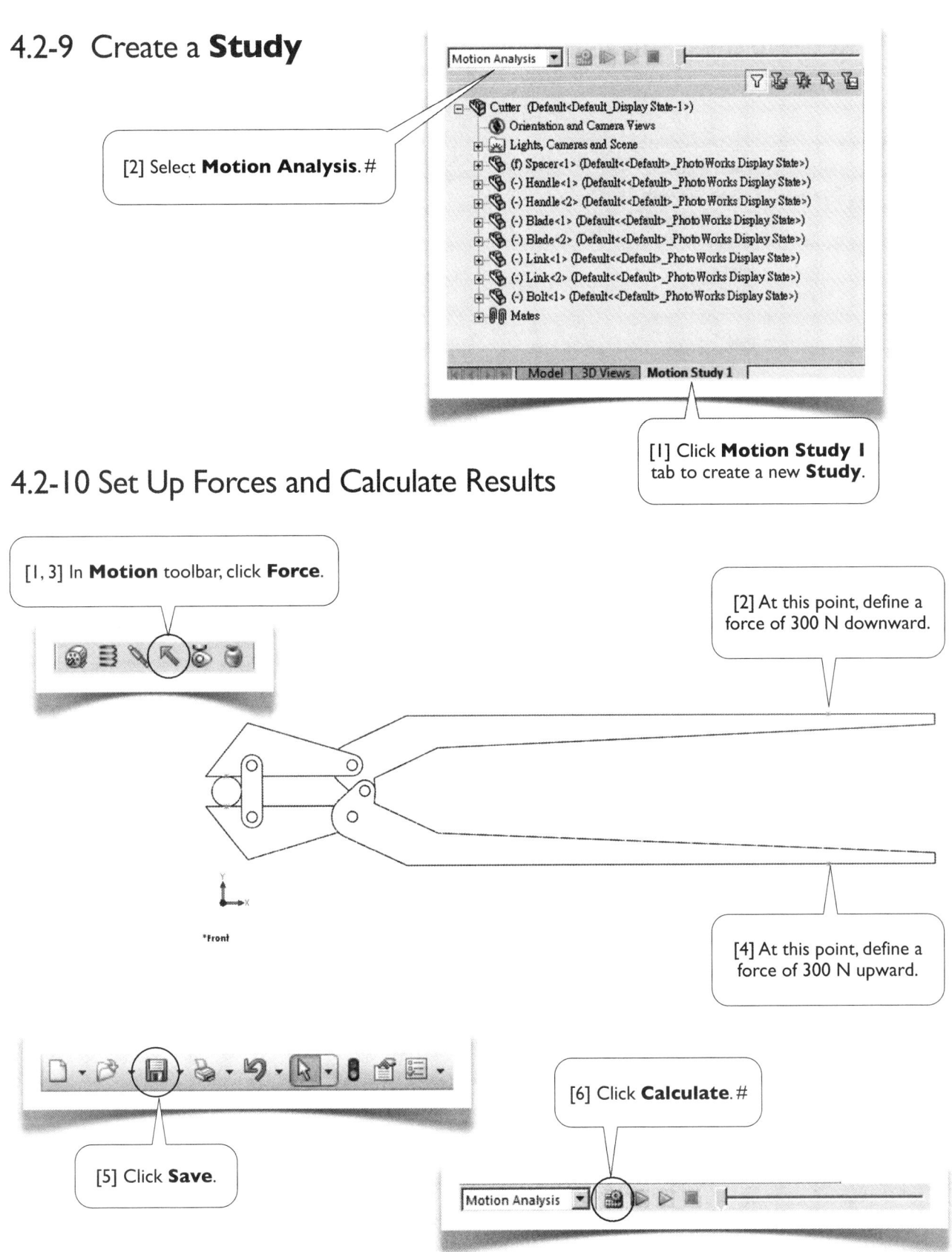

[2] Select **Motion Analysis**. #

[1] Click **Motion Study 1** tab to create a new **Study**.

4.2-10 Set Up Forces and Calculate Results

[1, 3] In **Motion** toolbar, click **Force**.

[2] At this point, define a force of 300 N downward.

[4] At this point, define a force of 300 N upward.

*Front

[5] Click **Save**.

[6] Click **Calculate**. #

4.2-11 Retrieve the Results

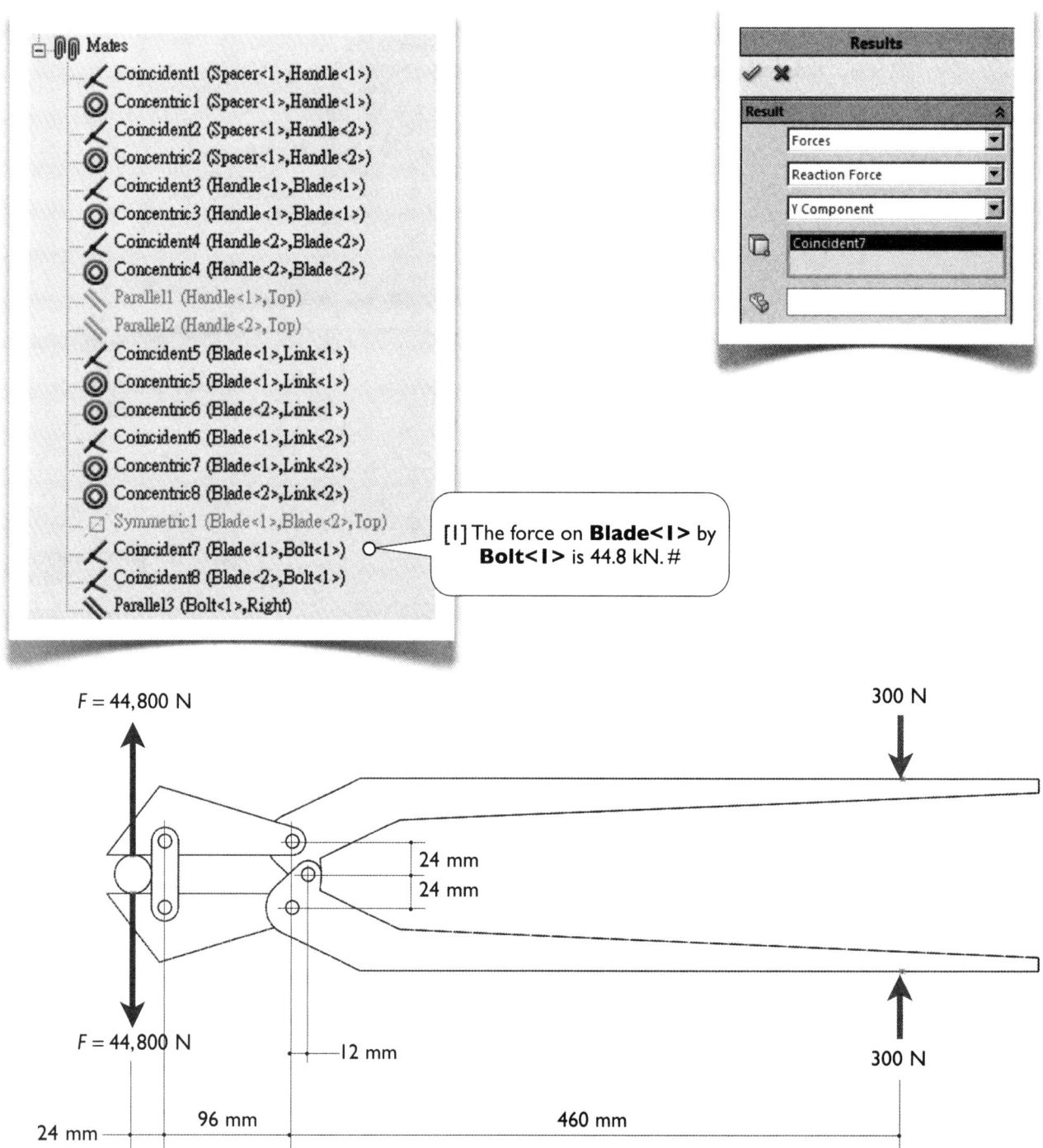

4.2-12 Do It Yourself: Forces on Components

Do It Yourself

[1] Retrieve forces acting on a Handle. Draw a free body diagram. Verify the results by checking the force and moment equilibria.

Do It Yourself

[2] Retrieve forces acting on a blade. Draw a free body diagram. Verify the results by checking the force and moment equilibria.

[3] Quit **SOLIDWORKS**. Click **Save all.** Click **Rebuild and save the document**. #

Chapter 5
Internal Forces in Beams

This chapter discusses the concepts of internal forces acting on cross-sections of a beam. To study the internal forces, we'll construct a beam by "gluing" blocks together. Internal forces then can be obtained by examining the reaction forces between blocks.

The reaction force normal to the cross-section is called the **axial force**, and the reaction force parallel to the cross-section is called the **shear force**. The reaction moment whose direction is normal to the cross-section is called the **torsion**, and the moment whose direction is parallel to the cross-section is called the **bending moment**.

In this chapter, we'll consider beams subject to transversal loads, and study **shear forces** and **bending moments** on cross-sections.

Section 5.1

Cantilever Beam

5.1-1 Introduction

[1] Consider a cantilever beam of length L = 4 m subject to a uniform load of 1000 Pa downward [2-4], and we want to study the internal shear forces and moments at each cross section of the beam [5]. The load per unit length of the beam is

$$w = 1000 \text{ Pa} \times 0.2 \text{ m} = 200 \text{ N/m}$$

The shear is distributed linearly and the maximum shear is (see sign convention in [6])

$$V_{max} = wL = (200 \text{ N/m}) \times (4 \text{ m}) = +800 \text{ N}$$

The moment is distributed quadratically and the maximum moment is (see sign convention in [7])

$$M_{max} = -\frac{wL^2}{2} = -\frac{(200 \text{ N/m})(4 \text{ m})^2}{2} = -1600 \text{ N-m}$$

[5] To study the internal forces and moments of the beam, we'll construct the beam by "gluing" 8 blocks together, each 0.5-m long. Internal forces and moments then can be obtained by examining the reaction forces and moments between blocks.

[4] A uniform load of 1000 Pa downward is applied on the top face of the beam.

[3] The beam has a cross section of 0.2 m x 0.4 m and a length of 4 m.

[2] The cantilever beam is fixed at this end.

[6] Sign convention for the shear force, positive when sheared clockwise; i.e., when the force on the right face is **downward**.

[7] Sign convention for the bending moment, positive when bent concave up; i.e., when the moment on the right face is **counter-clockwise**. #

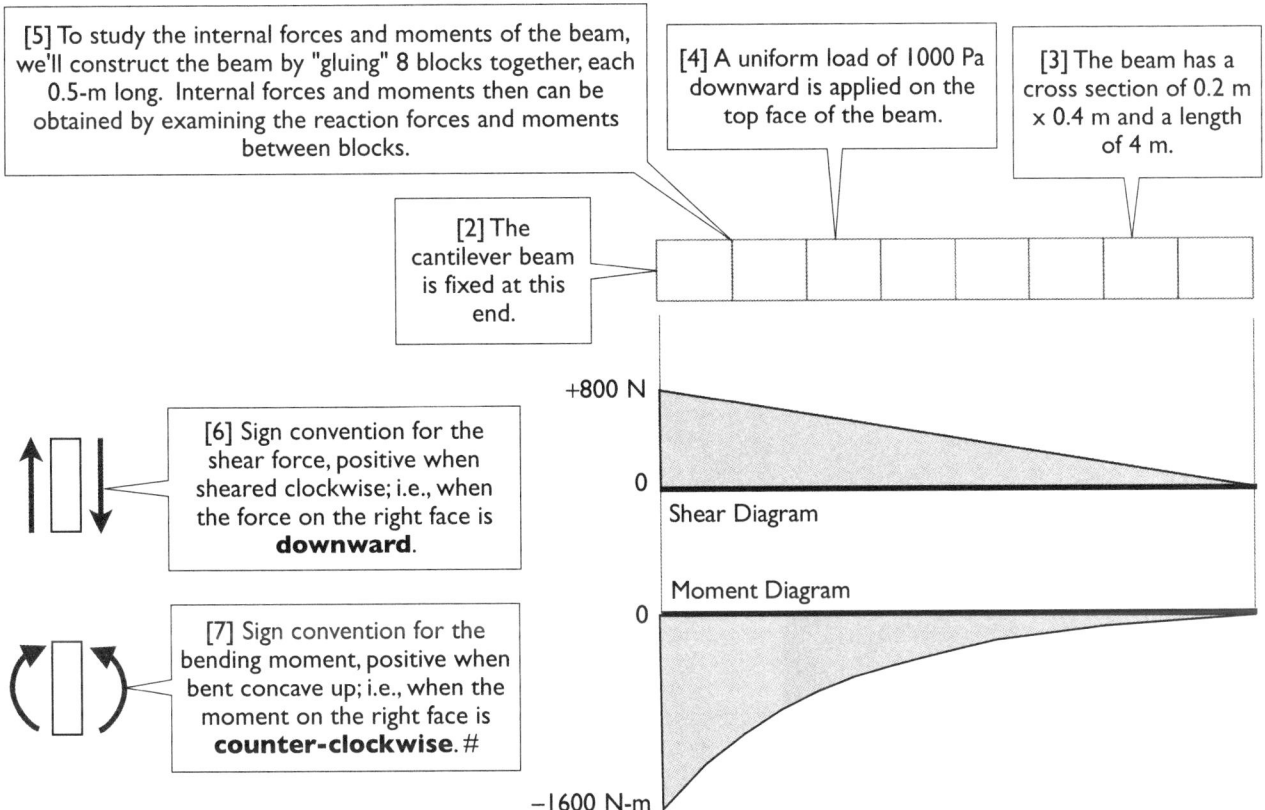

+800 N

0

Shear Diagram

Moment Diagram

0

−1600 N-m

5.1-2 Start Up and Create a Part: **Block**

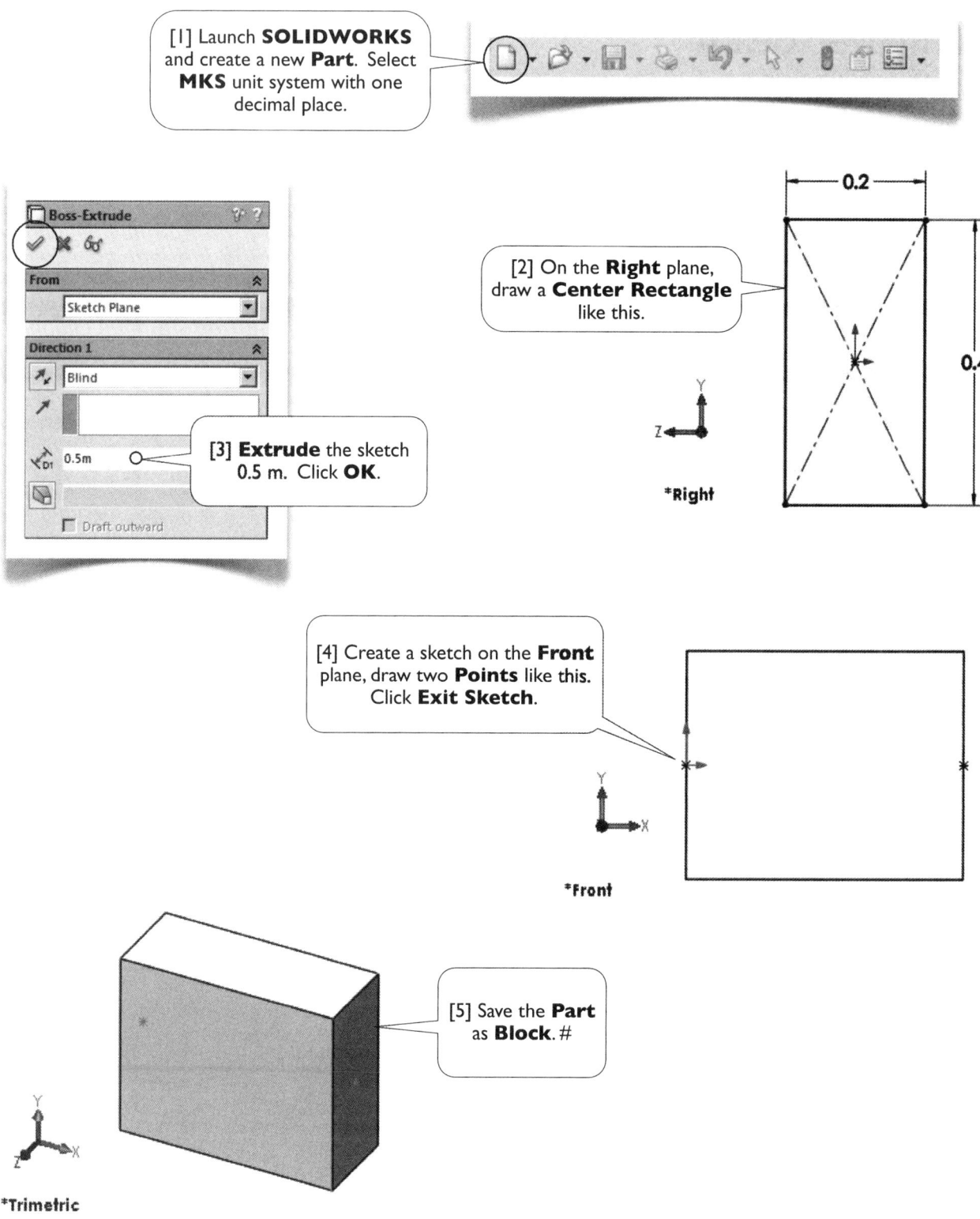

[1] Launch **SOLIDWORKS** and create a new **Part**. Select **MKS** unit system with one decimal place.

[2] On the **Right** plane, draw a **Center Rectangle** like this.

*Right

[3] **Extrude** the sketch 0.5 m. Click **OK**.

[4] Create a sketch on the **Front** plane, draw two **Points** like this. Click **Exit Sketch**.

*Front

[5] Save the **Part** as **Block**. #

*Trimetric

5.1-3 Create an Assembly: **Cantilever**

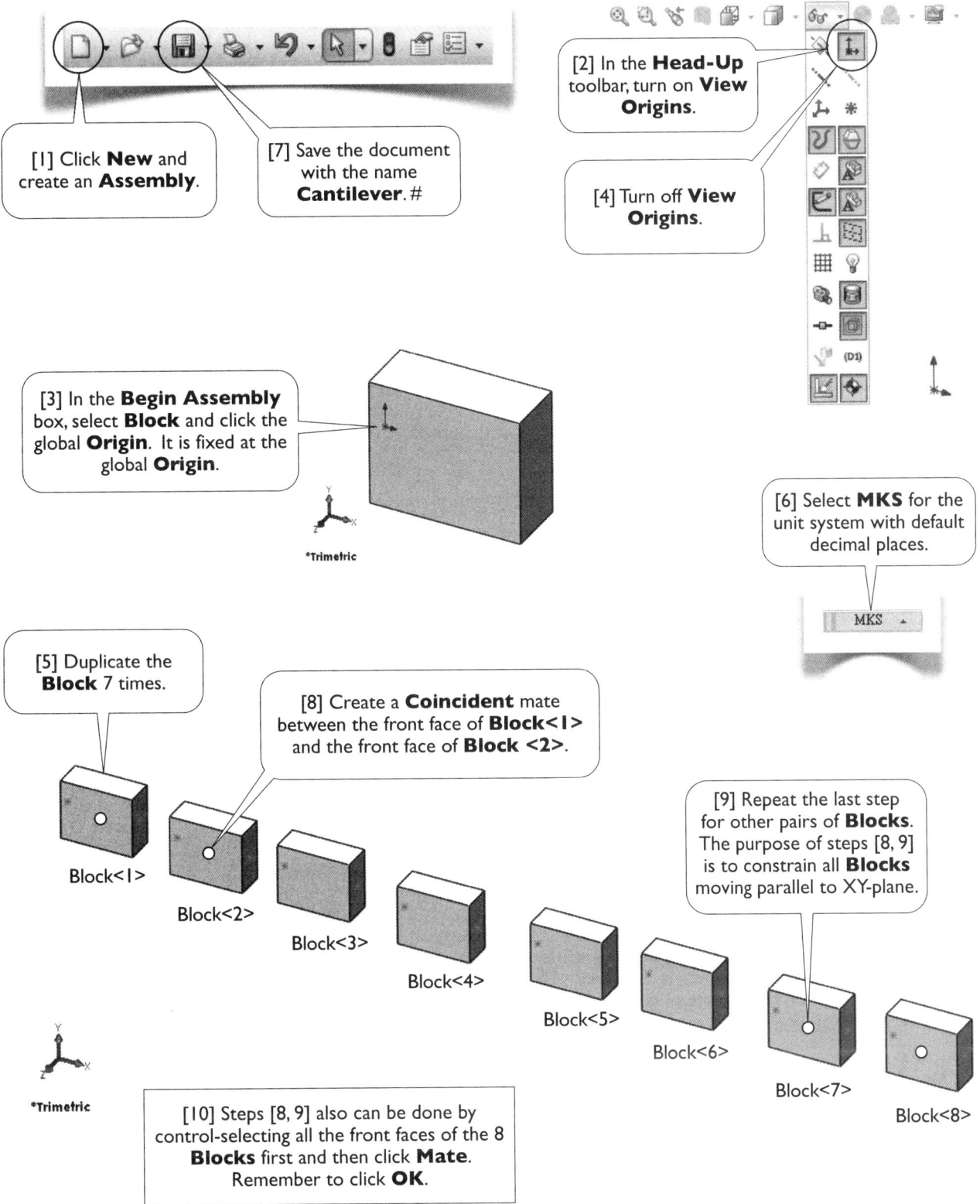

[2] In the **Head-Up** toolbar, turn on **View Origins**.

[1] Click **New** and create an **Assembly**.

[7] Save the document with the name **Cantilever**. #

[4] Turn off **View Origins**.

[3] In the **Begin Assembly** box, select **Block** and click the global **Origin**. It is fixed at the global **Origin**.

*Trimetric

[6] Select **MKS** for the unit system with default decimal places.

MKS

[5] Duplicate the **Block** 7 times.

[8] Create a **Coincident** mate between the front face of **Block<1>** and the front face of **Block <2>**.

[9] Repeat the last step for other pairs of **Blocks**. The purpose of steps [8, 9] is to constrain all **Blocks** moving parallel to XY-plane.

Block<1>

Block<2>

Block<3>

Block<4>

Block<5>

Block<6>

Block<7>

Block<8>

*Trimetric

[10] Steps [8, 9] also can be done by control-selecting all the front faces of the 8 **Blocks** first and then click **Mate**. Remember to click **OK**.

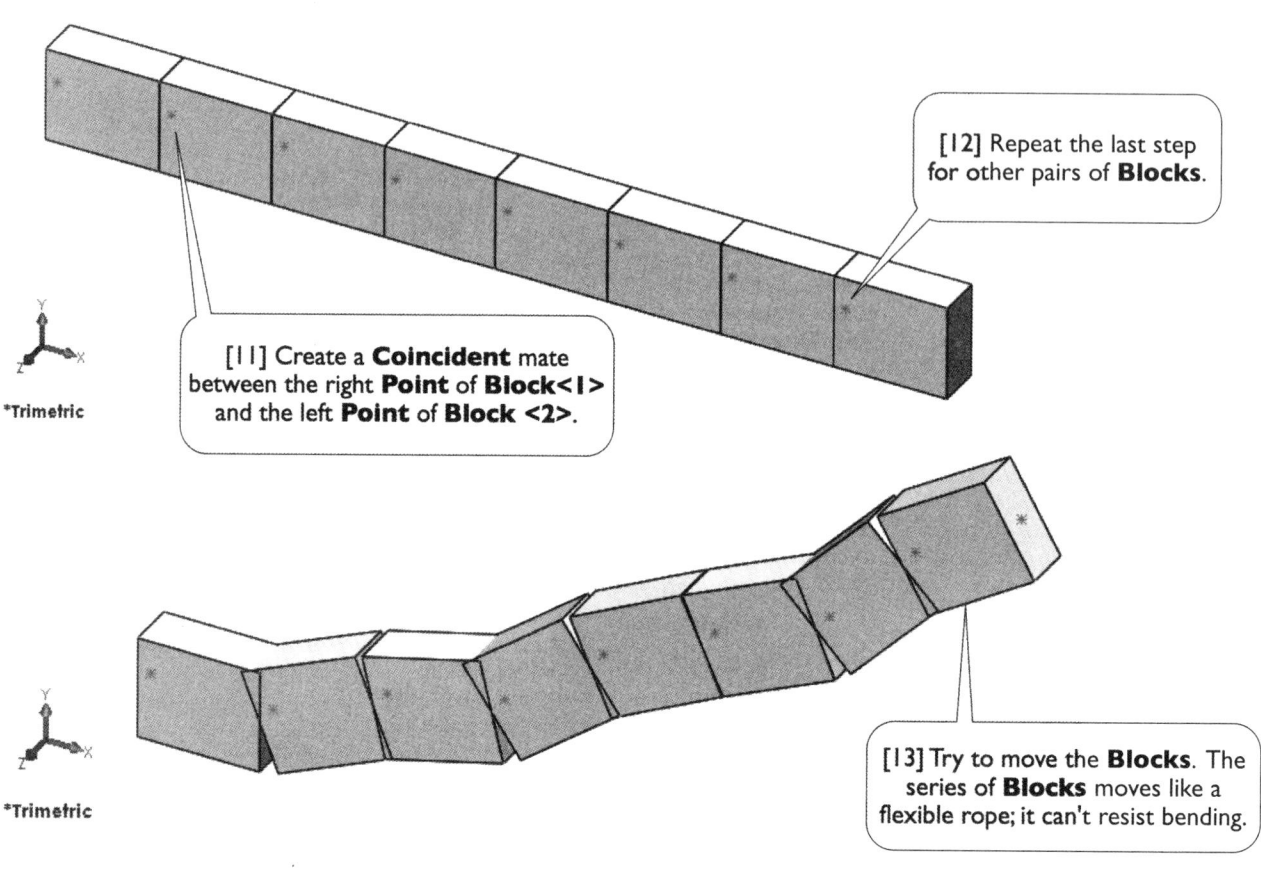

*Trimetric

[11] Create a **Coincident** mate between the right **Point** of **Block<1>** and the left **Point** of **Block <2>**.

[12] Repeat the last step for other pairs of **Blocks**.

*Trimetric

[13] Try to move the **Blocks**. The series of **Blocks** moves like a flexible rope; it can't resist bending.

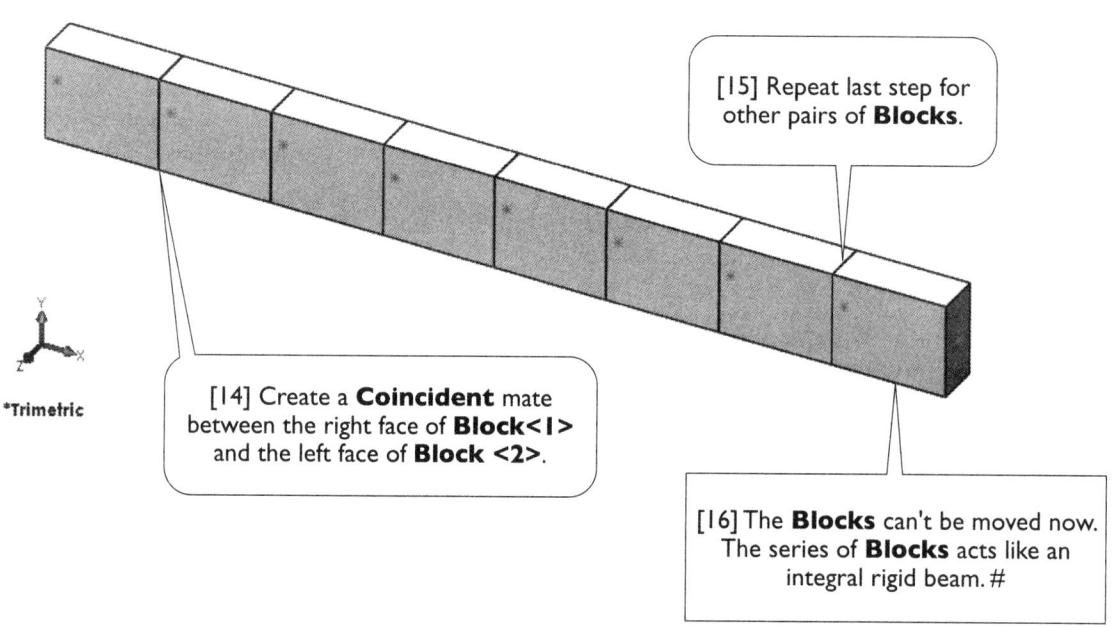

*Trimetric

[14] Create a **Coincident** mate between the right face of **Block<1>** and the left face of **Block <2>**.

[15] Repeat last step for other pairs of **Blocks**.

[16] The **Blocks** can't be moved now. The series of **Blocks** acts like an integral rigid beam. #

5.1-4 Set Up Supports

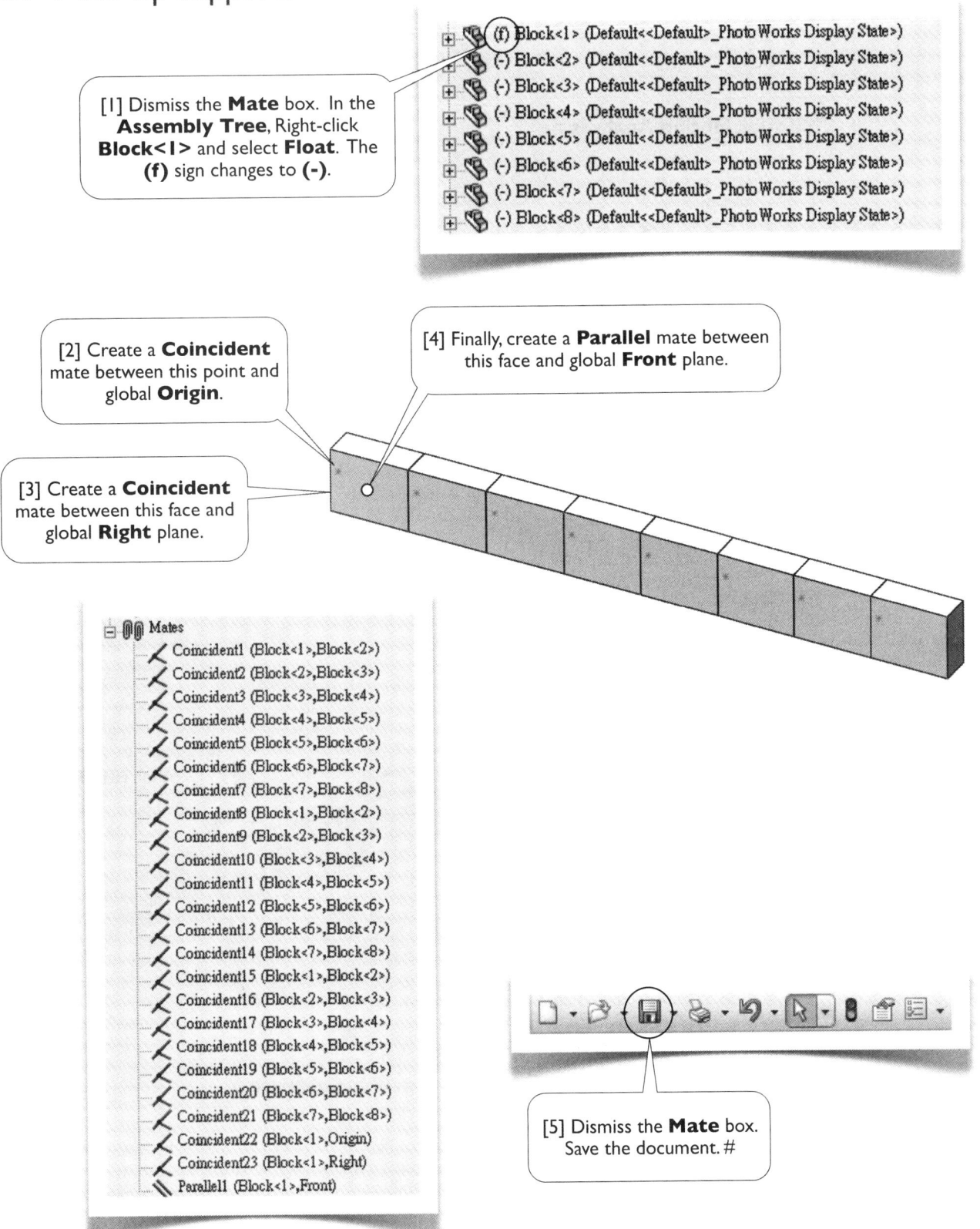

[1] Dismiss the **Mate** box. In the **Assembly Tree**, Right-click **Block<1>** and select **Float**. The **(f)** sign changes to **(-)**.

[2] Create a **Coincident** mate between this point and global **Origin**.

[3] Create a **Coincident** mate between this face and global **Right** plane.

[4] Finally, create a **Parallel** mate between this face and global **Front** plane.

[5] Dismiss the **Mate** box. Save the document. #

5.1-5 Create a **Study**

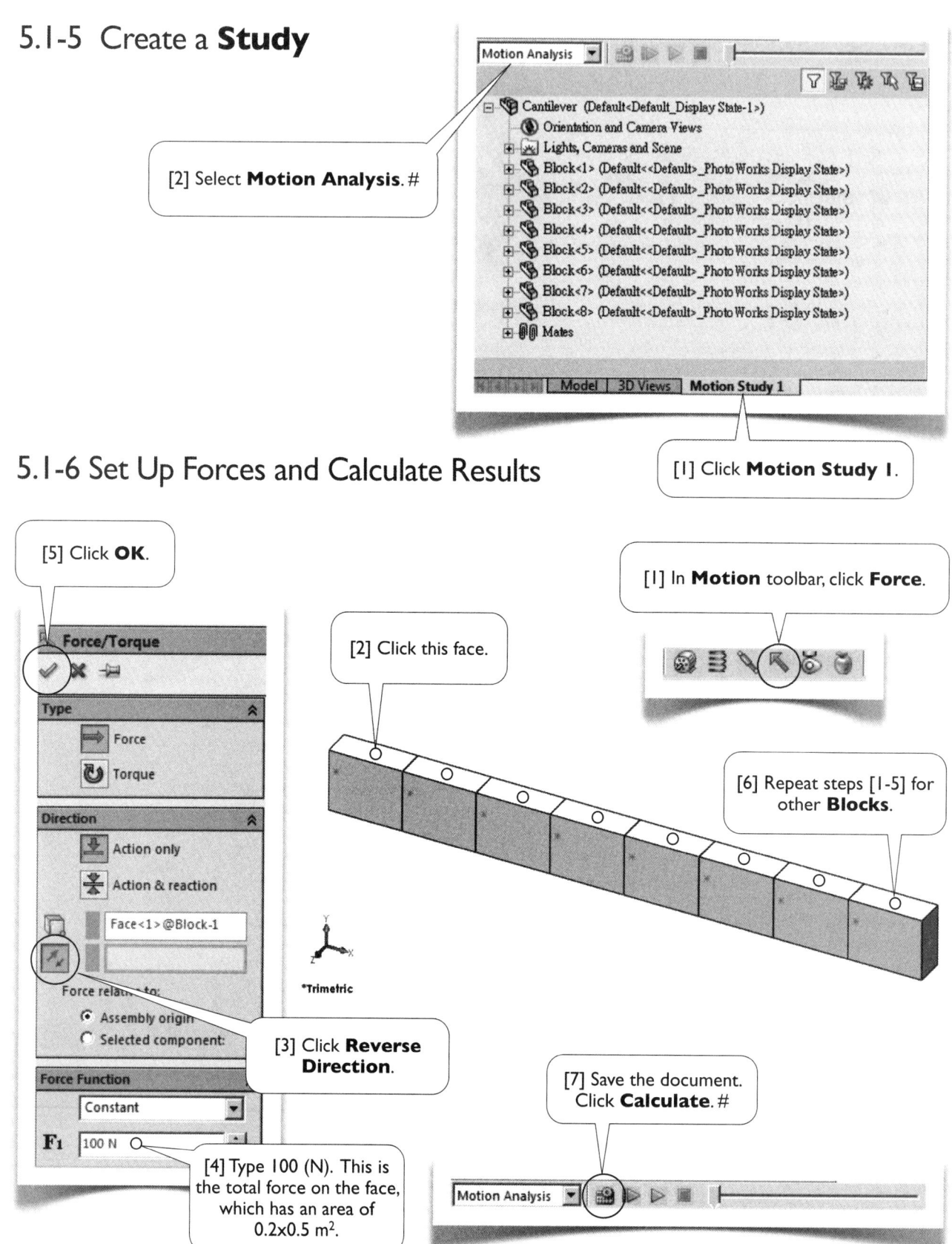

[2] Select **Motion Analysis**. #

[1] Click **Motion Study 1**.

5.1-6 Set Up Forces and Calculate Results

[5] Click **OK**.

[1] In **Motion** toolbar, click **Force**.

[2] Click this face.

[6] Repeat steps [1-5] for other **Blocks**.

[3] Click **Reverse Direction**.

[7] Save the document. Click **Calculate**. #

[4] Type 100 (N). This is the total force on the face, which has an area of 0.2x0.5 m².

Force/Torque

Type

Force
Torque

Direction

Action only
Action & reaction

Face<1>@Block-1

Force relative to:

○ Assembly origin
○ Selected component:

Force Function

Constant

F_1 100 N

*Trimetric

5.1-7 Retrieve Reaction Force and Moment at Fixed End

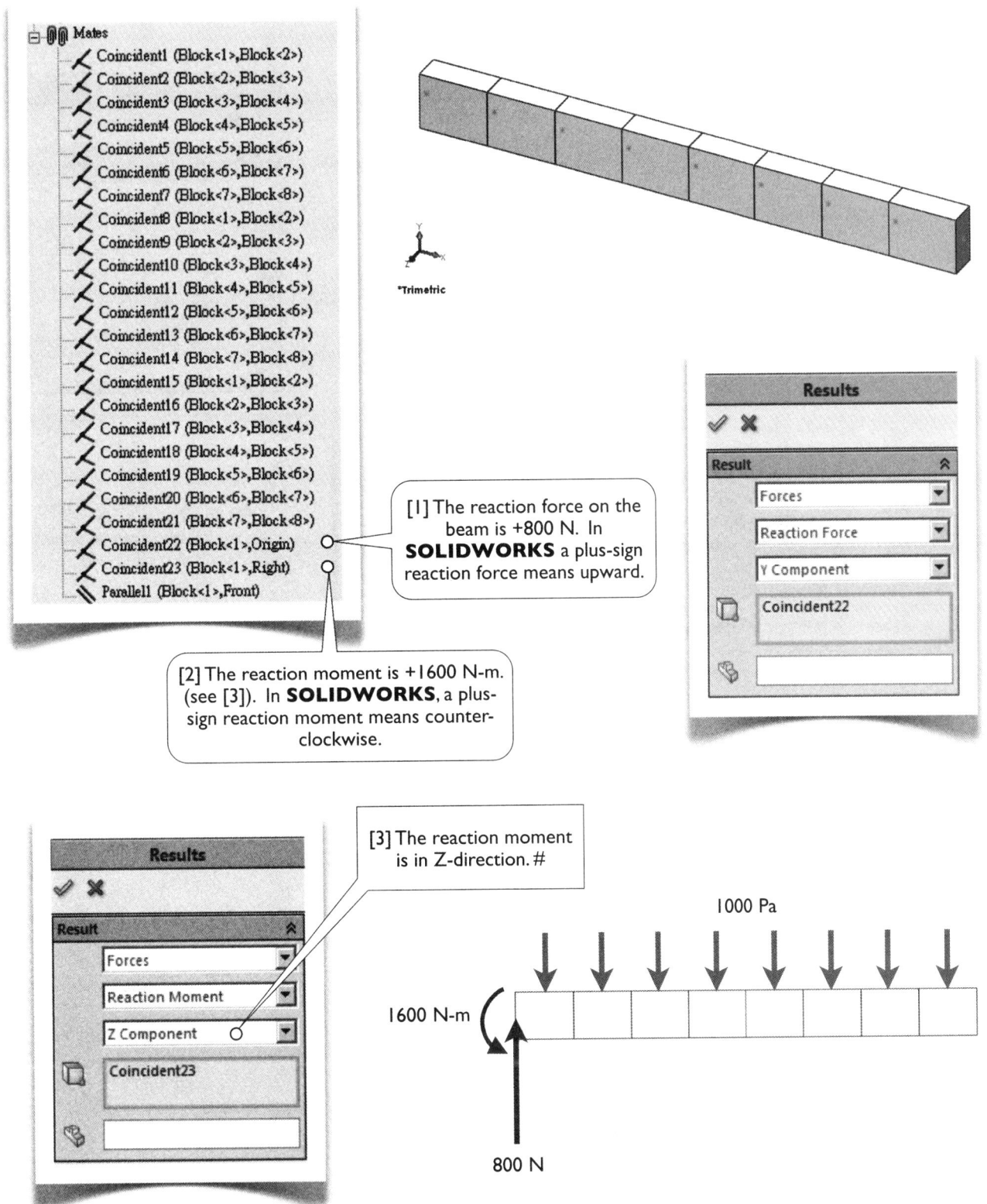

Mates
- Coincident1 (Block<1>,Block<2>)
- Coincident2 (Block<2>,Block<3>)
- Coincident3 (Block<3>,Block<4>)
- Coincident4 (Block<4>,Block<5>)
- Coincident5 (Block<5>,Block<6>)
- Coincident6 (Block<6>,Block<7>)
- Coincident7 (Block<7>,Block<8>)
- Coincident8 (Block<1>,Block<2>)
- Coincident9 (Block<2>,Block<3>)
- Coincident10 (Block<3>,Block<4>)
- Coincident11 (Block<4>,Block<5>)
- Coincident12 (Block<5>,Block<6>)
- Coincident13 (Block<6>,Block<7>)
- Coincident14 (Block<7>,Block<8>)
- Coincident15 (Block<1>,Block<2>)
- Coincident16 (Block<2>,Block<3>)
- Coincident17 (Block<3>,Block<4>)
- Coincident18 (Block<4>,Block<5>)
- Coincident19 (Block<5>,Block<6>)
- Coincident20 (Block<6>,Block<7>)
- Coincident21 (Block<7>,Block<8>)
- Coincident22 (Block<1>,Origin)
- Coincident23 (Block<1>,Right)
- Parallel1 (Block<1>,Front)

*Trimetric

[1] The reaction force on the beam is +800 N. In **SOLIDWORKS** a plus-sign reaction force means upward.

[2] The reaction moment is +1600 N-m. (see [3]). In **SOLIDWORKS**, a plus-sign reaction moment means counter-clockwise.

Results

Result

Forces

Reaction Force

Y Component

Coincident22

[3] The reaction moment is in Z-direction. #

Results

Result

Forces

Reaction Moment

Z Component

Coincident23

1000 Pa

1600 N-m

800 N

5.1-8 Retrieve Shear Forces

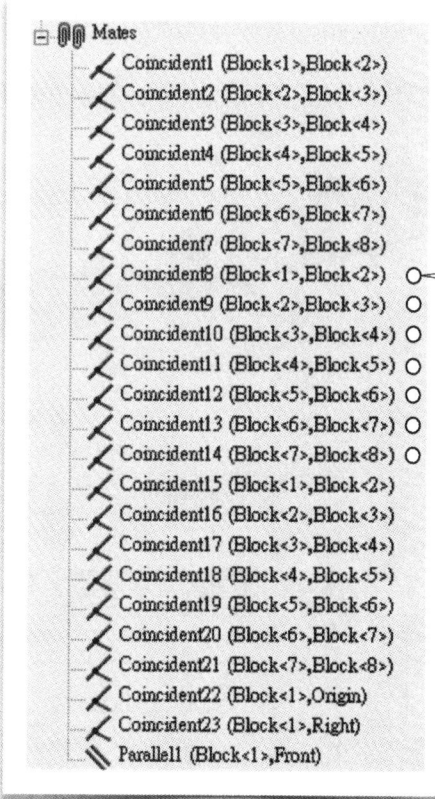

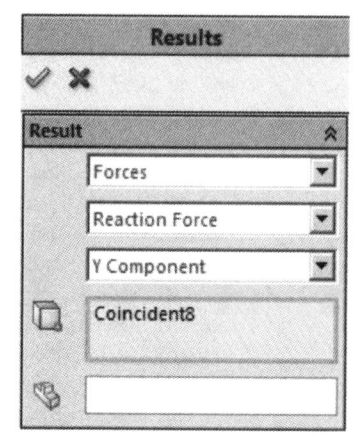

[1] The reaction force exerted on **Block<1>** by **Block<2>** is -700 N; the negative sign means downward. According to the sign convention in 5.1-1[6] (page 99), it is positive in a shear diagram.

Distance from left end	Shear Force
0.5 m	+700 N
1.0 m	+600 N
1.5 m	+500 N
2.0 m	+400 N
2.5 m	+300 N
3.0 m	+200 N
3.5 m	+100 N

[2] Shear forces on other sections can be retrieved and tabulated like this.

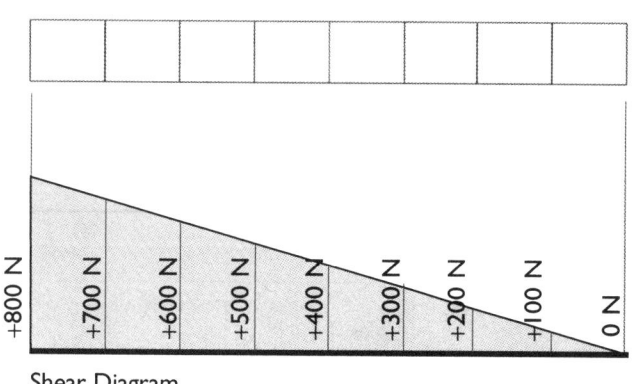

Shear Diagram

[3] The shear diagram. #

5.1-9 Retrieve Moments

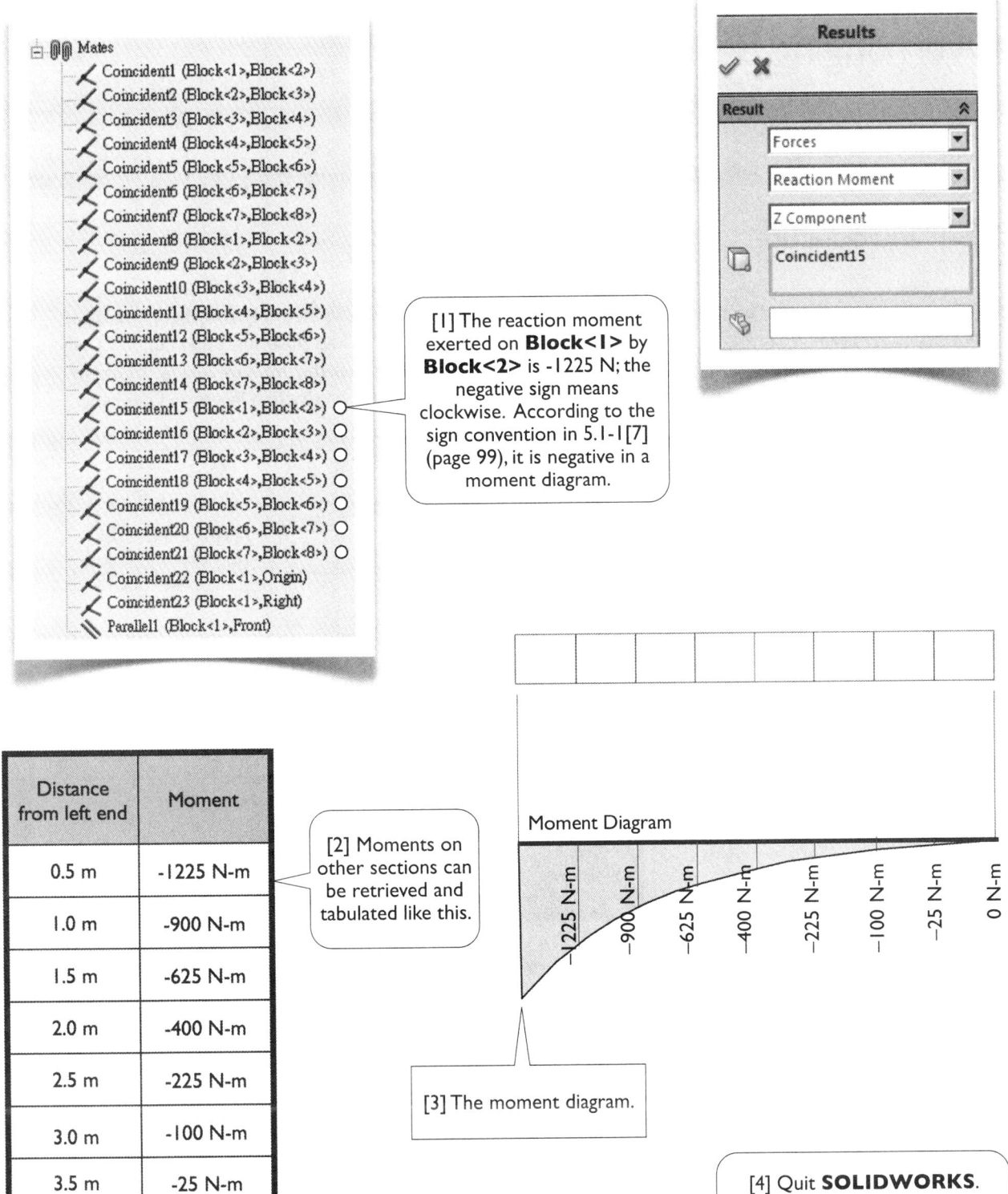

Mates
Coincident1 (Block<1>,Block<2>)
Coincident2 (Block<2>,Block<3>)
Coincident3 (Block<3>,Block<4>)
Coincident4 (Block<4>,Block<5>)
Coincident5 (Block<5>,Block<6>)
Coincident6 (Block<6>,Block<7>)
Coincident7 (Block<7>,Block<8>)
Coincident8 (Block<1>,Block<2>)
Coincident9 (Block<2>,Block<3>)
Coincident10 (Block<3>,Block<4>)
Coincident11 (Block<4>,Block<5>)
Coincident12 (Block<5>,Block<6>)
Coincident13 (Block<6>,Block<7>)
Coincident14 (Block<7>,Block<8>)
Coincident15 (Block<1>,Block<2>)
Coincident16 (Block<2>,Block<3>)
Coincident17 (Block<3>,Block<4>)
Coincident18 (Block<4>,Block<5>)
Coincident19 (Block<5>,Block<6>)
Coincident20 (Block<6>,Block<7>)
Coincident21 (Block<7>,Block<8>)
Coincident22 (Block<1>,Origin)
Coincident23 (Block<1>,Right)
Parallel1 (Block<1>,Front)

Results

Result

Forces

Reaction Moment

Z Component

Coincident15

[1] The reaction moment exerted on **Block<1>** by **Block<2>** is -1225 N; the negative sign means clockwise. According to the sign convention in 5.1-1[7] (page 99), it is negative in a moment diagram.

Distance from left end	Moment
0.5 m	-1225 N-m
1.0 m	-900 N-m
1.5 m	-625 N-m
2.0 m	-400 N-m
2.5 m	-225 N-m
3.0 m	-100 N-m
3.5 m	-25 N-m

[2] Moments on other sections can be retrieved and tabulated like this.

Moment Diagram

-1225 N-m   -900 N-m   -625 N-m   -400 N-m   -225 N-m   -100 N-m   -25 N-m   0 N-m

[3] The moment diagram.

[4] Quit **SOLIDWORKS**. Click **Save all.** Click **Rebuild and save the document**. #

Section 5.2

Overhanging Beam

5.2-1 Introduction

[1] Consider an overhanging beam subject to vertical loads [2-4] and we want to study internal shear forces and moments of the beam. Following the sign conventions in 5.1-1[6, 7] (page 99), the shear force and moments can be drawn as shown below. To study the internal forces of the beam, we'll construct the beam by "gluing" 15 blocks together, each 0.5-m long. Internal forces then can be obtained by examining the reaction forces and moments between blocks.

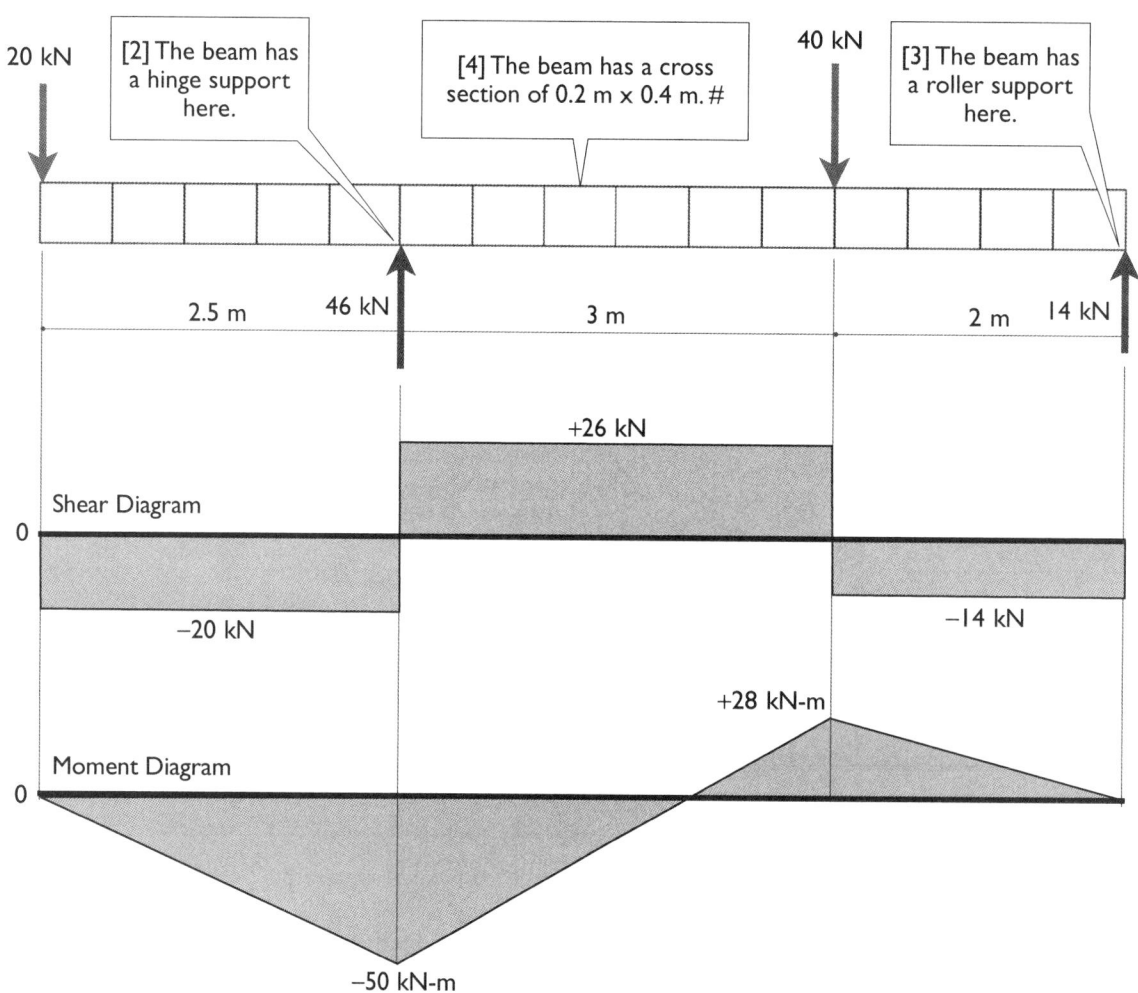

5.2-2 Start Up and Create a Part: **Block**

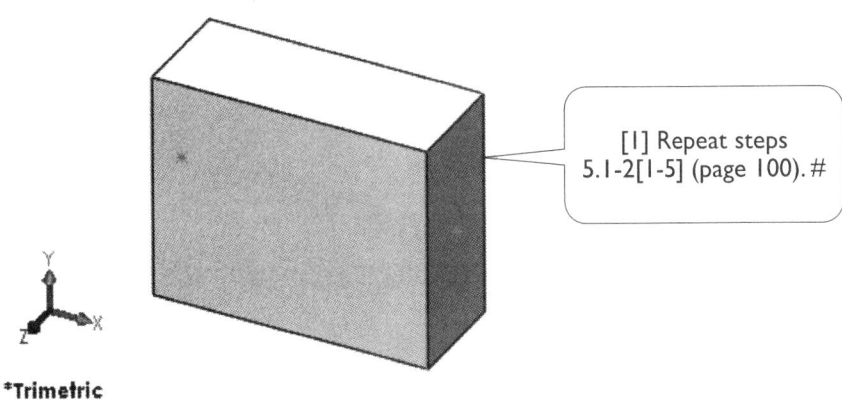

[1] Repeat steps 5.1-2[1-5] (page 100). #

*Trimetric

5.2-3 Create an Assembly: **OverhangingBeam**

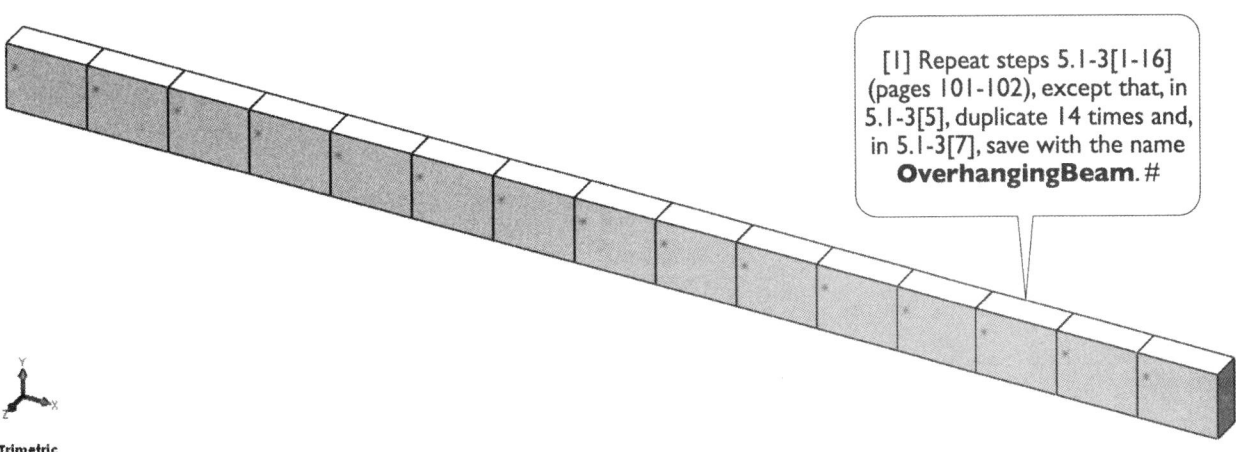

[1] Repeat steps 5.1-3[1-16] (pages 101-102), except that, in 5.1-3[5], duplicate 14 times and, in 5.1-3[7], save with the name **OverhangingBeam**. #

*Trimetric

5.2-4 Set Up Supports

[1] Dismiss **Mate** box. In the **Assembly Tree**, Right-click **Block<1>** and select **Float**. The **(f)** sign changes to **(-)**.

(f) Block<1> (Default<<Default>_PhotoWorks Display State>)
(-) Block<2> (Default<<Default>_PhotoWorks Display State>)
(-) Block<3> (Default<<Default>_PhotoWorks Display State>)
(-) Block<4> (Default<<Default>_PhotoWorks Display State>)
(-) Block<5> (Default<<Default>_PhotoWorks Display State>)
(-) Block<6> (Default<<Default>_PhotoWorks Display State>)
(-) Block<7> (Default<<Default>_PhotoWorks Display State>)
(-) Block<8> (Default<<Default>_PhotoWorks Display State>)
(-) Block<9> (Default<<Default>_PhotoWorks Display State>)
(-) Block<10> (Default<<Default>_PhotoWorks Display State>)
(-) Block<11> (Default<<Default>_PhotoWorks Display State>)
(-) Block<12> (Default<<Default>_PhotoWorks Display State>)
(-) Block<13> (Default<<Default>_PhotoWorks Display State>)
(-) Block<14> (Default<<Default>_PhotoWorks Display State>)
(-) Block<15> (Default<<Default>_PhotoWorks Display State>)

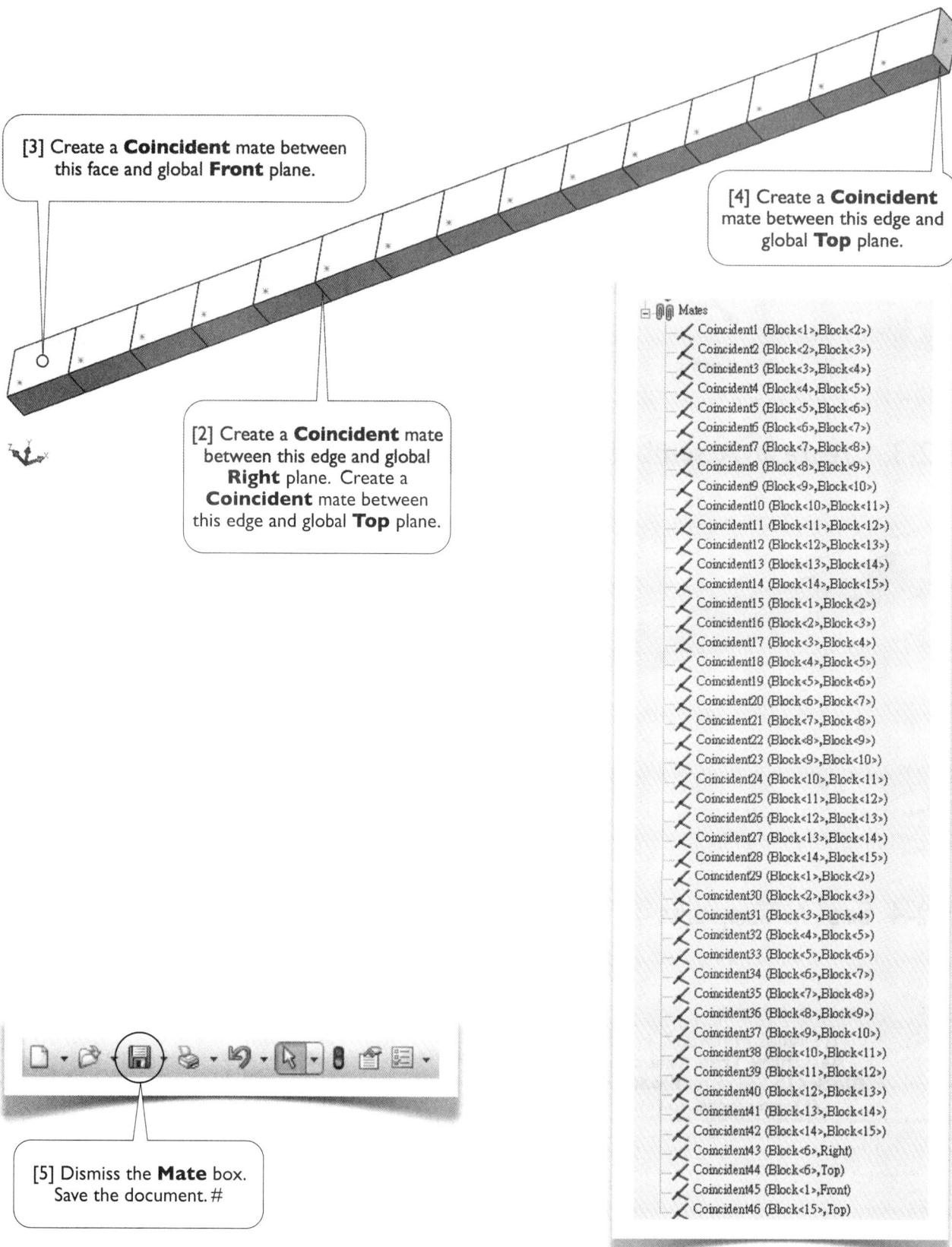

[3] Create a **Coincident** mate between this face and global **Front** plane.

[4] Create a **Coincident** mate between this edge and global **Top** plane.

[2] Create a **Coincident** mate between this edge and global **Right** plane. Create a **Coincident** mate between this edge and global **Top** plane.

Mates
- Coincident1 (Block<1>,Block<2>)
- Coincident2 (Block<2>,Block<3>)
- Coincident3 (Block<3>,Block<4>)
- Coincident4 (Block<4>,Block<5>)
- Coincident5 (Block<5>,Block<6>)
- Coincident6 (Block<6>,Block<7>)
- Coincident7 (Block<7>,Block<8>)
- Coincident8 (Block<8>,Block<9>)
- Coincident9 (Block<9>,Block<10>)
- Coincident10 (Block<10>,Block<11>)
- Coincident11 (Block<11>,Block<12>)
- Coincident12 (Block<12>,Block<13>)
- Coincident13 (Block<13>,Block<14>)
- Coincident14 (Block<14>,Block<15>)
- Coincident15 (Block<1>,Block<2>)
- Coincident16 (Block<2>,Block<3>)
- Coincident17 (Block<3>,Block<4>)
- Coincident18 (Block<4>,Block<5>)
- Coincident19 (Block<5>,Block<6>)
- Coincident20 (Block<6>,Block<7>)
- Coincident21 (Block<7>,Block<8>)
- Coincident22 (Block<8>,Block<9>)
- Coincident23 (Block<9>,Block<10>)
- Coincident24 (Block<10>,Block<11>)
- Coincident25 (Block<11>,Block<12>)
- Coincident26 (Block<12>,Block<13>)
- Coincident27 (Block<13>,Block<14>)
- Coincident28 (Block<14>,Block<15>)
- Coincident29 (Block<1>,Block<2>)
- Coincident30 (Block<2>,Block<3>)
- Coincident31 (Block<3>,Block<4>)
- Coincident32 (Block<4>,Block<5>)
- Coincident33 (Block<5>,Block<6>)
- Coincident34 (Block<6>,Block<7>)
- Coincident35 (Block<7>,Block<8>)
- Coincident36 (Block<8>,Block<9>)
- Coincident37 (Block<9>,Block<10>)
- Coincident38 (Block<10>,Block<11>)
- Coincident39 (Block<11>,Block<12>)
- Coincident40 (Block<12>,Block<13>)
- Coincident41 (Block<13>,Block<14>)
- Coincident42 (Block<14>,Block<15>)
- Coincident43 (Block<6>,Right)
- Coincident44 (Block<6>,Top)
- Coincident45 (Block<1>,Front)
- Coincident46 (Block<15>,Top)

[5] Dismiss the **Mate** box. Save the document. #

5.2-5 Create a **Study**

[2] Select **Motion Analysis**. #

[1] Click **Motion Study 1**.

5.2-6 Set Up Forces and Calculate Results

[1] On this edge, apply a force of 20,000 N downward.

[2] On this edge, apply a force of 40,000 N downward.

[3] Save the document. Click **Calculate**. #

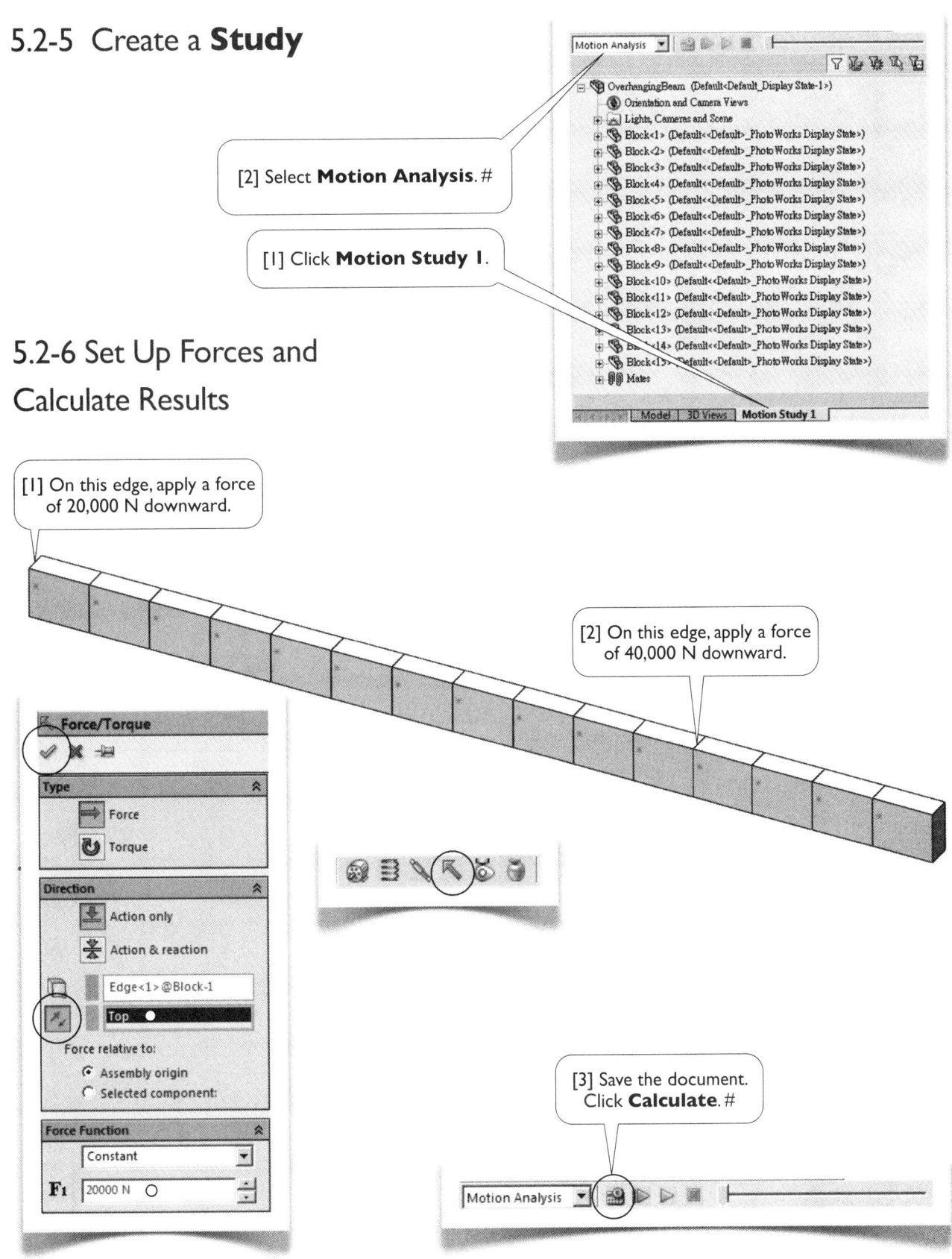

5.2-7 Retrieve Reaction Forces at Supports

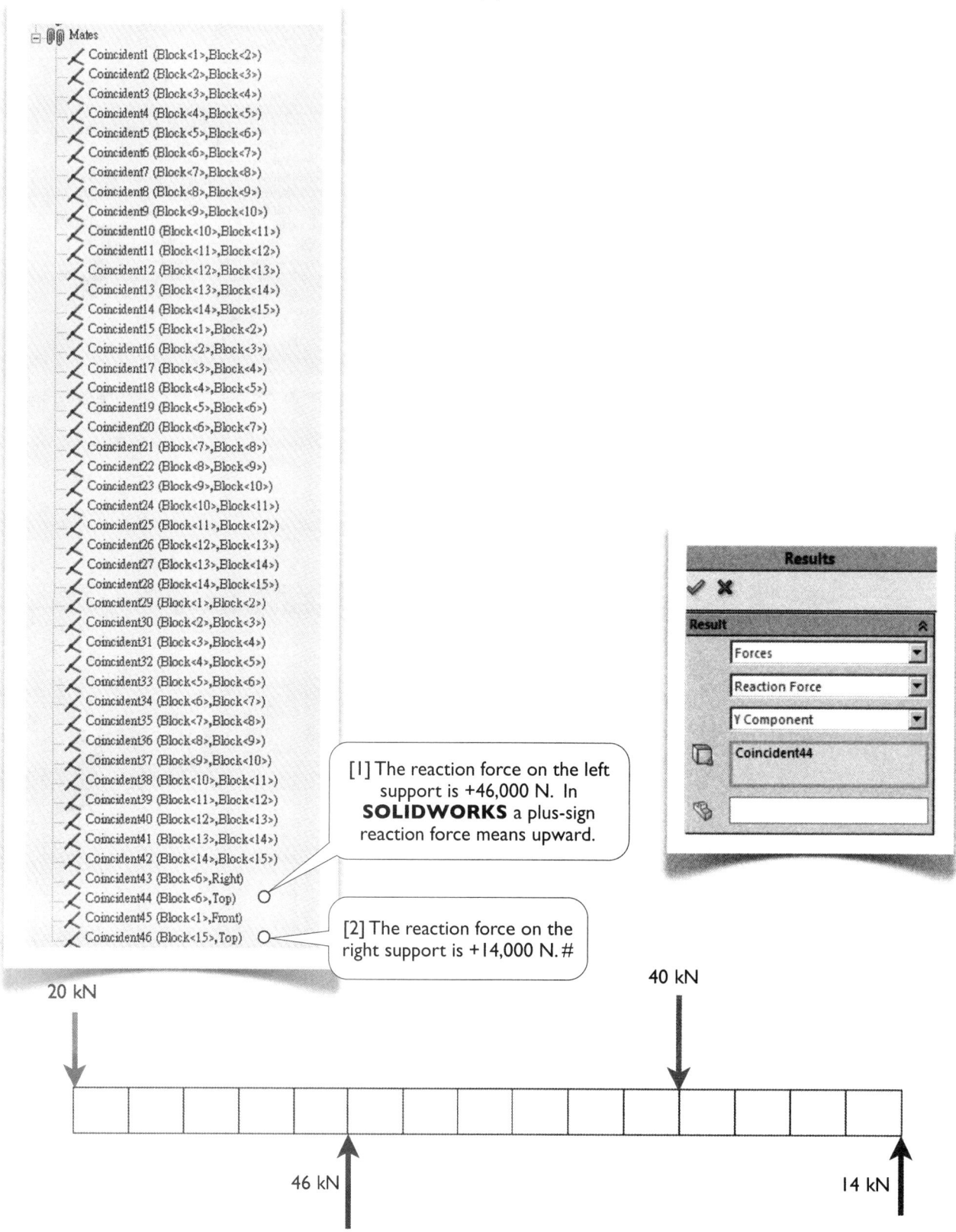

Mates
- Coincident1 (Block<1>,Block<2>)
- Coincident2 (Block<2>,Block<3>)
- Coincident3 (Block<3>,Block<4>)
- Coincident4 (Block<4>,Block<5>)
- Coincident5 (Block<5>,Block<6>)
- Coincident6 (Block<6>,Block<7>)
- Coincident7 (Block<7>,Block<8>)
- Coincident8 (Block<8>,Block<9>)
- Coincident9 (Block<9>,Block<10>)
- Coincident10 (Block<10>,Block<11>)
- Coincident11 (Block<11>,Block<12>)
- Coincident12 (Block<12>,Block<13>)
- Coincident13 (Block<13>,Block<14>)
- Coincident14 (Block<14>,Block<15>)
- Coincident15 (Block<1>,Block<2>)
- Coincident16 (Block<2>,Block<3>)
- Coincident17 (Block<3>,Block<4>)
- Coincident18 (Block<4>,Block<5>)
- Coincident19 (Block<5>,Block<6>)
- Coincident20 (Block<6>,Block<7>)
- Coincident21 (Block<7>,Block<8>)
- Coincident22 (Block<8>,Block<9>)
- Coincident23 (Block<9>,Block<10>)
- Coincident24 (Block<10>,Block<11>)
- Coincident25 (Block<11>,Block<12>)
- Coincident26 (Block<12>,Block<13>)
- Coincident27 (Block<13>,Block<14>)
- Coincident28 (Block<14>,Block<15>)
- Coincident29 (Block<1>,Block<2>)
- Coincident30 (Block<2>,Block<3>)
- Coincident31 (Block<3>,Block<4>)
- Coincident32 (Block<4>,Block<5>)
- Coincident33 (Block<5>,Block<6>)
- Coincident34 (Block<6>,Block<7>)
- Coincident35 (Block<7>,Block<8>)
- Coincident36 (Block<8>,Block<9>)
- Coincident37 (Block<9>,Block<10>)
- Coincident38 (Block<10>,Block<11>)
- Coincident39 (Block<11>,Block<12>)
- Coincident40 (Block<12>,Block<13>)
- Coincident41 (Block<13>,Block<14>)
- Coincident42 (Block<14>,Block<15>)
- Coincident43 (Block<6>,Right)
- Coincident44 (Block<6>,Top)
- Coincident45 (Block<1>,Front)
- Coincident46 (Block<15>,Top)

Results

Result

Forces

Reaction Force

Y Component

Coincident44

[1] The reaction force on the left support is +46,000 N. In **SOLIDWORKS** a plus-sign reaction force means upward.

[2] The reaction force on the right support is +14,000 N. #

20 kN

40 kN

46 kN

14 kN

5.2-8 Retrieve Shear Forces

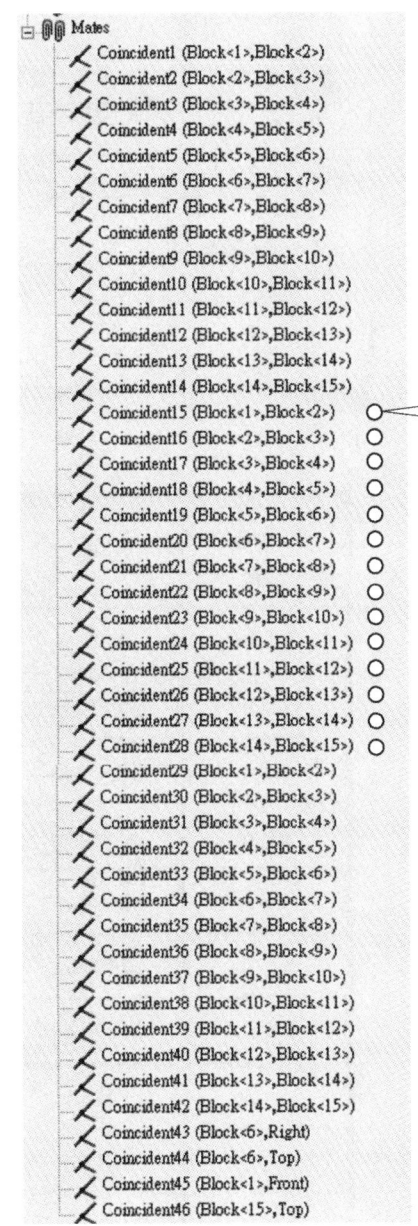

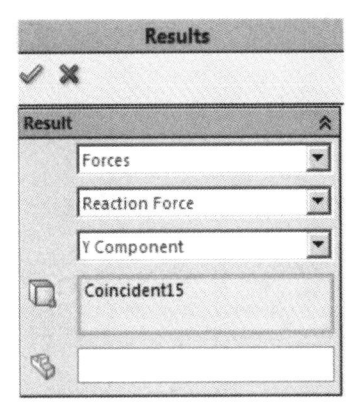

[1] The reaction force exerted on **Block<1>** by **Block<2>** is +20,000 N; the positive sign means upward. According to the sign convention in 5.1-1[6] (page 99), it is negative in a shear diagram.

[2] Shear forces on other sections can be retrieved and tabulated like this. A shear diagram can be drawn as shown in 5.2-1 (page 108). #

Distance from left end	Shear Force
0.5 m	-20 kN
1.0 m	-20 kN
1.5 m	-20 kN
2.0 m	-20 kN
2.5 m	-20 kN
3.0 m	+26 kN
3.5 m	+26 kN
4.0 m	+26 kN
4.5 m	+26 kN
5,0 m	+26 kN
5.5 m	+26 kN
6.0 m	-14 kN
6.5 m	-14 kN
7.0 m	-14 kN

5.2-9 Retrieve Moments

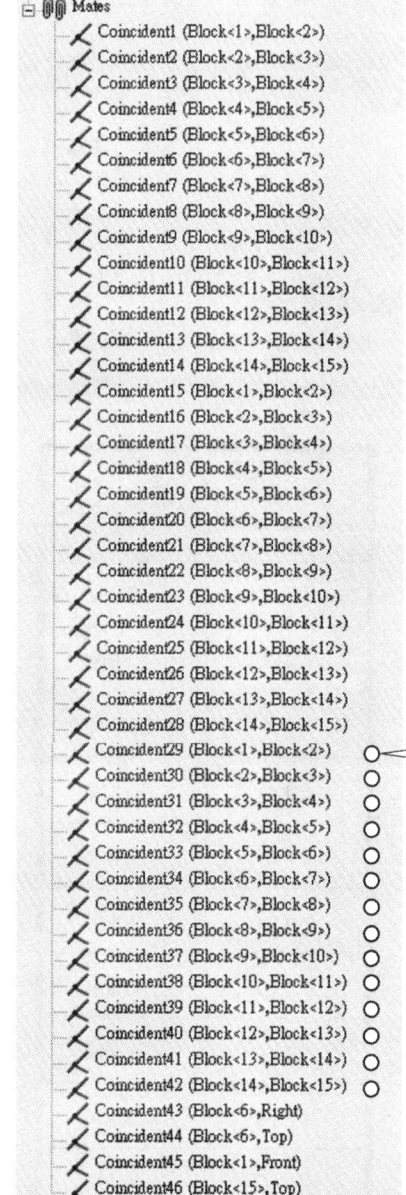

Results

Result	
Forces	
Reaction Moment	
Z Component	
Coincident29	

[1] The reaction moment exerted on **Block<1>** by **Block<2>** is -10,000 N-m; the negative sign means clockwise. According to the sign convention in 5.1-1[7] (page 99), it is negative in a moment diagram.

[2] Moments on other sections can be retrieved and tabulated like this. A moment diagram can be drawn as shown in 5.2-1 (page 108).

[3] Quit **SOLIDWORKS**. Click **Save all.** Click **Rebuild and save the document**. #

Distance from left end	Moment
0.5 m	-10 kN-m
1.0 m	-20 kN-m
1.5 m	-30 kN-m
2.0 m	-40 kN-m
2.5 m	-50 kN-m
3.0 m	-37 kN-m
3.5 m	-24 kN-m
4.0 m	-11 kN-m
4.5 m	+2 kN-m
5.0 m	+15 kN-m
5.5 m	+28 kN-m
6.0 m	+21 kN-m
6.5 m	+14 kN-m
7.0 m	+7 kN-m

Chapter 6
Cables

Cables are structural components that can resist tensile only; they are not capable of resisting compression or bending. They are also called tension-only components. Use of cables as structural components is very cost-effective if the components are subject to tension only.

To model cables in **SOLIDWORKS Motion**, which performs rigid-body mechanical analysis, we'll construct a cable system by connecting sections of "rigid cables." The connections allow two adjacent "rigid cables" to rotate each other. The cable system then behaves like a flexible cable.

Section 6.1

Cable with Point Loads

6.1-1 Introduction

[1] Consider a cable of 10 m long, its two ends *A* and *B* fixed in the space, its intermediate points *C* and *D* subject to downward loads of 10 kN respectively. We want to find the positions of *C* and *D* and the tensions in *AC*, *CD*, and *BD*. Since **SOLIDWORKS Motion** assumes all bodies are rigid, we'll construct the cable system by connecting 20 sections of cable, each 0.5-m long. The connections allow two adjacent sections of cable to rotate each other. The cable system then becomes "flexible." Tensions of the cable can be obtained by examining the reaction forces between adjacent sections of the cable system.

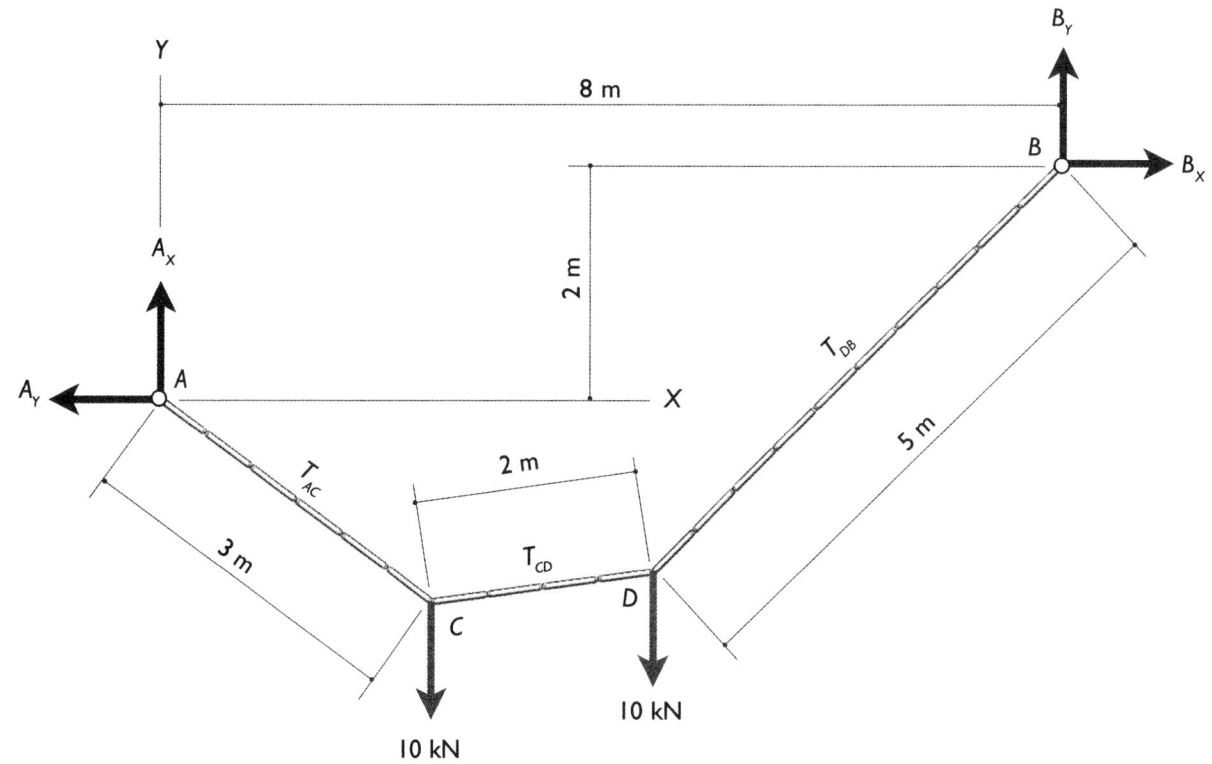

6.1-2 Start Up and Create a Part: **Cable**

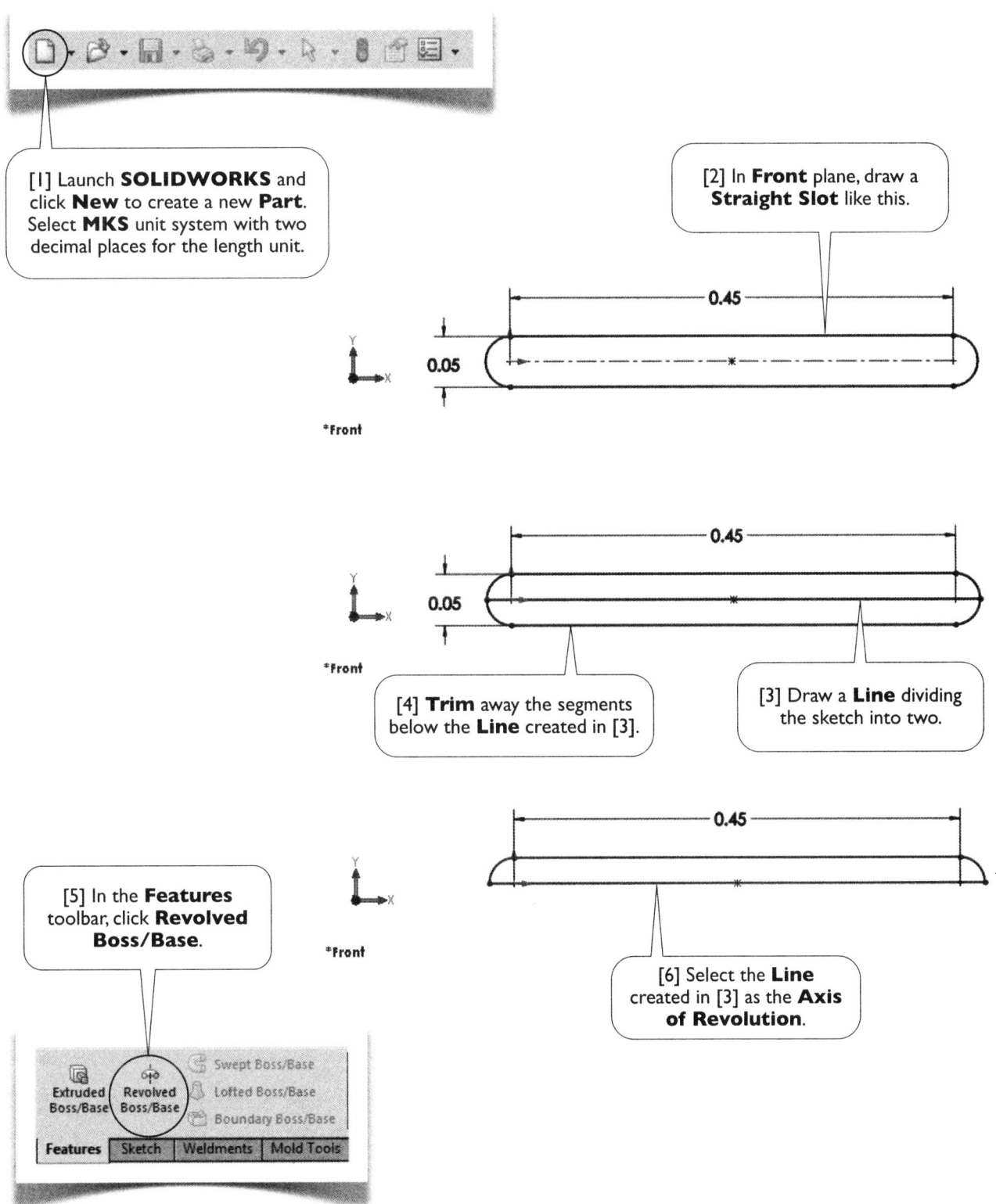

[1] Launch **SOLIDWORKS** and click **New** to create a new **Part**. Select **MKS** unit system with two decimal places for the length unit.

[2] In **Front** plane, draw a **Straight Slot** like this.

[4] **Trim** away the segments below the **Line** created in [3].

[3] Draw a **Line** dividing the sketch into two.

[5] In the **Features** toolbar, click **Revolved Boss/Base**.

[6] Select the **Line** created in [3] as the **Axis of Revolution**.

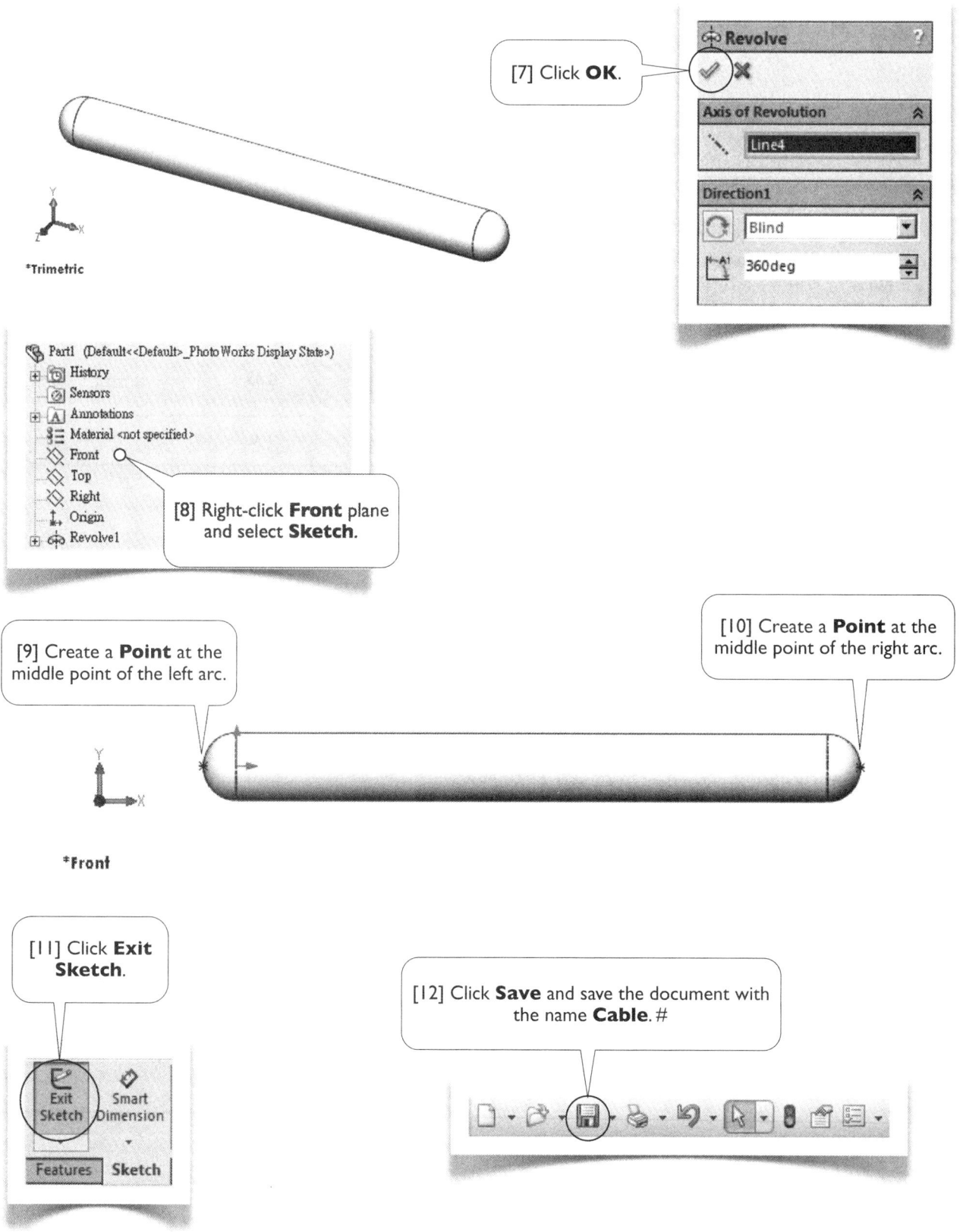

*Trimetric

[7] Click **OK**.

Revolve

Axis of Revolution

Line4

Direction1

Blind

360deg

Part1 (Default<<Default>_PhotoWorks Display State>)
- History
- Sensors
- Annotations
- Material <not specified>
- Front
- Top
- Right
- Origin
- Revolve1

[8] Right-click **Front** plane and select **Sketch**.

[9] Create a **Point** at the middle point of the left arc.

[10] Create a **Point** at the middle point of the right arc.

*Front

[11] Click **Exit Sketch**.

Exit
Sketch

Smart
Dimension

Features Sketch

[12] Click **Save** and save the document with the name **Cable**. #

6.1-3 Create an Assembly: **CableSystem**

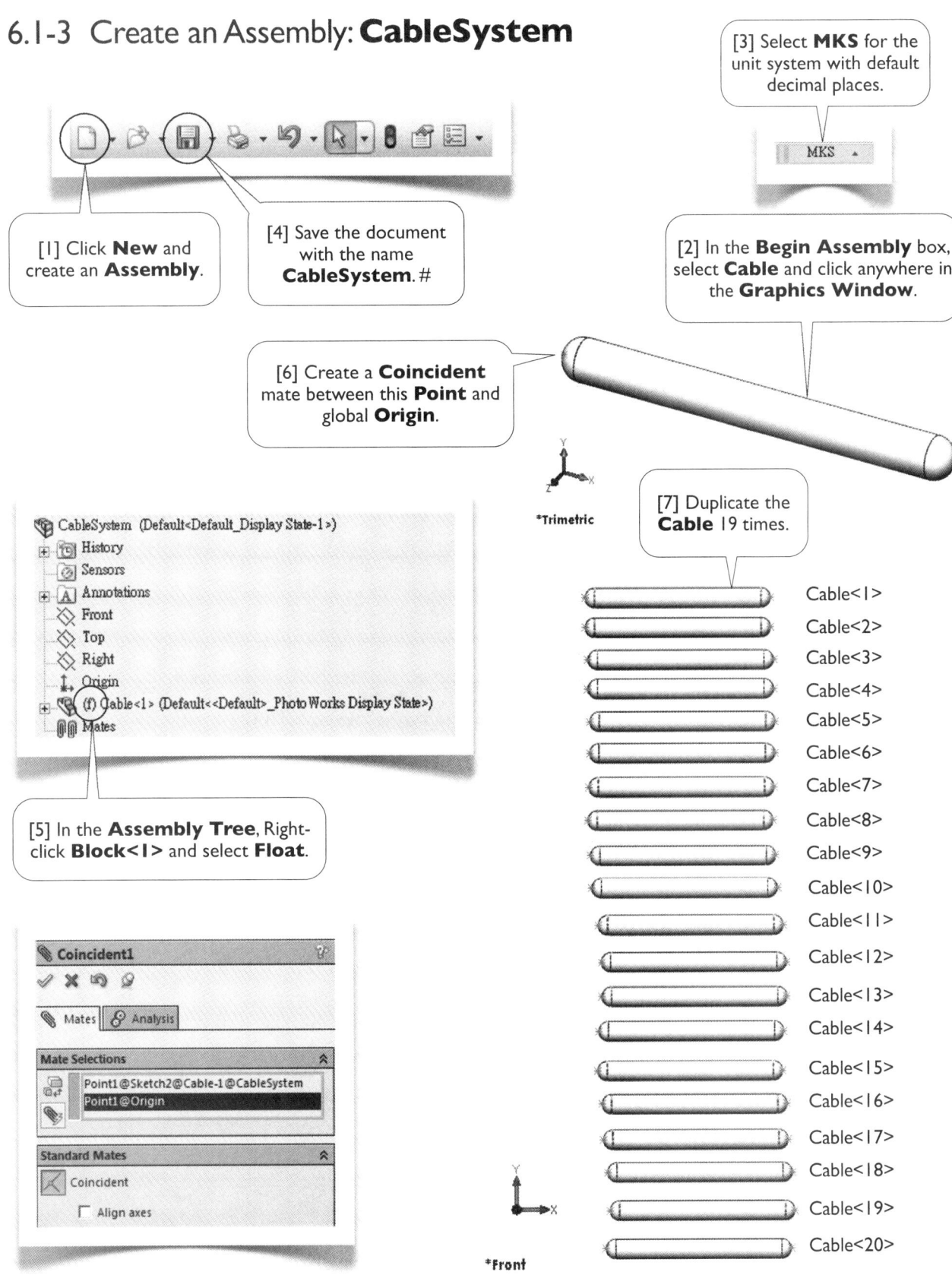

[1] Click **New** and create an **Assembly**.

[4] Save the document with the name **CableSystem**. #

[3] Select **MKS** for the unit system with default decimal places.

MKS

[2] In the **Begin Assembly** box, select **Cable** and click anywhere in the **Graphics Window**.

[6] Create a **Coincident** mate between this **Point** and global **Origin**.

*Trimetric

[7] Duplicate the **Cable** 19 times.

CableSystem (Default<Default_Display State-1>)
  History
  Sensors
  Annotations
  Front
  Top
  Right
  Origin
  (f) Cable<1> (Default<<Default>_PhotoWorks Display State>)
  Mates

[5] In the **Assembly Tree**, Right-click **Block<1>** and select **Float**.

Coincident1

Mates   Analysis

Mate Selections
  Point1@Sketch2@Cable-1@CableSystem
  Point1@Origin

Standard Mates
  Coincident
  Align axes

*Front

Cable<1>
Cable<2>
Cable<3>
Cable<4>
Cable<5>
Cable<6>
Cable<7>
Cable<8>
Cable<9>
Cable<10>
Cable<11>
Cable<12>
Cable<13>
Cable<14>
Cable<15>
Cable<16>
Cable<17>
Cable<18>
Cable<19>
Cable<20>

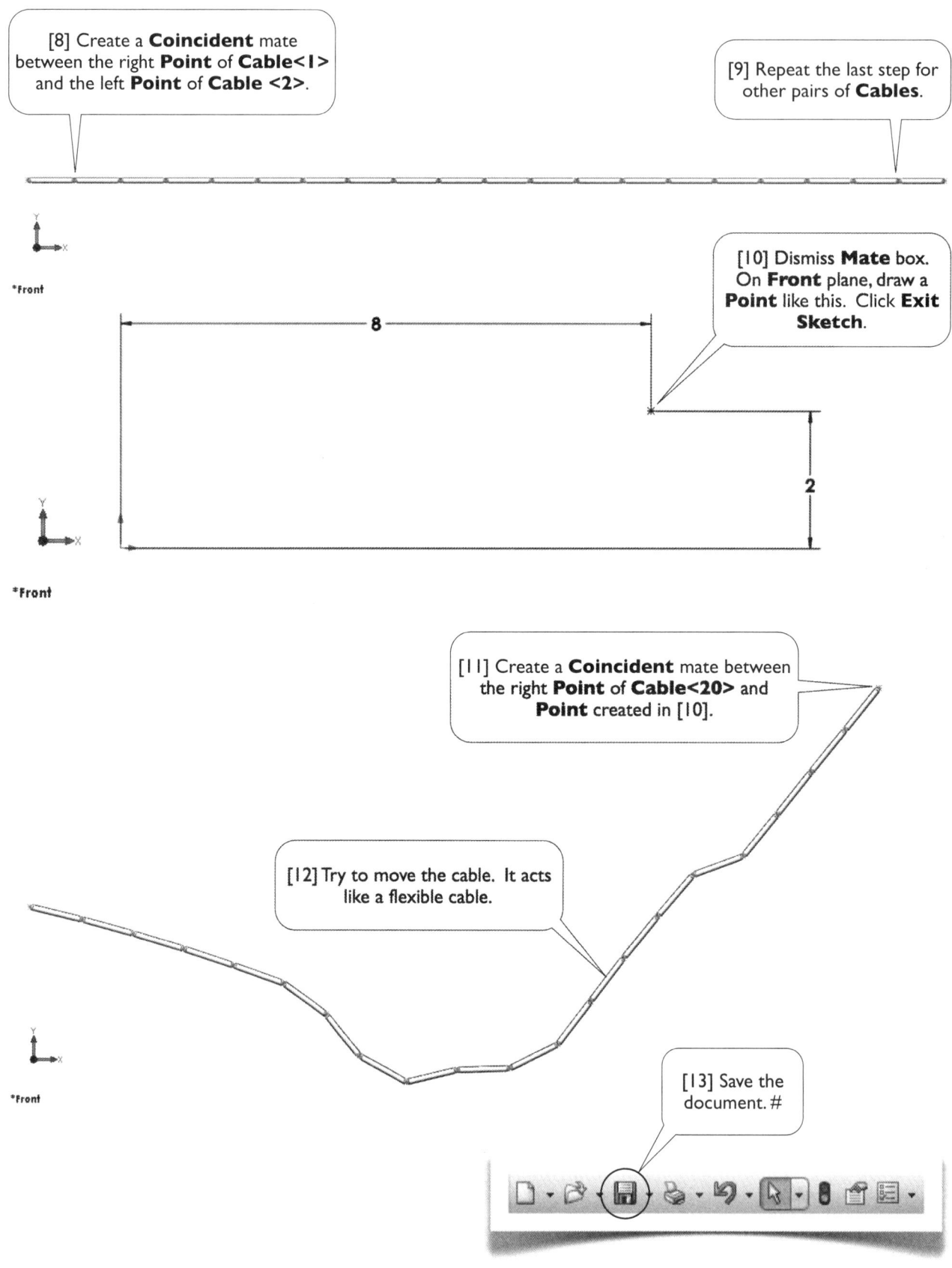

[8] Create a **Coincident** mate between the right **Point** of Cable<1> and the left **Point** of **Cable <2>**.

[9] Repeat the last step for other pairs of **Cables**.

[10] Dismiss **Mate** box. On **Front** plane, draw a **Point** like this. Click **Exit Sketch**.

[11] Create a **Coincident** mate between the right **Point** of **Cable<20>** and **Point** created in [10].

[12] Try to move the cable. It acts like a flexible cable.

[13] Save the document. #

6.1-4 Create a **Study**

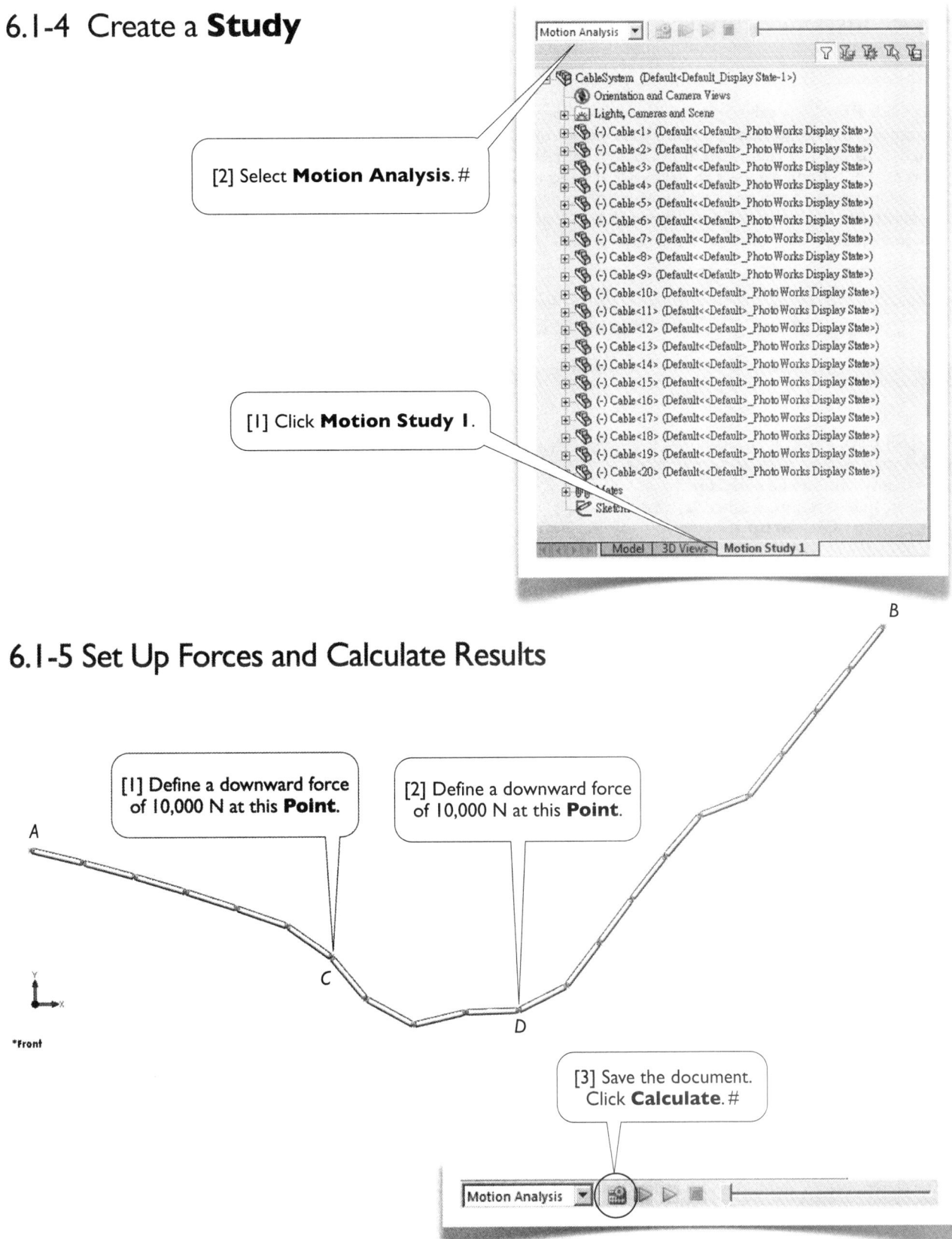

[2] Select **Motion Analysis**. #

[1] Click **Motion Study 1**.

6.1-5 Set Up Forces and Calculate Results

[1] Define a downward force of 10,000 N at this **Point**.

[2] Define a downward force of 10,000 N at this **Point**.

[3] Save the document. Click **Calculate**. #

6.1-6 Install Dampers

[1] Click **Play**. The cable system vibrates and does not cease in 5 seconds. Actually, without any energy dissipation mechanism (e.g., friction, damping) a cable system will vibrate forever after applying forces. Let's install dampers to damp out vibrations. The number and locations of dampers can be arbitrary as long as they can effectively eliminate the vibrations; they do not affect the final configuration and tensions of the cable system. Click **Stop**.

[9] Save the document. Click **Calculate**.

[2] Drag the **Time Slider** to the beginning. We want to install the dampers at $t = 0$.

[10] Click **Play**. In 5 seconds, the vibrations become small enough that the cable system can be considered as a *steady-state* system; i.e., a *static* system. Click **Stop**. #

[7] Click **OK**.

[3] In **Motion** toolbar, click **Damper**.

[6] Type an arbitrary large damping coefficient, say 1000 (N/(m/s)).

[4] Click the **Point** at A.

[5] Click the **Point** at C.

[8] Repeat similar procedure in steps [3-7] to install three more dampers connecting (1) B and C, (2) A and D, and (3) B and D.

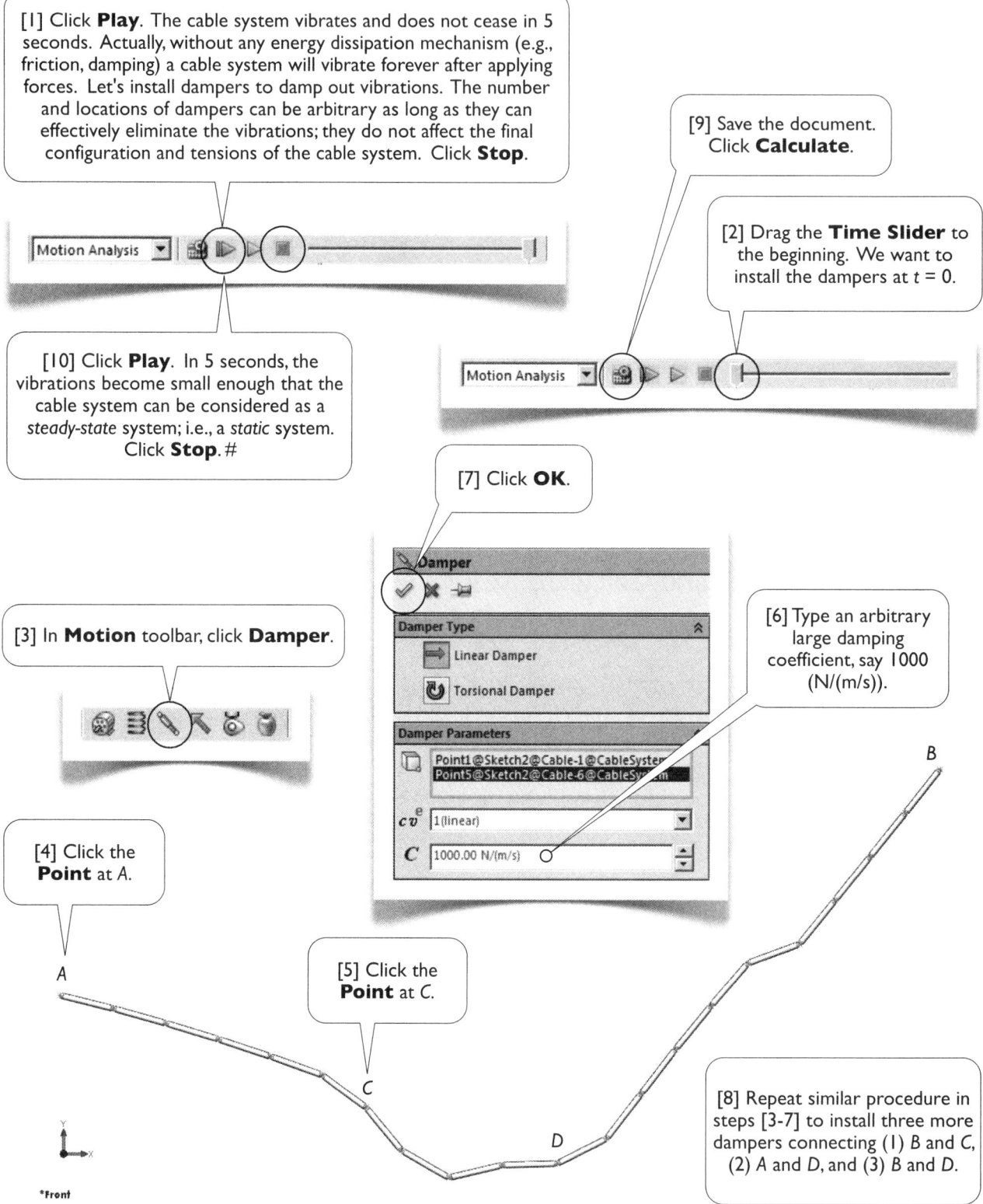

6.1-7 Retrieve the Positions of *C* and *D*

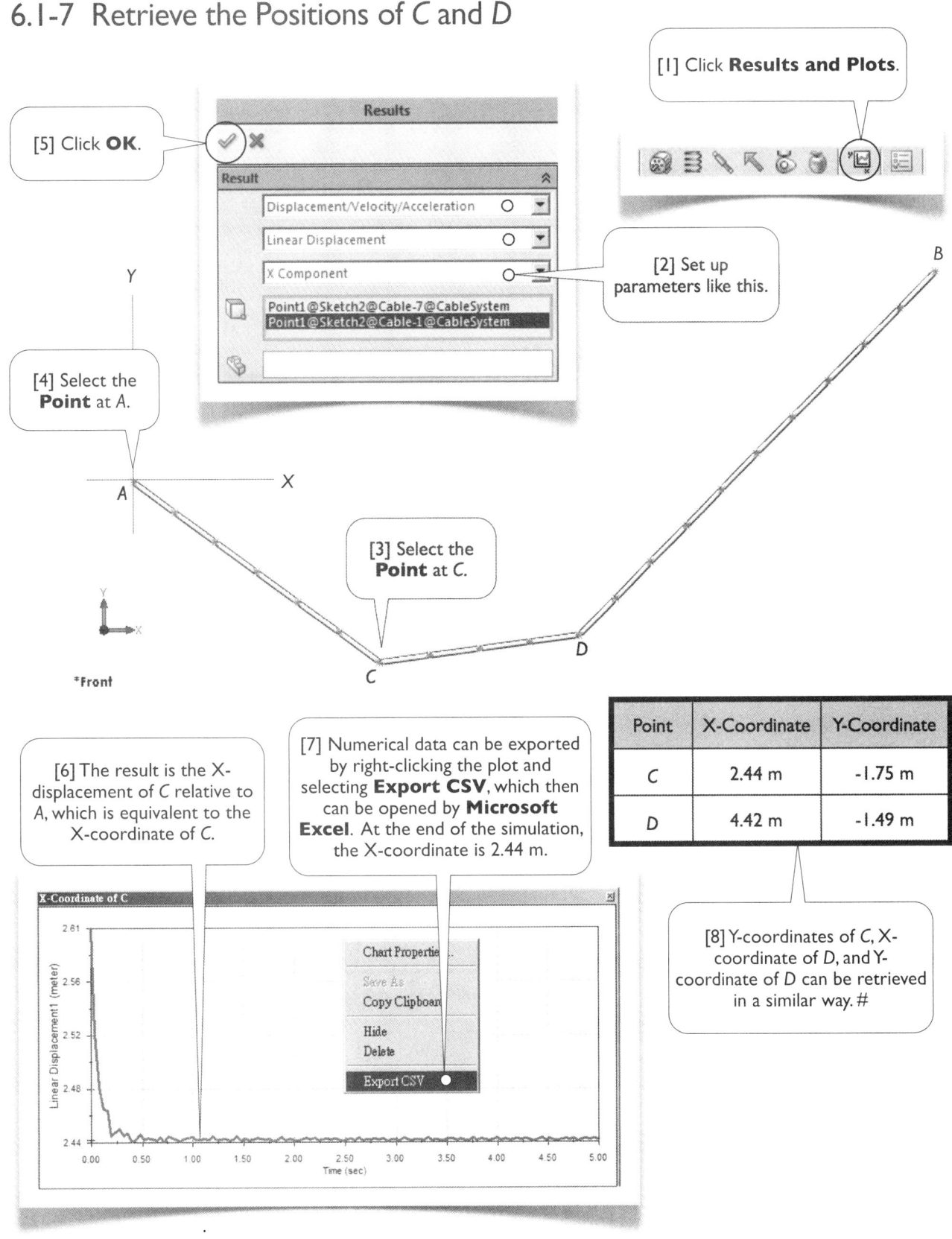

[1] Click **Results and Plots**.

[5] Click **OK**.

Results

Result

Displacement/Velocity/Acceleration

Linear Displacement

X Component

[2] Set up parameters like this.

Point1@Sketch2@Cable-7@CableSystem
Point1@Sketch2@Cable-1@CableSystem

[4] Select the **Point** at *A*.

[3] Select the **Point** at *C*.

*Front

[6] The result is the X-displacement of *C* relative to *A*, which is equivalent to the X-coordinate of *C*.

[7] Numerical data can be exported by right-clicking the plot and selecting **Export CSV**, which then can be opened by **Microsoft Excel**. At the end of the simulation, the X-coordinate is 2.44 m.

Point	X-Coordinate	Y-Coordinate
C	2.44 m	-1.75 m
D	4.42 m	-1.49 m

[8] Y-coordinates of *C*, X-coordinate of *D*, and Y-coordinate of *D* can be retrieved in a similar way. #

X-Coordinate of C

Chart Properties
Save As
Copy Clipboard
Hide
Delete
Export CSV

6.1-8 Retrieve Tensions in the Cable System

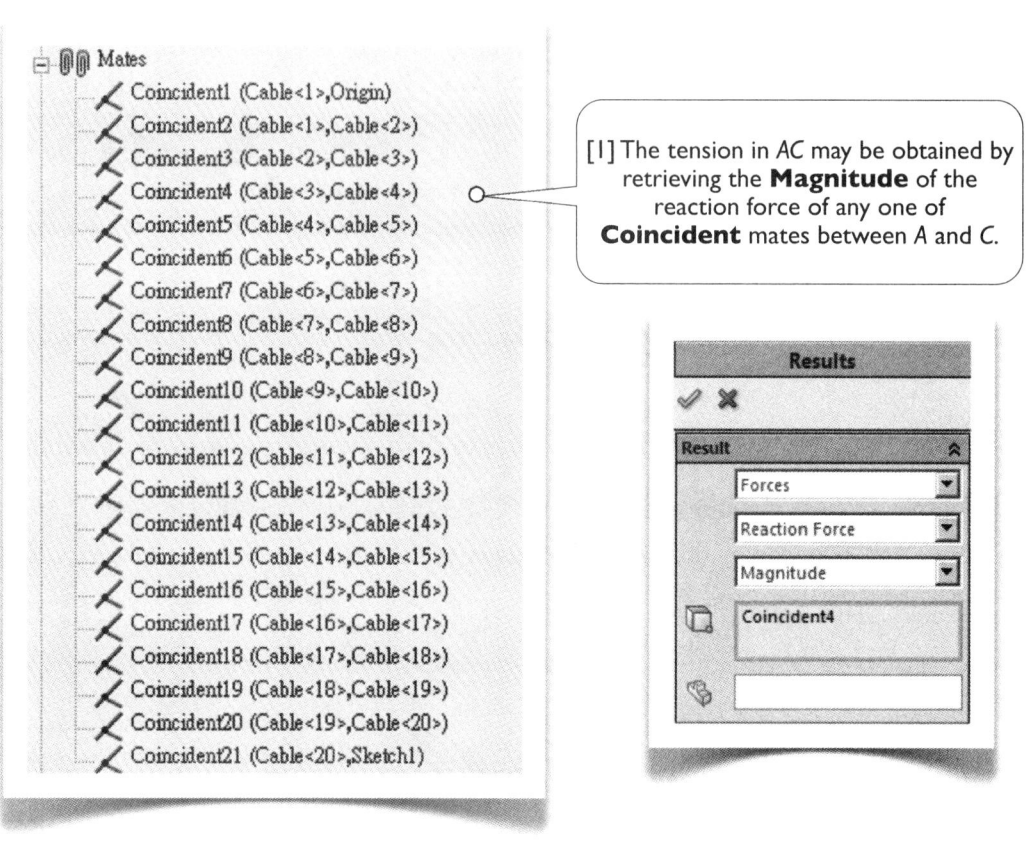

[1] The tension in *AC* may be obtained by retrieving the **Magnitude** of the reaction force of any one of **Coincident** mates between *A* and *C*.

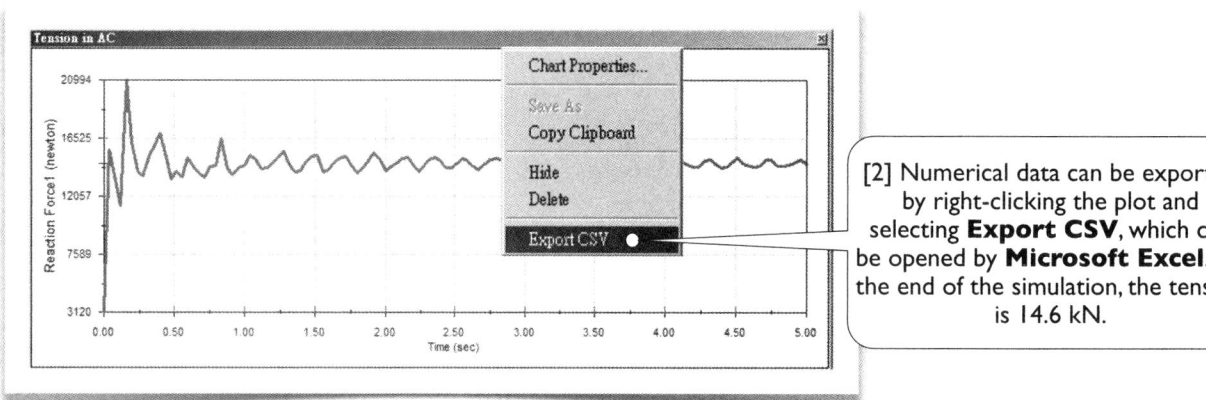

[2] Numerical data can be exported by right-clicking the plot and selecting **Export CSV**, which can be opened by **Microsoft Excel**. At the end of the simulation, the tension is 14.6 kN.

Section	Y-Coordinate
AC	14.6 kN
CD	11.9 kN
DB	16.6 kN

[3] Tension in *CD* and tension in *DB* can be retrieved in a similar way. #

6.1-9 Do It Yourself

Do It Yourself

[1] Verify the equilibria at C and at D.

Do It Yourself

[2] Retrieve the reaction forces at A and B, and verify the force equilibria at A and at B.

[3] Quit **SOLIDWORKS**. Click **Save all.** Click **Rebuild and save the document.** #

Section 6.2

Suspending Cable

6.2-1 Introduction

[1] Consider the cable in Section 6.1 again. Now, we remove the concentrated loads at C and at D, and consider the gravitational forces on the cable. We want to know the maximum tension force in the cable. Note that, by default, **SOLIDWORKS** assumes the material has a mass density of 1000 kg/m^3.

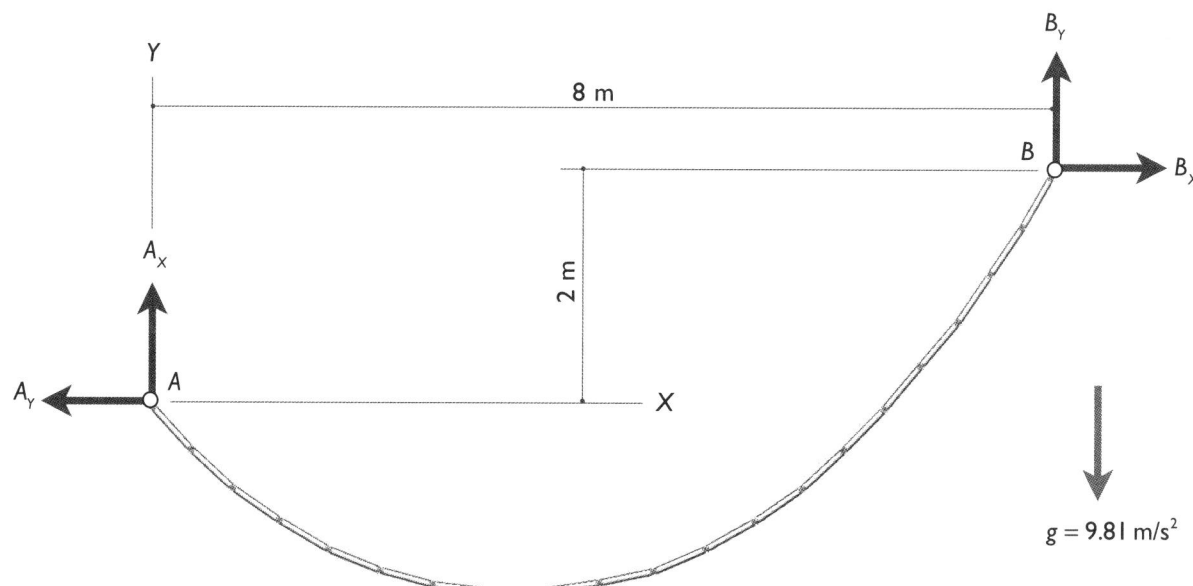

6.2-2 Start Up

[1] Launch **SOLIDWORKS** and click **Open** to open the file **CableSystem**, which was saved at the end of Section 6.1.

6.2-3 Create a New **Study** and Apply Gravitational Force

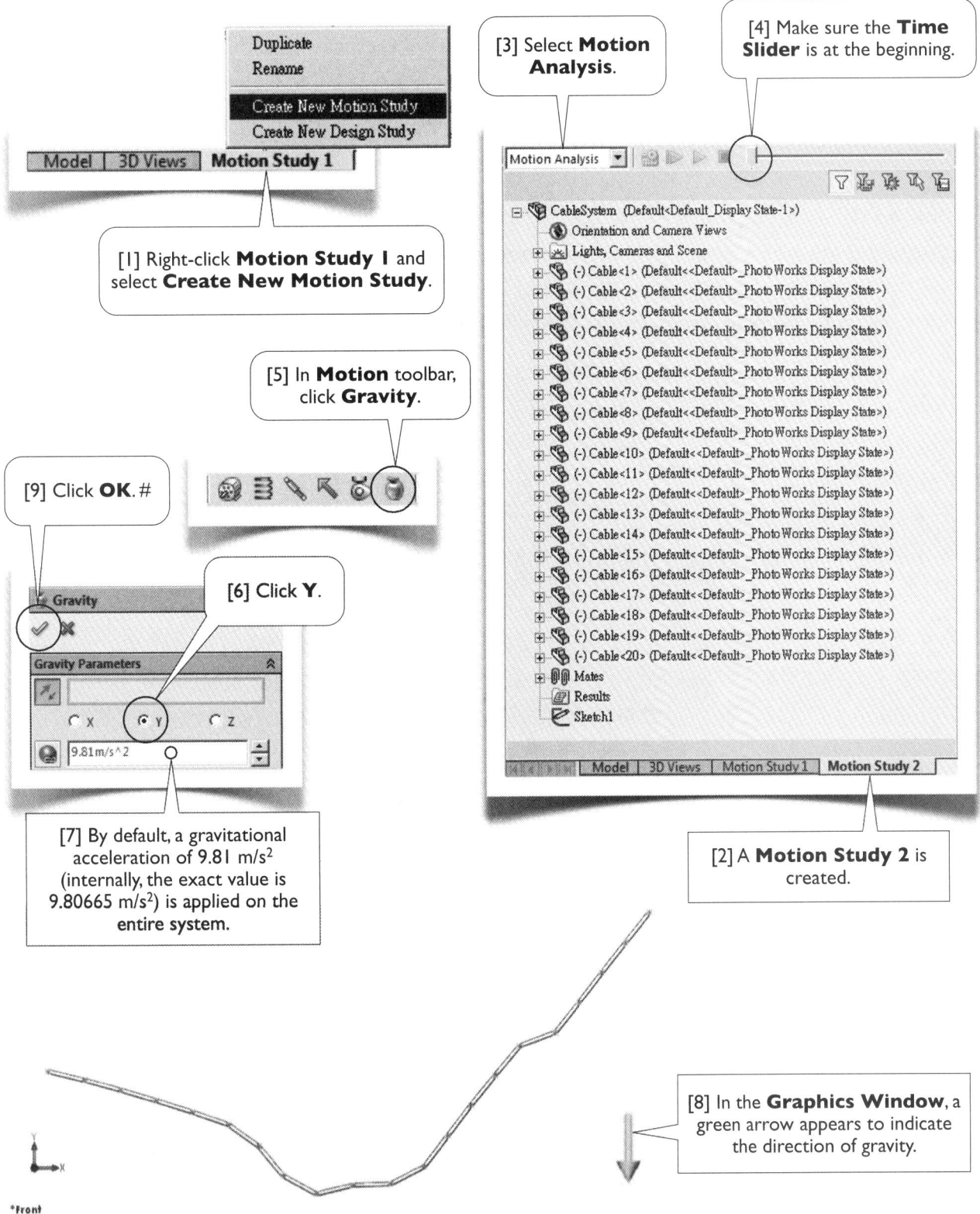

Duplicate
Rename
Create New Motion Study
Create New Design Study

Model | 3D Views | **Motion Study 1** |

[1] Right-click **Motion Study 1** and select **Create New Motion Study**.

[3] Select **Motion Analysis**.

[4] Make sure the **Time Slider** is at the beginning.

Motion Analysis

CableSystem (Default<Default_Display State-1>)
  Orientation and Camera Views
  Lights, Cameras and Scene
  (-) Cable<1> (Default<<Default>_PhotoWorks Display State>)
  (-) Cable<2> (Default<<Default>_PhotoWorks Display State>)
  (-) Cable<3> (Default<<Default>_PhotoWorks Display State>)
  (-) Cable<4> (Default<<Default>_PhotoWorks Display State>)
  (-) Cable<5> (Default<<Default>_PhotoWorks Display State>)
  (-) Cable<6> (Default<<Default>_PhotoWorks Display State>)
  (-) Cable<7> (Default<<Default>_PhotoWorks Display State>)
  (-) Cable<8> (Default<<Default>_PhotoWorks Display State>)
  (-) Cable<9> (Default<<Default>_PhotoWorks Display State>)
  (-) Cable<10> (Default<<Default>_PhotoWorks Display State>)
  (-) Cable<11> (Default<<Default>_PhotoWorks Display State>)
  (-) Cable<12> (Default<<Default>_PhotoWorks Display State>)
  (-) Cable<13> (Default<<Default>_PhotoWorks Display State>)
  (-) Cable<14> (Default<<Default>_PhotoWorks Display State>)
  (-) Cable<15> (Default<<Default>_PhotoWorks Display State>)
  (-) Cable<16> (Default<<Default>_PhotoWorks Display State>)
  (-) Cable<17> (Default<<Default>_PhotoWorks Display State>)
  (-) Cable<18> (Default<<Default>_PhotoWorks Display State>)
  (-) Cable<19> (Default<<Default>_PhotoWorks Display State>)
  (-) Cable<20> (Default<<Default>_PhotoWorks Display State>)
  Mates
  Results
  Sketch1

Model | 3D Views | Motion Study 1 | **Motion Study 2** |

[5] In **Motion** toolbar, click **Gravity**.

[9] Click **OK**. #

[6] Click **Y**.

Gravity

Gravity Parameters

C x   C Y   C Z

9.81m/s^2

[7] By default, a gravitational acceleration of 9.81 m/s^2 (internally, the exact value is 9.80665 m/s^2) is applied on the entire system.

[2] A **Motion Study 2** is created.

[8] In the **Graphics Window**, a green arrow appears to indicate the direction of gravity.

Y
X

*Front

6.2-4 Install Dampers

[1, 9] Save the document. Click **Calculate**.

[2] Click **Play**. The cable system vibrates. Let's install dampers to damp out vibrations. Click **Stop**.

Motion Analysis

[10] Click **Play**. At the end of the 5th second, the vibration is still large. Let's extend the simulation time. Click **Stop**. #

[3] Move the **Time Slider** to the beginning (see 6.2-3[4], last page). In **Motion** toolbar, click **Damper**.

[7] Click **OK**.

Damper

Damper Type

Linear Damper

Torsional Damper

[6] Type an arbitrary large damping coefficient, say 100 (N/(m/s)).

Damper Parameters

Point1@Sketch2@Cable-1@CableSystem
Point5@Sketch2@Cable-8@CableSystem

cv^e 1(linear)

C 100.00 N/(m/s)

B

[4] Click the **Point** at A.

[5] Click the **Point** that, by estimation, would be the lowest **Point**. Here, we select the point between **Cable<8>** and **Cable<9>**.

A

[8] Repeat similar procedure in steps [3-7] to install one more damper connecting B and the **Point** mentioned in [5].

*Front

6.2-5 Extend Simulation Time

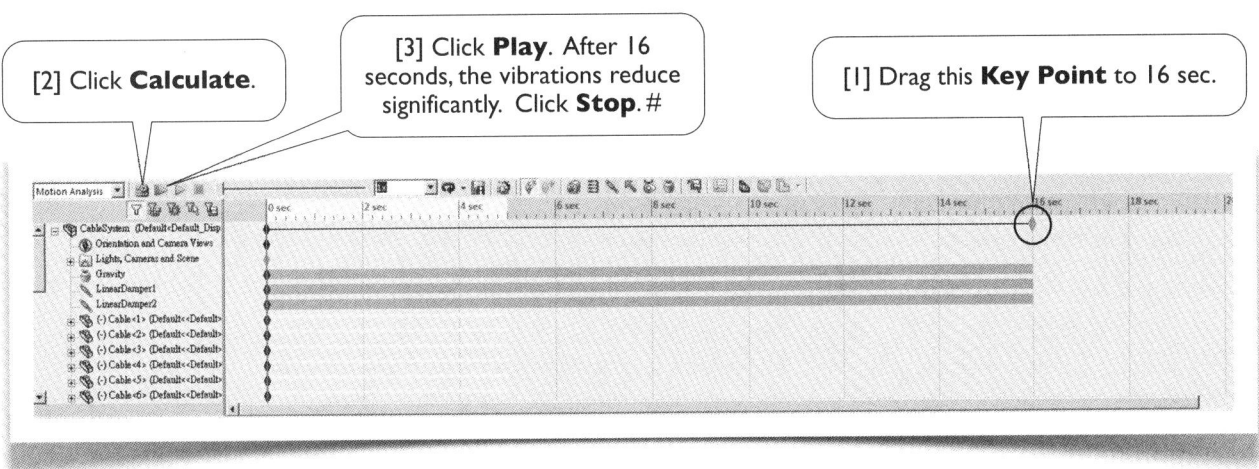

[2] Click **Calculate**.

[3] Click **Play**. After 16 seconds, the vibrations reduce significantly. Click **Stop**. #

[1] Drag this **Key Point** to 16 sec.

6.2-6 Retrieve Maximum Tensions

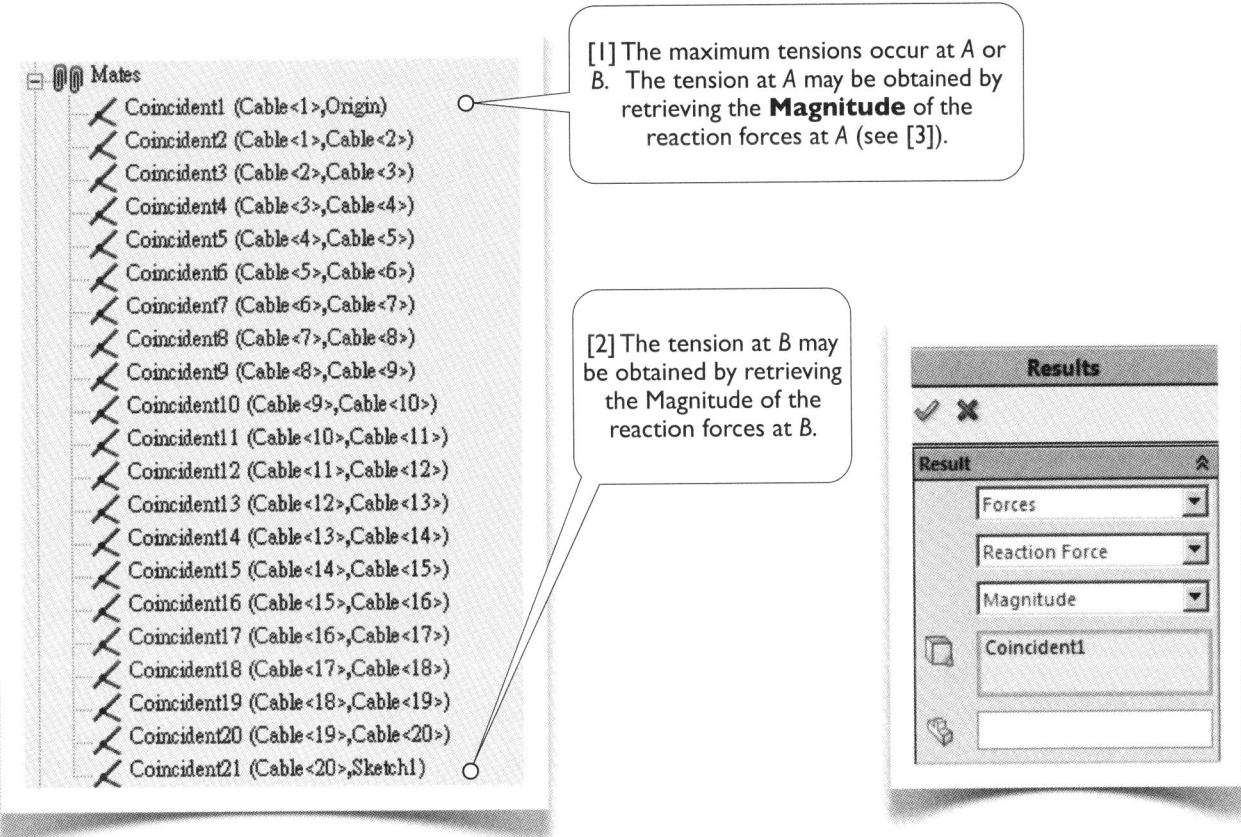

[1] The maximum tensions occur at A or B. The tension at A may be obtained by retrieving the **Magnitude** of the reaction forces at A (see [3]).

[2] The tension at B may be obtained by retrieving the Magnitude of the reaction forces at B.

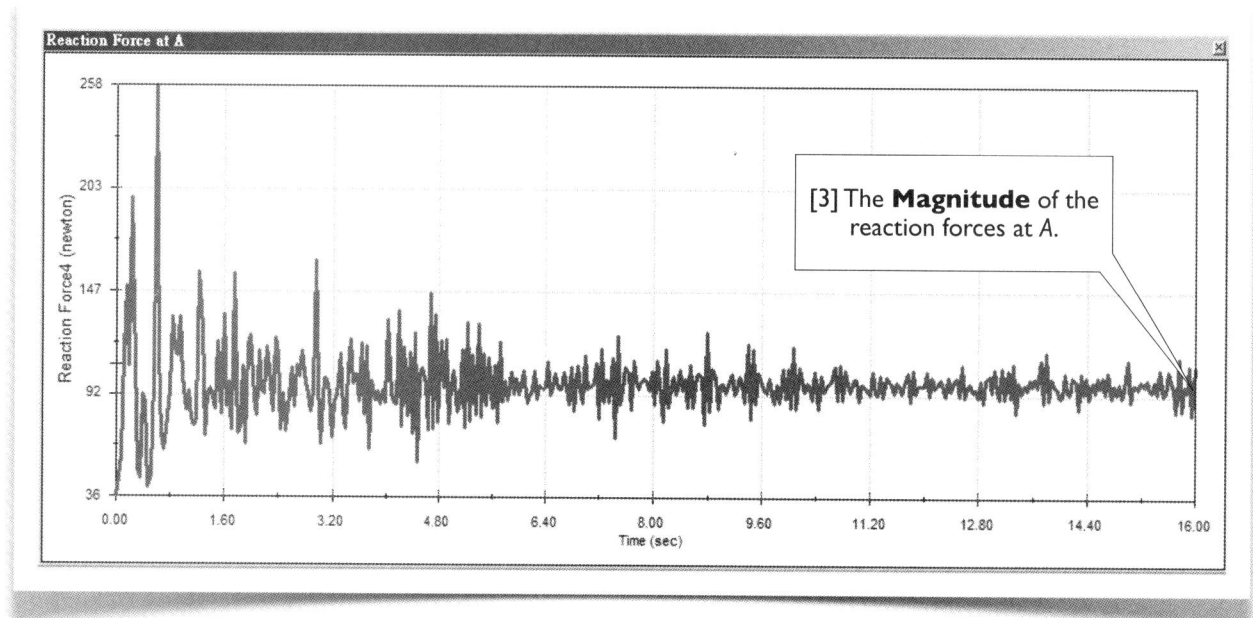

[3] The **Magnitude** of the reaction forces at *A*.

Discussion

[4] We're solving a static case using dynamic approach. In cases like this, we need to add some energy dissipation mechanism to obtain a steady-state solution. Obviously, in this section, the vibration is still large and the solution is not accurate enough. To obtain a more accurate solution, you may try to add more dampers or extend more simulation time. Friction is also a way to dissipate energy. We'll discuss friction in the next chapter.

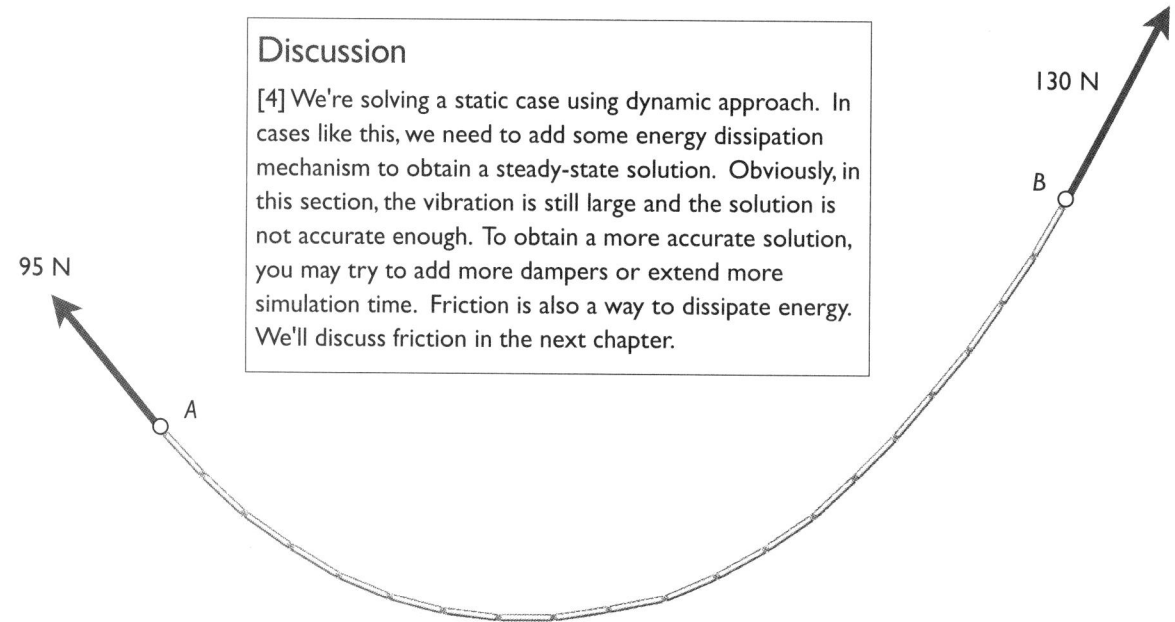

130 N

B

95 N

A

Do It Yourself

[5] At any point of the cable system, the horizontal component of the tension must be the same. Verify this.

[6] Quit **SOLIDWORKS**. Click **Save all.** Click **Rebuild and save the document**. #

Chapter 7

Friction

In the first five chapters, we assumed that no frictions were involved in structures or machines. In reality, frictions exist in a structural or mechanical system.

Dampers can be used to dissipate mechanical energy, similar to that electrical resistors are used to dissipate electrical energy. In Chapter 6, we've added dampers to dissipate mechanical energy, to stabilize a cable system. Without dampers, the cable system would vibrate forever. Typically, a damper is made by utilizing "viscous friction," in which the friction force increases as the relative speed between solid bodies and surrounding fluid increases.

Besides the viscous frictions (frictions between solid bodies and surrounding fluid), two additional types of frictions can be identified in a mechanical system: dry frictions and material frictions. Dry frictions are the friction between solid bodies. Material frictions are the internal frictions between the grains or molecules of a material.

In this chapter, we discuss the **dry friction**.

Section 7.1

Block on Floor

7.1-1 Introduction

[1] This section is designed to reinforce the concepts of *friction force* and *coefficient of friction*.

Consider a **Block** on a **Floor** [2, 3]. A constant vertical force $V = 100$ N and a lateral force P, increasing gradually from 0 to 50 N, are applied on the **Block** [4, 5]. Reaction force on the **Block** has a normal component N and a shear component F, called a friction force [6]. Before the **Block** starts to move, it is in a static equilibrium state, therefore $F = P$ [6, 7]. When F reaches a value $\mu_s N$, the **Block** starts to move. μ_s is called the *coefficient of static friction* [8, 9]. As soon as the **Block** starts to move, F reduces to a constant $\mu_k N$. μ_k is called the *coefficient of dynamic friction* [10].

In the following simulation, we'll neglect the difference between μ_s and μ_k and assume $\mu_s = \mu_k = 0.3$.

[9] When F reaches a value $\mu_s N$, the **Block** starts to move. μ_s is called the coefficient of static friction.

[10] As soon as the **Block** starts to move, F reduces to a constant $\mu_k N$. μ_k is called the coefficient of dynamic friction. #

[7] Before the **Block** starts to move, $F = P$.

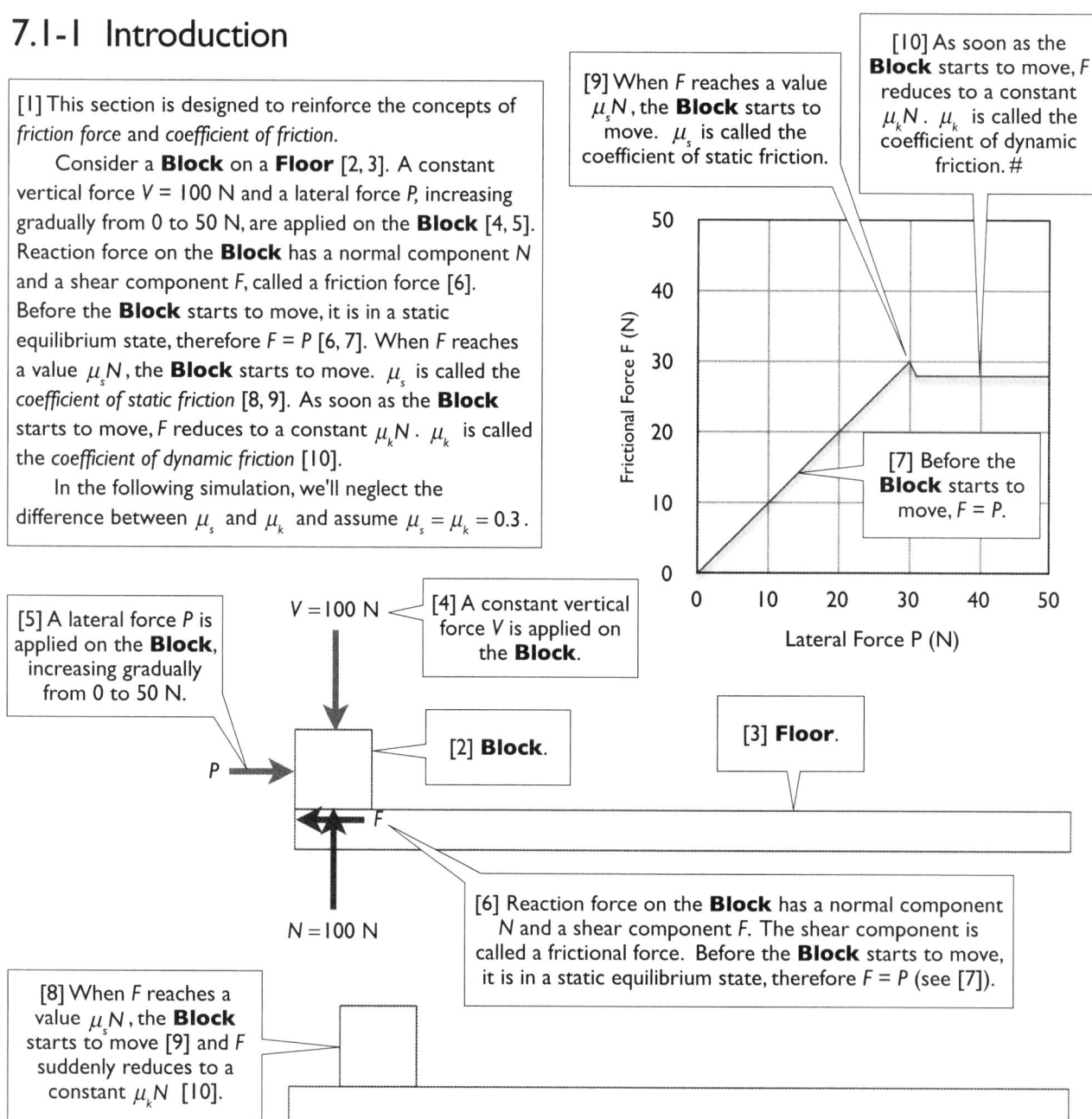

[5] A lateral force P is applied on the **Block**, increasing gradually from 0 to 50 N.

$V = 100$ N

[4] A constant vertical force V is applied on the **Block**.

[2] **Block**.

[3] **Floor**.

P

F

$N = 100$ N

[6] Reaction force on the **Block** has a normal component N and a shear component F. The shear component is called a frictional force. Before the **Block** starts to move, it is in a static equilibrium state, therefore $F = P$ (see [7]).

[8] When F reaches a value $\mu_s N$, the **Block** starts to move [9] and F suddenly reduces to a constant $\mu_k N$ [10].

7.1-2 Start Up and Create a Part: **Block**

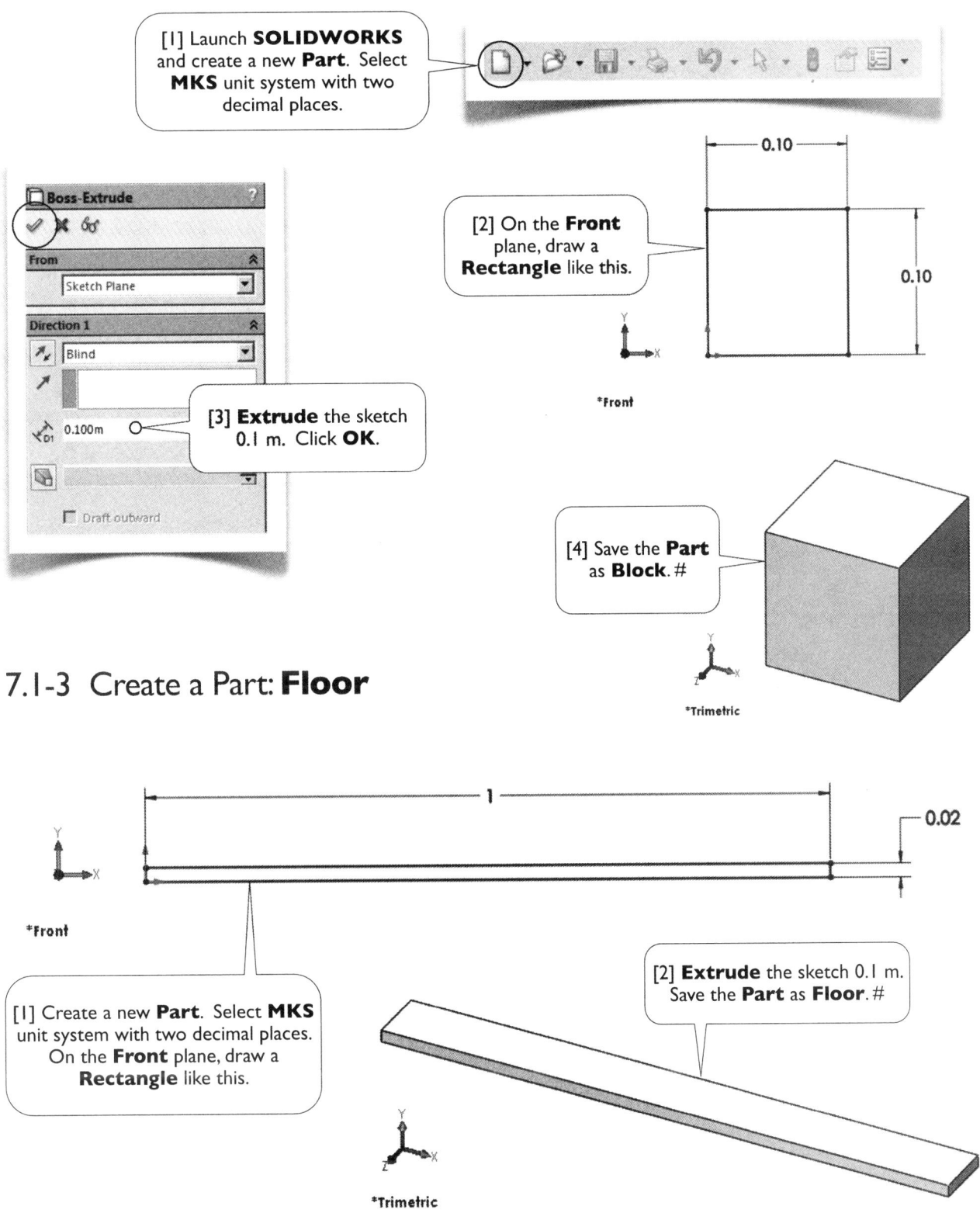

[1] Launch **SOLIDWORKS** and create a new **Part**. Select **MKS** unit system with two decimal places.

[2] On the **Front** plane, draw a **Rectangle** like this.

0.10

0.10

*Front

Boss-Extrude

From

Sketch Plane

Direction 1

Blind

0.100m

[3] **Extrude** the sketch 0.1 m. Click **OK**.

Draft outward

[4] Save the **Part** as **Block**. #

*Trimetric

7.1-3 Create a Part: **Floor**

1

0.02

*Front

[1] Create a new **Part**. Select **MKS** unit system with two decimal places. On the **Front** plane, draw a **Rectangle** like this.

[2] **Extrude** the sketch 0.1 m. Save the **Part** as **Floor**. #

*Trimetric

7.1-4 Create an Assembly: **Block-On-Floor**

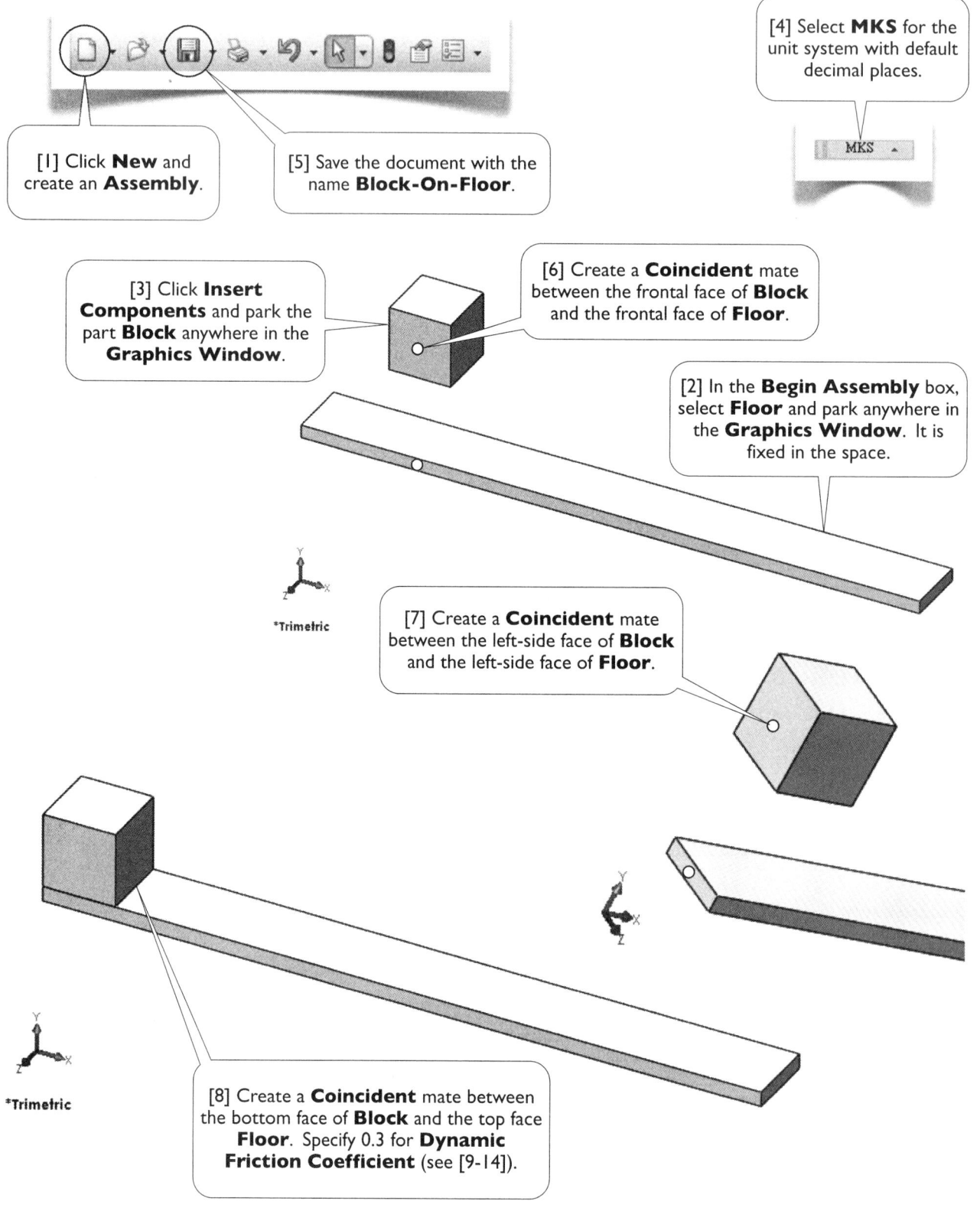

[4] Select **MKS** for the unit system with default decimal places.

[1] Click **New** and create an **Assembly**.

[5] Save the document with the name **Block-On-Floor**.

[3] Click **Insert Components** and park the part **Block** anywhere in the **Graphics Window**.

[6] Create a **Coincident** mate between the frontal face of **Block** and the frontal face of **Floor**.

[2] In the **Begin Assembly** box, select **Floor** and park anywhere in the **Graphics Window**. It is fixed in the space.

[7] Create a **Coincident** mate between the left-side face of **Block** and the left-side face of **Floor**.

*Trimetric

[8] Create a **Coincident** mate between the bottom face of **Block** and the top face **Floor**. Specify 0.3 for **Dynamic Friction Coefficient** (see [9-14]).

*Trimetric

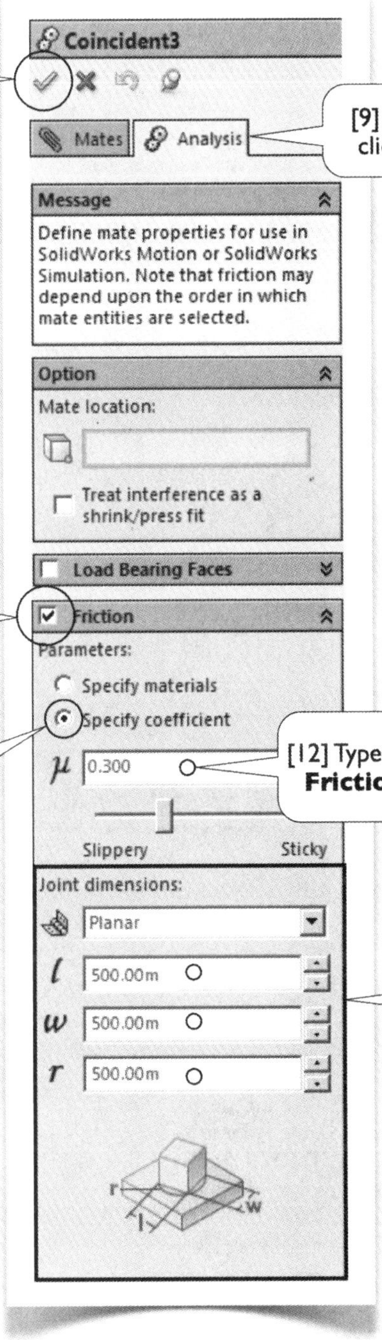

Coincident3

[14] Click **OK**.

Mates | Analysis

[9] To specify friction, click **Analysis** tab.

Message

Define mate properties for use in SolidWorks Motion or SolidWorks Simulation. Note that friction may depend upon the order in which mate entities are selected.

Option

Mate location:

☐ Treat interference as a shrink/press fit

Load Bearing Faces

[10] Click **Friction**.

☑ **Friction**

Parameters:

○ Specify materials
⦿ Specify coefficient

[11] Click **Specify coefficient**.

μ 0.300

[12] Type 0.3 for **Dynamic Friction Coefficient**.

Slippery   Sticky

Joint dimensions:

Planar

l 500.00m
w 500.00m
r 500.00m

[13] **SOLIDWORKS Motion** uses a friction model that includes **Joint dimensions** effects. To disable the **Joint dimensions** effects, simply type large dimensions. Here, we type 500 (m) for all the dimensions.

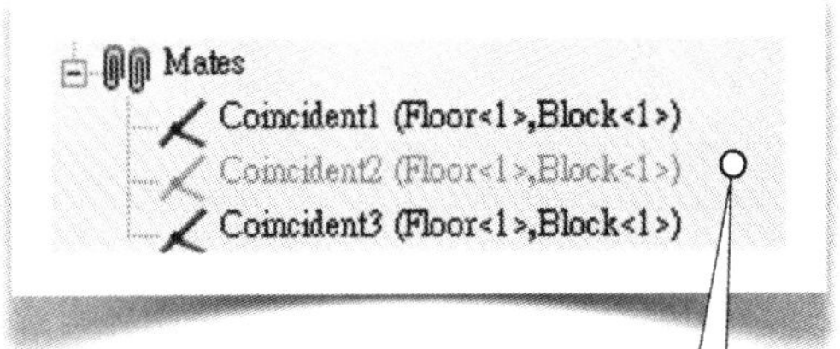

Mates
  Coincident1 (Floor<1>,Block<1>)
  Coincident2 (Floor<1>,Block<1>)
  Coincident3 (Floor<1>,Block<1>)

[15] Dismiss **Mate** box. Right-click **Coincident2** and select **Suppress**.

[16] Click to save the document. #

7.1-5 Create a **Study**

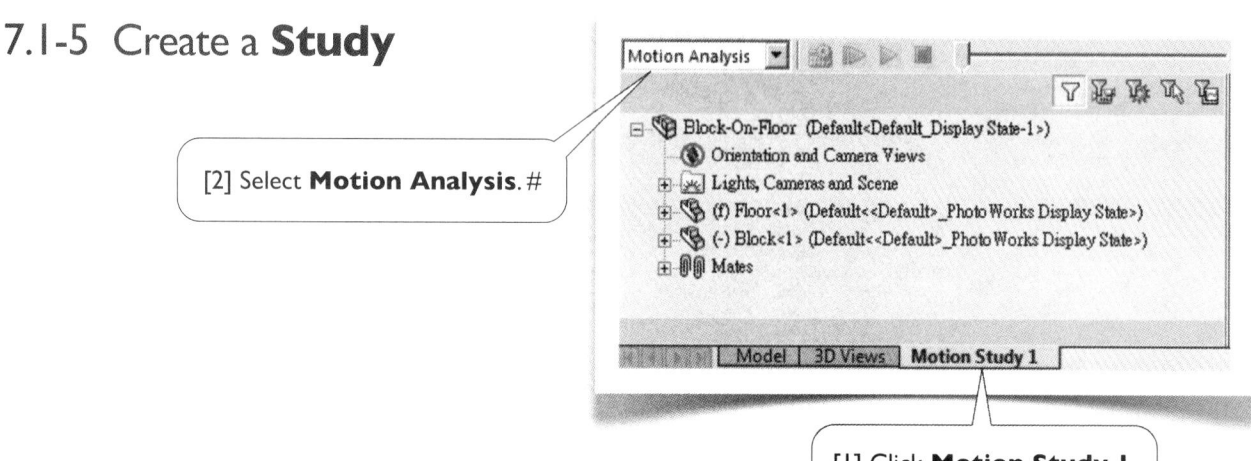

[2] Select **Motion Analysis**. #

[1] Click **Motion Study 1**.

7.1-6 Set Up Forces and Calculate Results

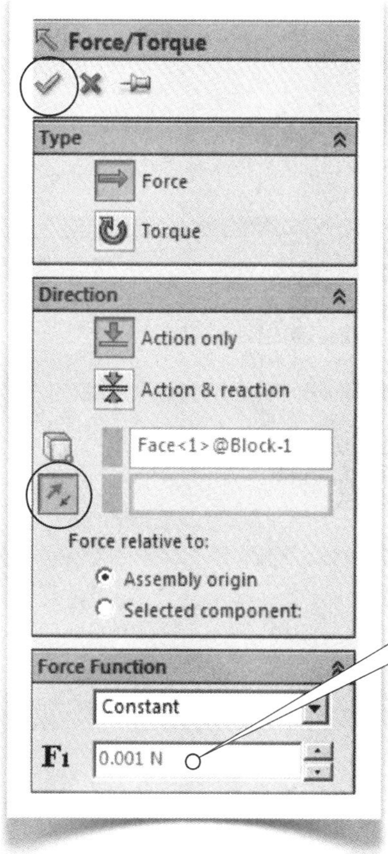

[1] Make sure the **Time Slider** is at the beginning (Time = 0 sec).

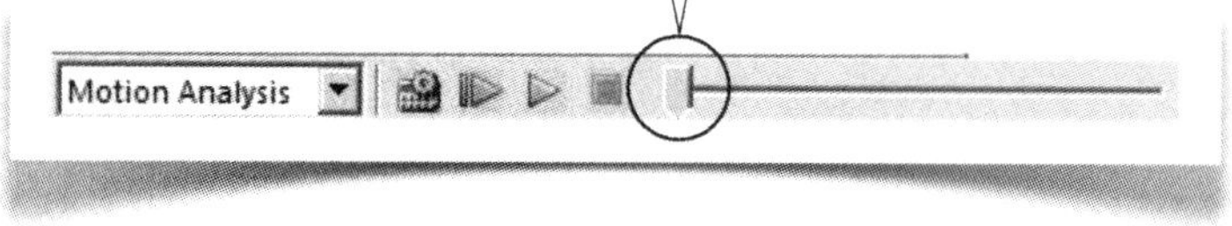

[2, 4] In **Motion** toolbar, click **Force**.

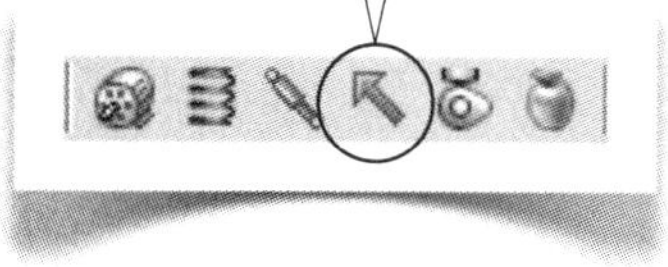

[6] Type 0.001 (N). We intend to specify a zero force value at Time = 0. However, **SOLIDWORKS** doesn't allow a zero value for a force. Therefore we type a very small value. This number essentially equals zero.

[3] Apply 100 N downward on the top face of the **Block**.

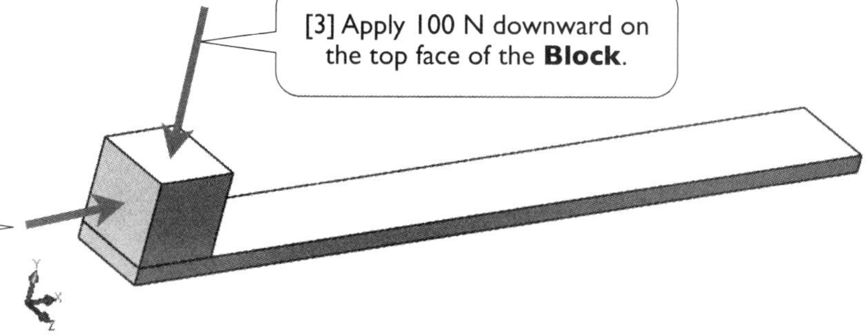

[5] Apply 0.001 N rightward on the left-side face of the **Block** (see [6]).

Force/Torque

Type

→ Force

↻ Torque

Direction

↧ Action only

⥮ Action & reaction

Face<1>@Block-1

Force relative to:
⦿ Assembly origin
○ Selected component:

Force Function

Constant

F_1 0.001 N

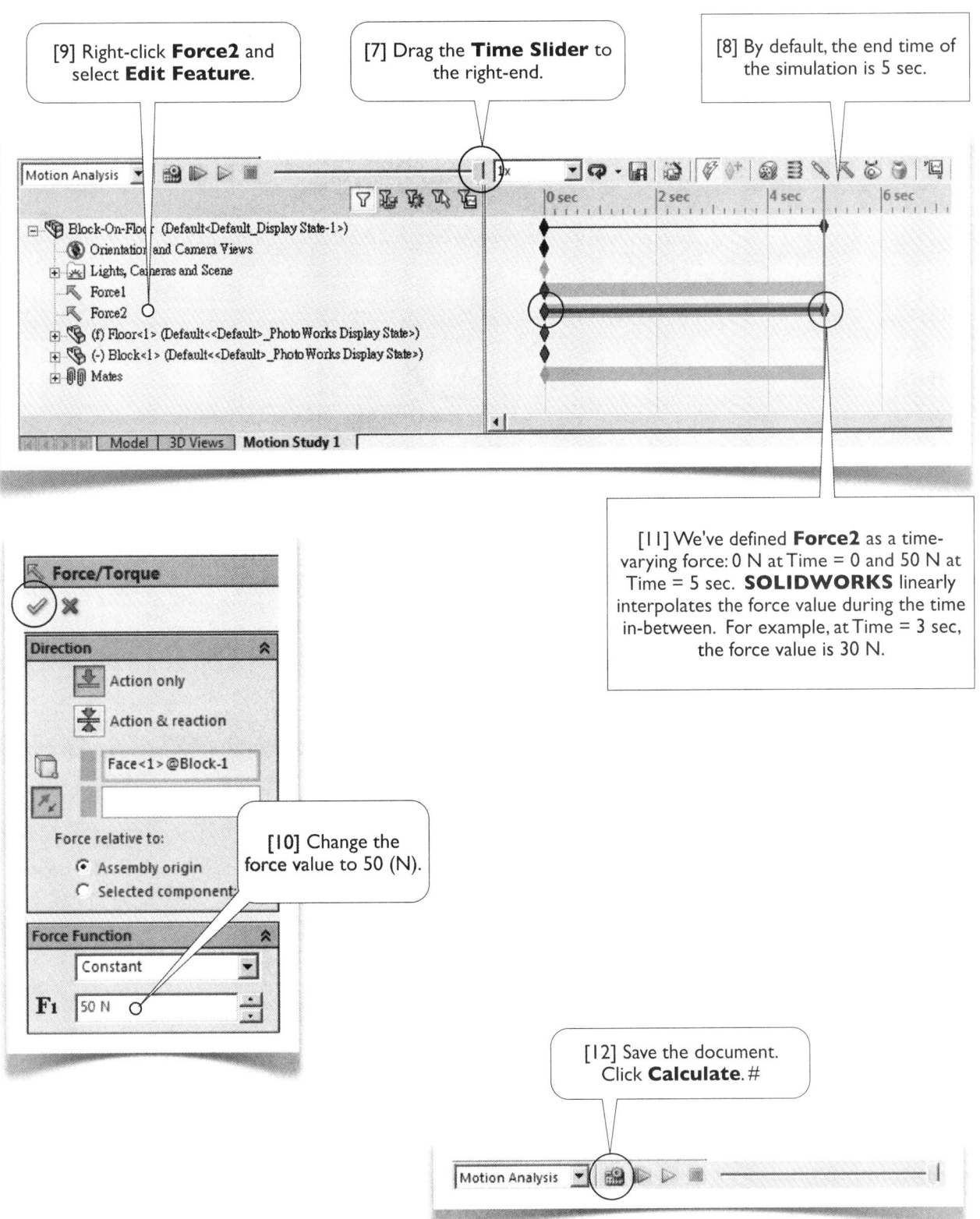

[9] Right-click **Force2** and select **Edit Feature**.

[7] Drag the **Time Slider** to the right-end.

[8] By default, the end time of the simulation is 5 sec.

[11] We've defined **Force2** as a time-varying force: 0 N at Time = 0 and 50 N at Time = 5 sec. **SOLIDWORKS** linearly interpolates the force value during the time in-between. For example, at Time = 3 sec, the force value is 30 N.

[10] Change the force value to 50 (N).

[12] Save the document. Click **Calculate**. #

7.1-7 Retrieve Friction Force

[1] Click **Play**.

[2] Click **Stop**.

[3] In **Motion** toolbar, click **Results and Plots**.

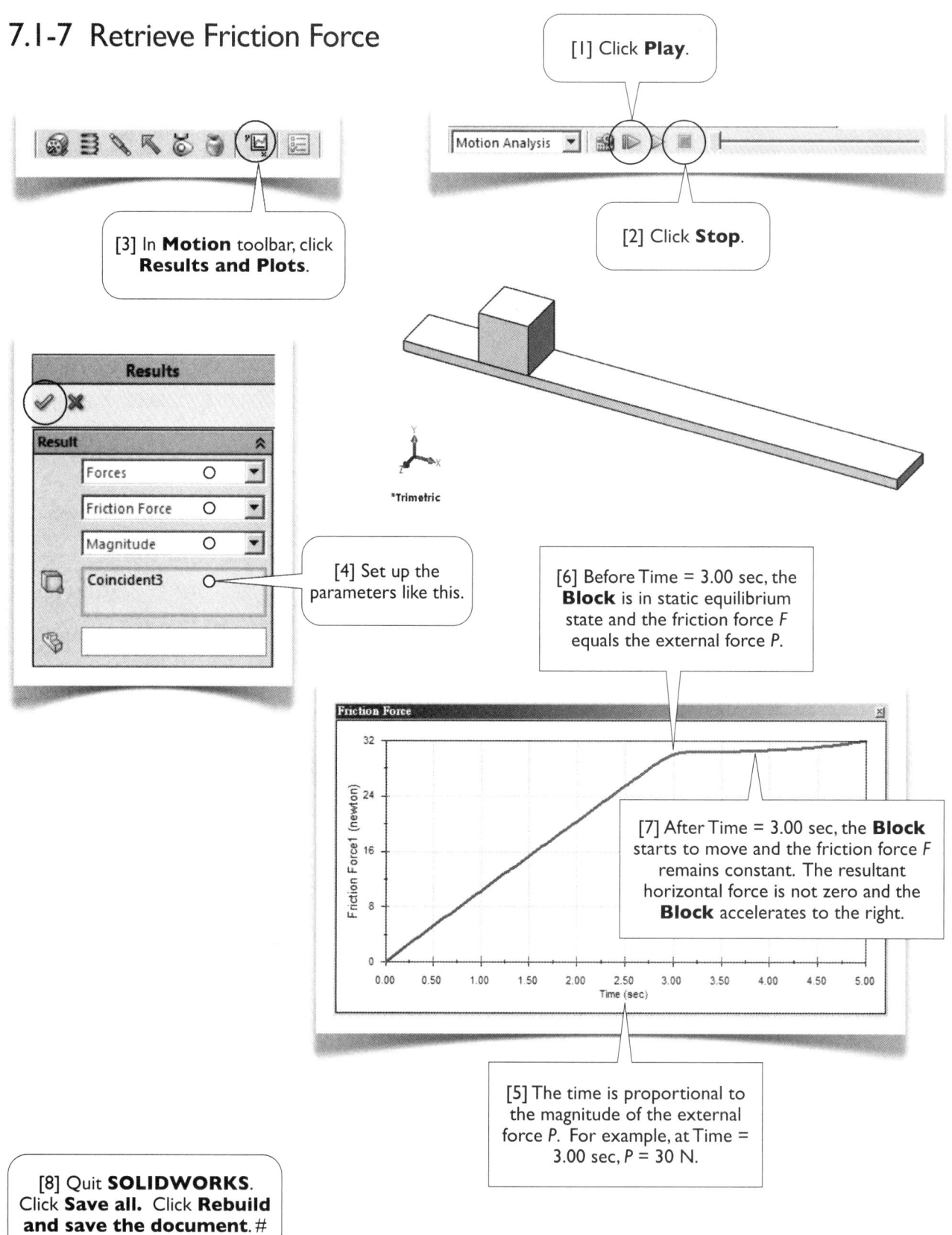

*Trimetric

Results

Result

Forces

Friction Force

Magnitude

Coincident3

[4] Set up the parameters like this.

[6] Before Time = 3.00 sec, the **Block** is in static equilibrium state and the friction force *F* equals the external force *P*.

[7] After Time = 3.00 sec, the **Block** starts to move and the friction force *F* remains constant. The resultant horizontal force is not zero and the **Block** accelerates to the right.

[5] The time is proportional to the magnitude of the external force *P*. For example, at Time = 3.00 sec, *P* = 30 N.

[8] Quit **SOLIDWORKS**. Click **Save all.** Click **Rebuild and save the document**. #

Section 7.2

Wedge

7.2-1 Introduction

[1] The position of a **Block** can be adjusted by moving a **Wedge** [2-7]. Knowing that the coefficient of static friction is 0.35 between all surfaces of contact, including the vertical walls guiding the **Block**. We want to find the force *P* required to raise the **Block** [8].

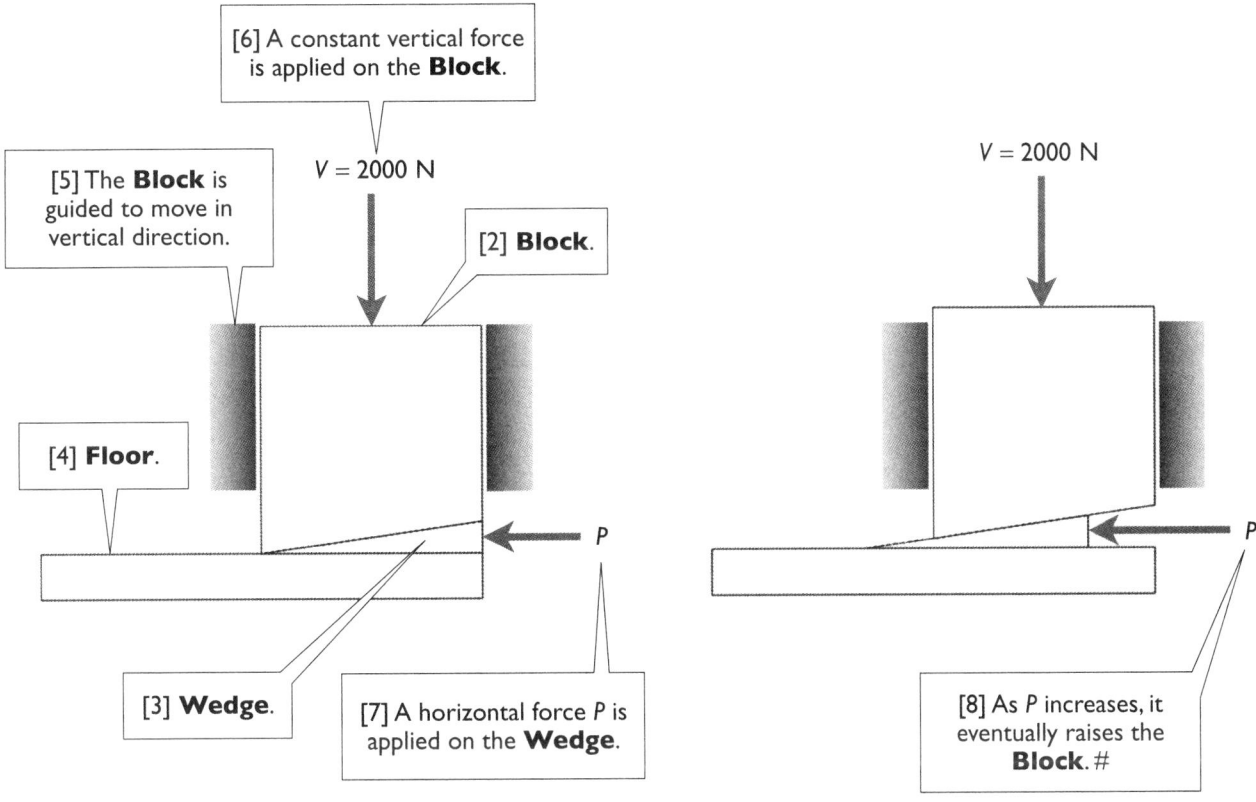

[6] A constant vertical force is applied on the **Block**.

[5] The **Block** is guided to move in vertical direction.

[2] **Block**.

[4] **Floor**.

$V = 2000$ N

[3] **Wedge**.

[7] A horizontal force *P* is applied on the **Wedge**.

$V = 2000$ N

[8] As *P* increases, it eventually raises the **Block**. #

7.2-2 Start Up and Create a Part: **Block**

[1] Launch **SOLIDWORKS** and create a new **Part**. Select **MKS** unit system with two decimal places.

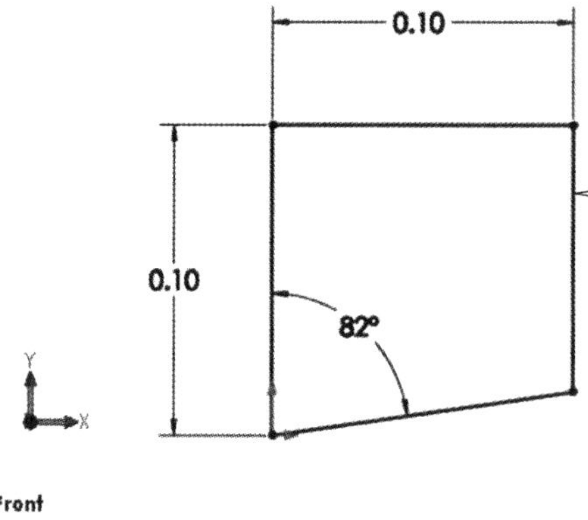

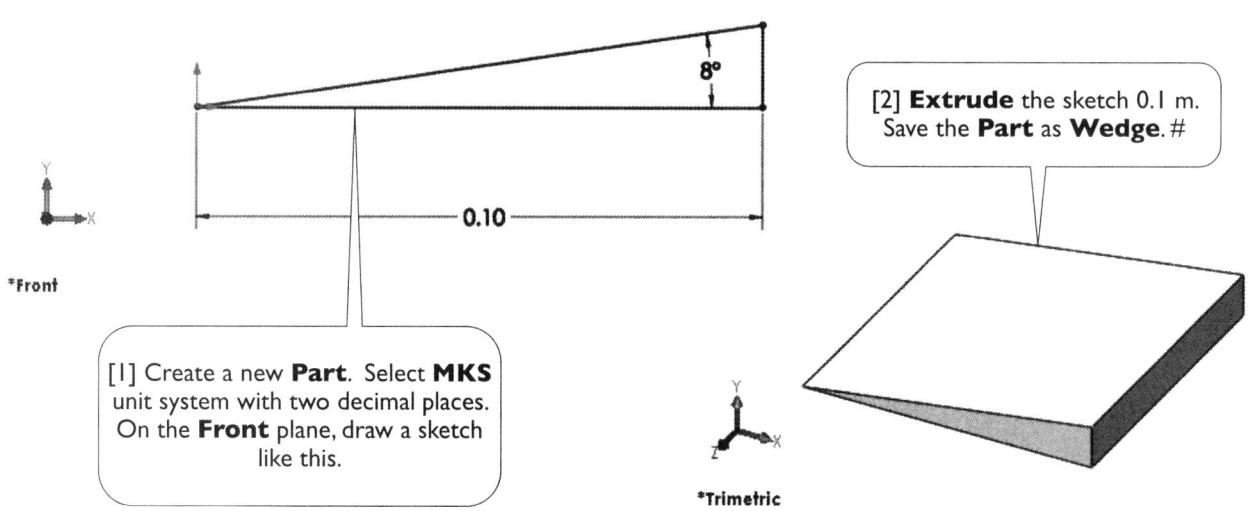

[2] On the **Front** plane, draw a sketch like this.

0.10

0.10

82°

*Front

Boss-Extrude

From

Sketch Plane

Direction 1

Blind

0.100m

☐ Draft outward

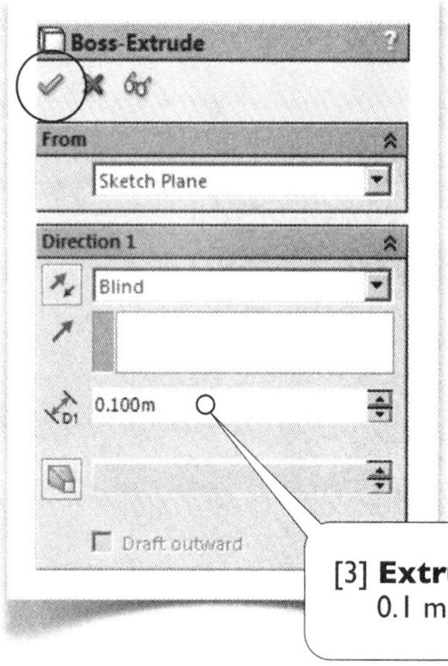

[3] **Extrude** the sketch 0.1 m. Click **OK**.

[4] Save the **Part** as **Block**. #

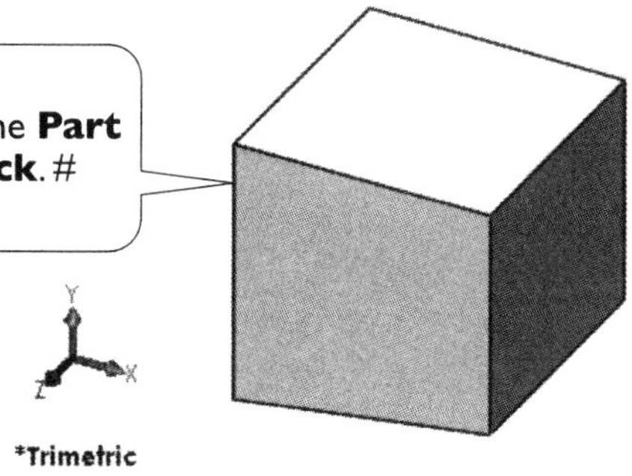

*Trimetric

7.2-3 Create a Part: **Wedge**

8°

0.10

*Front

[2] **Extrude** the sketch 0.1 m. Save the **Part** as **Wedge**. #

[1] Create a new **Part**. Select **MKS** unit system with two decimal places. On the **Front** plane, draw a sketch like this.

*Trimetric

7.2-4 Create a Part: **Floor**

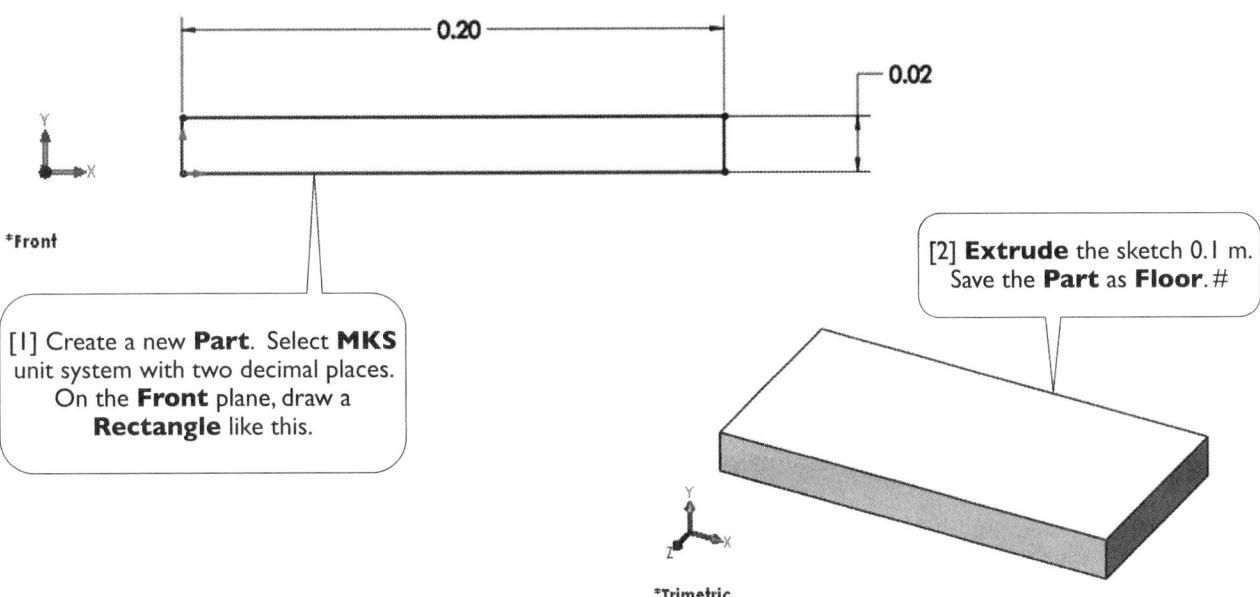

0.20

0.02

Y
X
*Front

[1] Create a new **Part**. Select **MKS** unit system with two decimal places. On the **Front** plane, draw a **Rectangle** like this.

[2] **Extrude** the sketch 0.1 m. Save the **Part** as **Floor**. #

*Trimetric

7.2-5 Create an Assembly: **WedgeSystem**

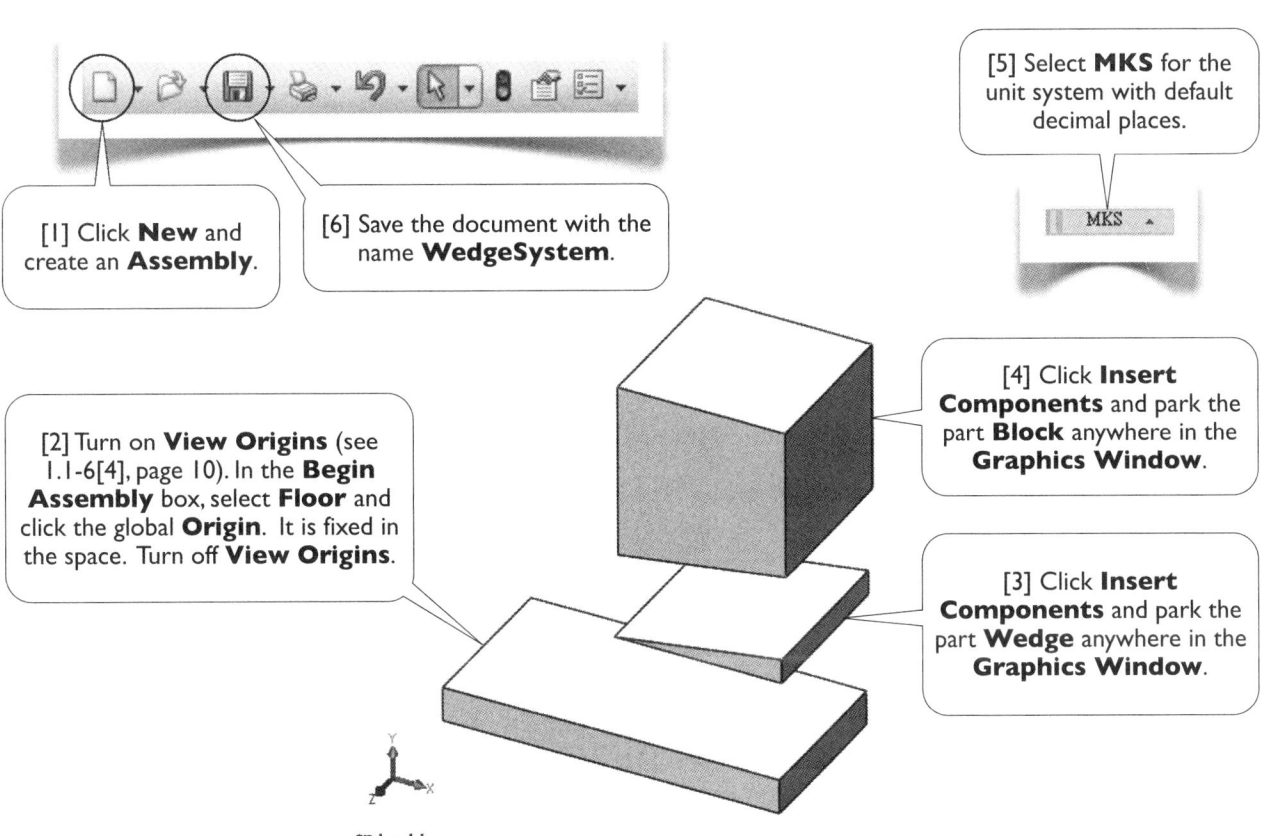

[5] Select **MKS** for the unit system with default decimal places.

MKS

[1] Click **New** and create an **Assembly**.

[6] Save the document with the name **WedgeSystem**.

[4] Click **Insert Components** and park the part **Block** anywhere in the **Graphics Window**.

[2] Turn on **View Origins** (see 1.1-6[4], page 10). In the **Begin Assembly** box, select **Floor** and click the global **Origin**. It is fixed in the space. Turn off **View Origins**.

[3] Click **Insert Components** and park the part **Wedge** anywhere in the **Graphics Window**.

*Trimetric

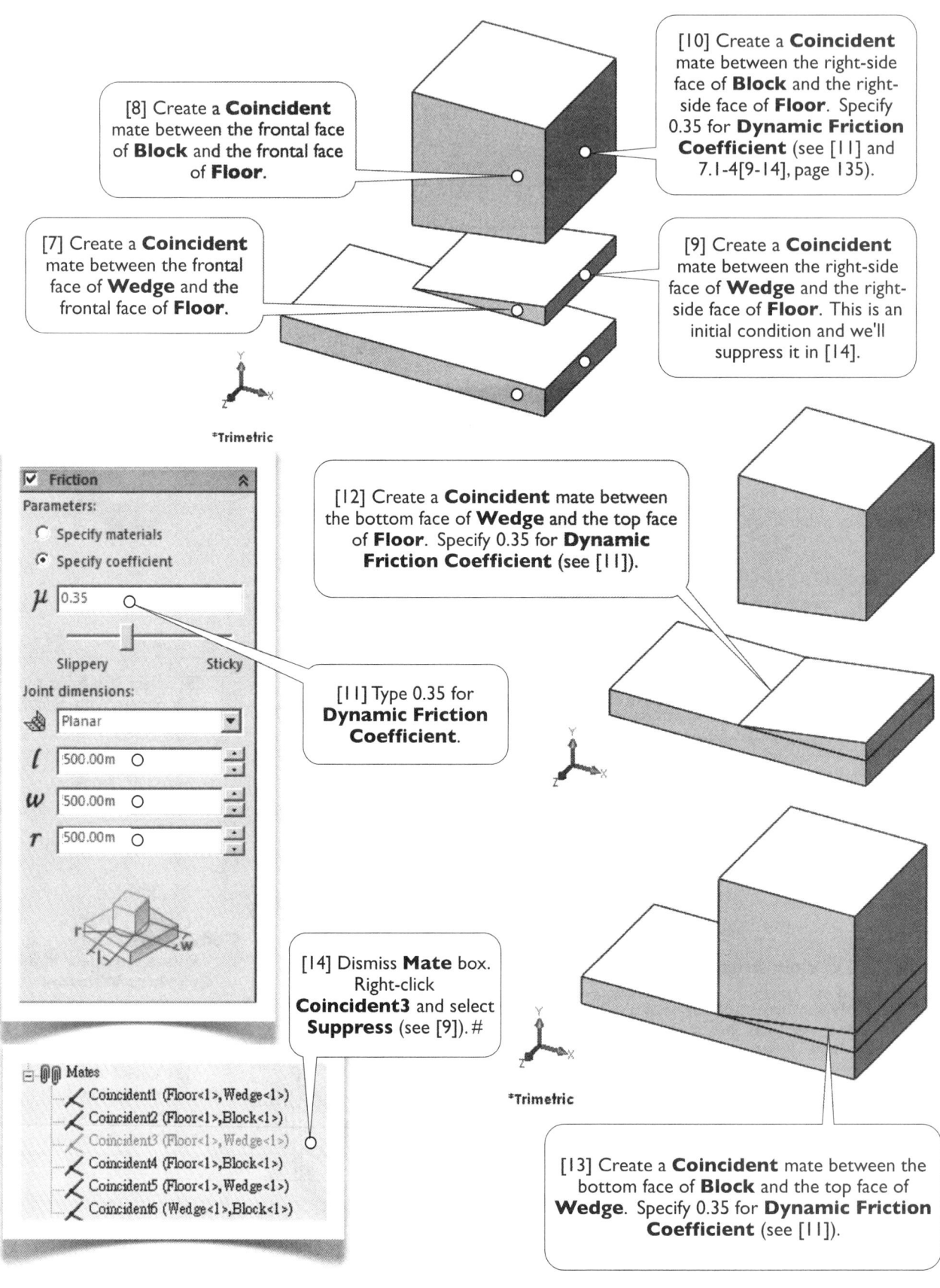

[8] Create a **Coincident** mate between the frontal face of **Block** and the frontal face of **Floor**.

[10] Create a **Coincident** mate between the right-side face of **Block** and the right-side face of **Floor**. Specify 0.35 for **Dynamic Friction Coefficient** (see [11] and 7.1-4[9-14], page 135).

[7] Create a **Coincident** mate between the frontal face of **Wedge** and the frontal face of **Floor**.

[9] Create a **Coincident** mate between the right-side face of **Wedge** and the right-side face of **Floor**. This is an initial condition and we'll suppress it in [14].

*Trimetric

[12] Create a **Coincident** mate between the bottom face of **Wedge** and the top face of **Floor**. Specify 0.35 for **Dynamic Friction Coefficient** (see [11]).

Friction

Parameters:
- Specify materials
- Specify coefficient

μ 0.35

Slippery   Sticky

Joint dimensions:

Planar

l 500.00m

w 500.00m

r 500.00m

[11] Type 0.35 for **Dynamic Friction Coefficient**.

[14] Dismiss **Mate** box. Right-click **Coincident3** and select **Suppress** (see [9]). #

*Trimetric

[13] Create a **Coincident** mate between the bottom face of **Block** and the top face of **Wedge**. Specify 0.35 for **Dynamic Friction Coefficient** (see [11]).

Mates
- Coincident1 (Floor<1>,Wedge<1>)
- Coincident2 (Floor<1>,Block<1>)
- Coincident3 (Floor<1>,Wedge<1>)
- Coincident4 (Floor<1>,Block<1>)
- Coincident5 (Floor<1>,Wedge<1>)
- Coincident6 (Wedge<1>,Block<1>)

7.2-6 Create a **Study**

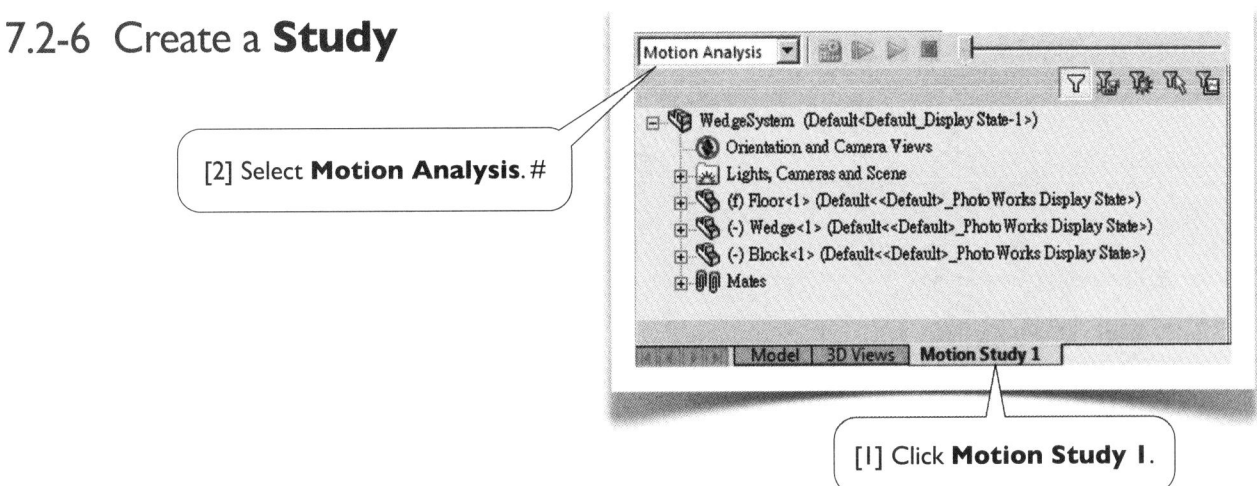

[2] Select **Motion Analysis**. #

[1] Click **Motion Study 1**.

7.2-7 Set Up Forces and Calculate Results

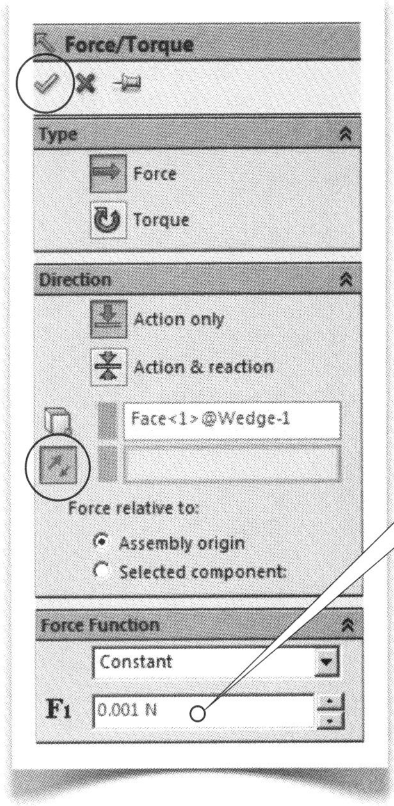

Force/Torque

Type

Force

Torque

Direction

Action only

Action & reaction

Face<1>@Wedge-1

Force relative to:
- Assembly origin
- Selected component

Force Function

Constant

F_1 0.001 N

[1] Make sure the **Time Slider** is at the beginning (Time = 0 sec).

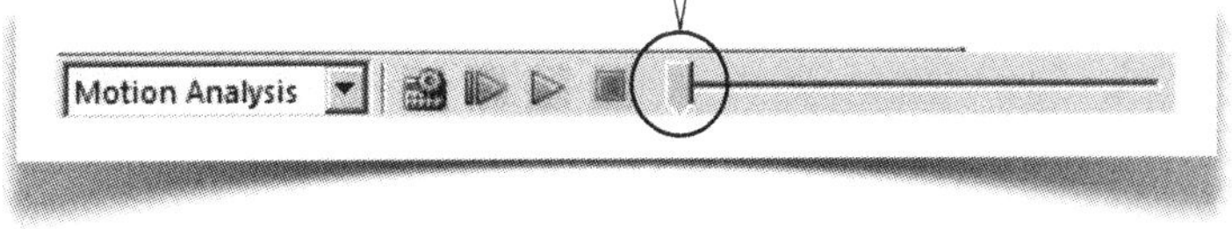

Motion Analysis

[2, 4] In **Motion** toolbar, click **Force**.

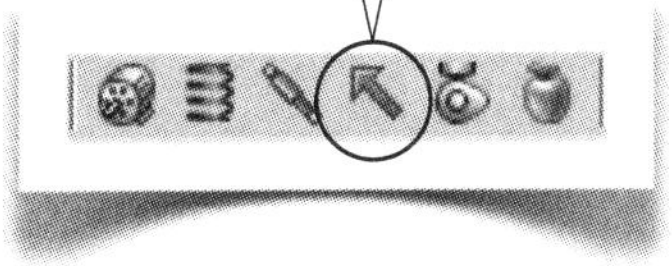

[6] Type 0.001 (N). We intend to specify a zero force value at Time = 0. However, **SOLIDWORKS** doesn't allow a zero value for a force. Therefore we type a very small value. This number essentially equals zero.

[3] Apply 2000 N downward on the top face of the **Block**.

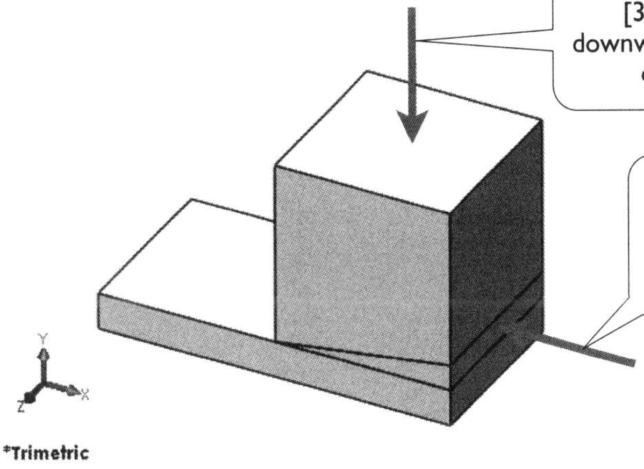

[5] Apply 0.001 N leftward on the right-side face of the **Wedge** (see [6]).

*Trimetric

[8] Right-click **Force2** and select **Edit Feature**.

[7] Drag the **Time Slider** to the right-end.

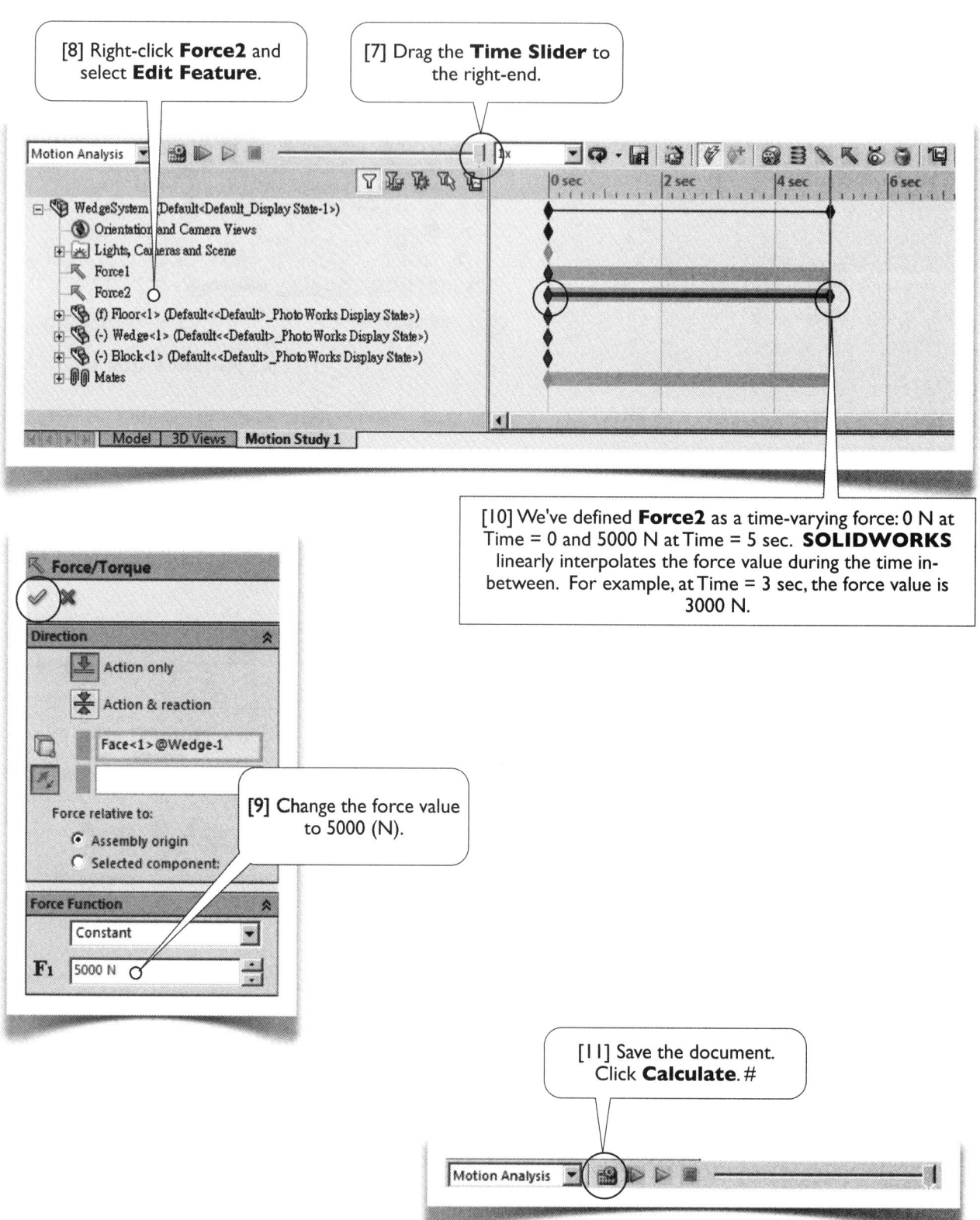

[10] We've defined **Force2** as a time-varying force: 0 N at Time = 0 and 5000 N at Time = 5 sec. **SOLIDWORKS** linearly interpolates the force value during the time in-between. For example, at Time = 3 sec, the force value is 3000 N.

[9] Change the force value to 5000 (N).

[11] Save the document. Click **Calculate**. #

7.2-8 Retrieve Friction Force

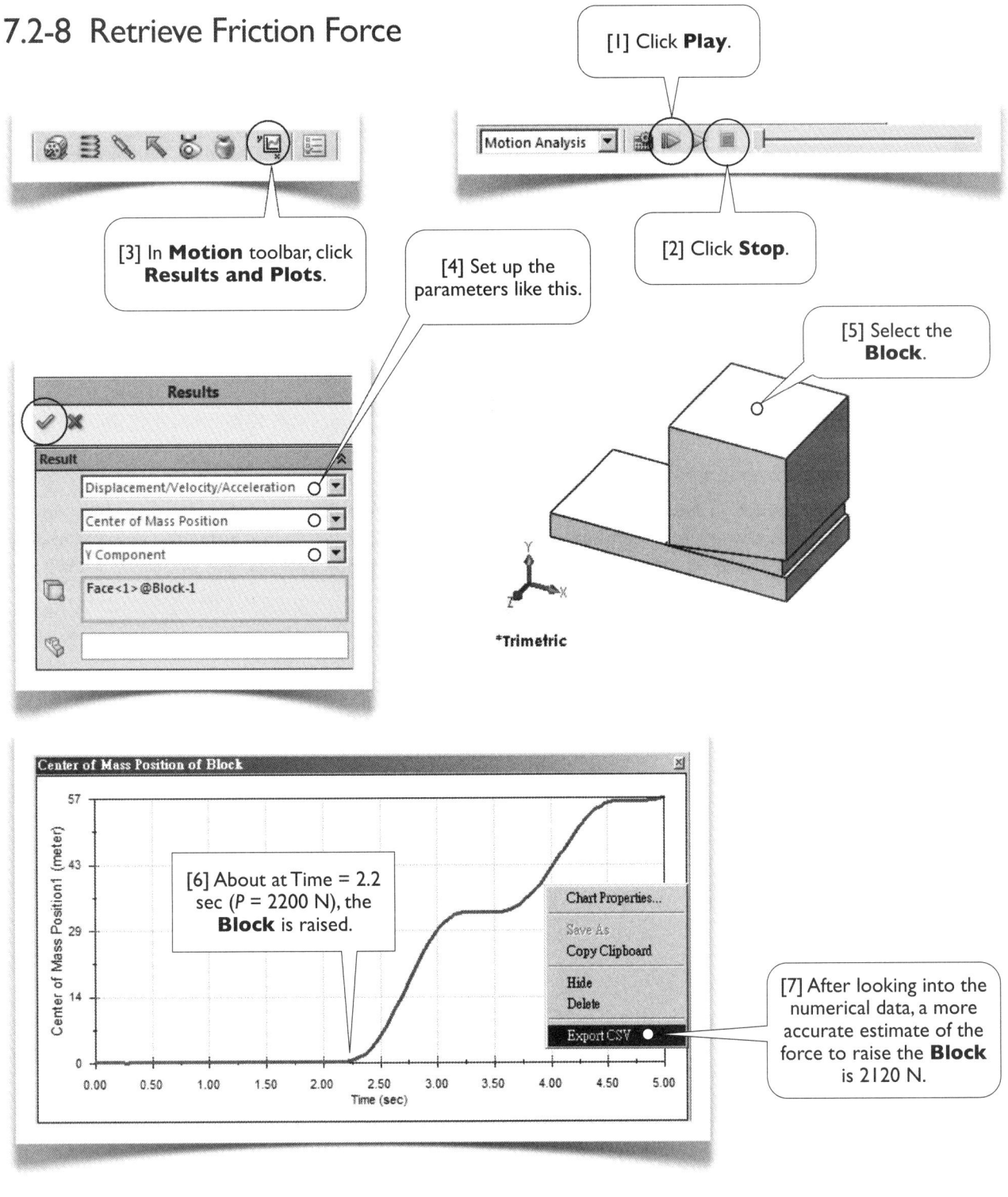

[1] Click **Play**.

[2] Click **Stop**.

[3] In **Motion** toolbar, click **Results and Plots**.

[4] Set up the parameters like this.

[5] Select the **Block**.

Results

Result

Displacement/Velocity/Acceleration

Center of Mass Position

Y Component

Face<1>@Block-1

*Trimetric

Center of Mass Position of Block

[6] About at Time = 2.2 sec (*P* = 2200 N), the **Block** is raised.

Chart Properties...
Save As
Copy Clipboard
Hide
Delete
Export CSV

[7] After looking into the numerical data, a more accurate estimate of the force to raise the **Block** is 2120 N.

[8] Quit **SOLIDWORKS**. Click **Save all**. Click **Rebuild and save the document**. #

Chapter 8
Moments of Inertia

In the study of engineering mechanics, many formulas contain cross-sectional area properties, such as moment of inertia, polar moment of inertia, and product of inertia. For example, in Mechanics of Materials, the bending stress σ is inversely proportional to the area moment of inertia I of the cross section,

$$\sigma = \frac{My}{I}$$

where M is the bending moment on the cross section and y is the distance from the neutral axis of the cross section.

Likewise, many formulas contain mass moments of inertia and products of inertia of a body. For example, in plane motions, the angular acceleration α of a free body is inversely proportional to the mass moment of inertia I of the body,

$$M = I\alpha$$

where M is the moment applied on the body.

This chapter introduces these area properties and mass properties, and their related theorems such as the parallel-axis theorem and Mohr's circles.

Section 8.1

Moments of Inertia of an Area

8.1-1 Introduction

[1] In the study of engineering mechanics, many formulas contain cross-sectional area properties, such as moment of inertia, polar moment of inertia, product of inertia, etc. This section introduces these area properties and their related theorems, such as parallel-axis theorem and Mohr's circle for moments and products of inertia.

Given a cross-sectional area and a coordinate system Oxy, the moments of inertia I_x and I_y are defined as

$$I_x = \int y^2 \, dA$$

$$I_y = \int x^2 \, dA$$

The polar moment of inertia J_o is defined as

$$J_o = \int r^2 \, dA = \int (x^2 + y^2) \, dA = I_x + I_y$$

where r is the distance from the origin O to the differential area dA. The product of inertia I_{xy} and I_{yx} are defined as

$$I_{xy} = \int xy \, dA$$

$$I_{yx} = \int yx \, dA$$

Obviously,

$$I_{xy} = I_{yx}$$

The parallel-axis theorem states that, knowing the moment of inertia $\overline{I}$ with respect to an axis passing through the centroid, the moment of inertia I with respect to an axis parallel to the centroidal axis is

$$I = \overline{I} + Ad^2$$

where A is the area and d is the distance between the two axes.

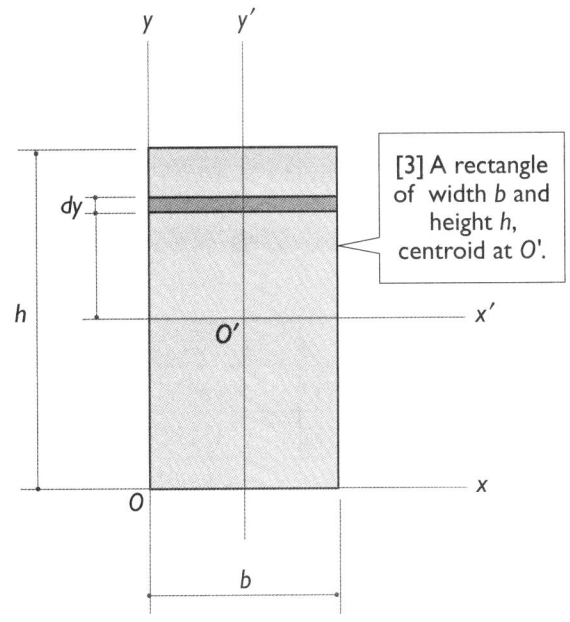

[3] A rectangle of width b and height h, centroid at O'.

[2] As an example, consider a rectangle of width b and height h [3]. The moment of inertia with respect to the x'-axis, which passes through the centroid, is

$$\overline{I}_{x'} = \int y^2 \, dA = \int_{-h/2}^{h/2} y^2 (b \, dy) = \frac{bh^3}{12}$$

The moment of inertia with respect to the x-axis is

$$I_x = \overline{I}_{x'} + Ad^2 = \frac{bh^3}{12} + (bh)(\frac{h}{2})^2 = \frac{bh^3}{3}$$

Similarly, $\overline{I}_{y'} = \frac{hb^3}{12}$ and the polar moments of inertia

$$J_{o'} = \overline{I}_{x'} + \overline{I}_{y'} = \frac{bh^3}{12} + \frac{hb^3}{12} = \frac{bh}{12}(b^2 + h^2)$$

The product of inertia $\overline{I}_{x'y'} = 0$ and

$$I_{xy} = \int xy \, dA = \overline{xy}A = \left(\frac{b}{2}\right)\left(\frac{h}{2}\right)(bh) = \frac{b^2 h^2}{4}$$

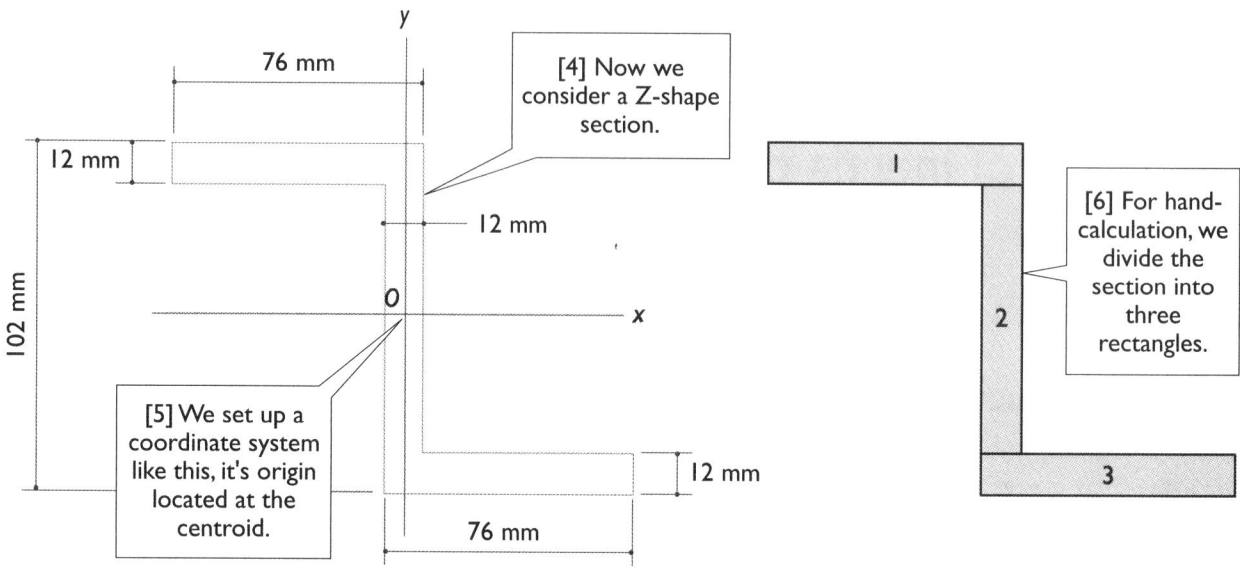

Rectangle	$\overline{I}_x$ (mm⁴)	$\overline{I}_y$ (mm⁴)	A (mm²)	$\overline{x}$ (mm)	$\overline{y}$ (mm)	$A\overline{x}^2$ (mm⁴)	$A\overline{y}^2$ (mm⁴)	$A\overline{xy}$ (mm⁴)
1	10944	438976	912	-32	45	933888	1846800	-1313280
2	474552	11232	936	0	0	0	0	0
3	10944	438976	912	32	-45	933888	1846800	-1313280
Total	496440	889184	2760			1867776	3693600	-2626560

[7] Properties of each rectangle can be calculated and totalled.

Summary of Section Properties

[8] The moments of inertia

$$I_x = 496440 + 3693600 = 4190040 \text{ mm}^4$$

$$I_y = 889184 + 1867776 = 2756960 \text{ mm}^4$$

The polar moments of inertia

$$J_O = I_x + I_y = 4190040 + 2756960 = 6947000 \text{ mm}^4$$

The product of inertia

$$I_{xy} = -2626560 \text{ mm}^4$$

The section area

$$A = 2760 \text{ mm}^2$$

Purpose of This Section

[9] In this section, we'll show you how to obtain these section properties with **SOLIDWORKS**. We'll also show you that the value of moment of inertia changes when the direction of referencing axis changes. In many applications, it is desirable to find the maximum or minimum values of moment of inertia and locate the corresponding axes. These extrema of moment of inertia are called **principal moments of inertia** and the corresponding axes are called **principal axes** of the section. These concepts can be easily understood by using a **Mohr's circle**. #

8.1-2 Start Up and Create a Part: **Z-Section**

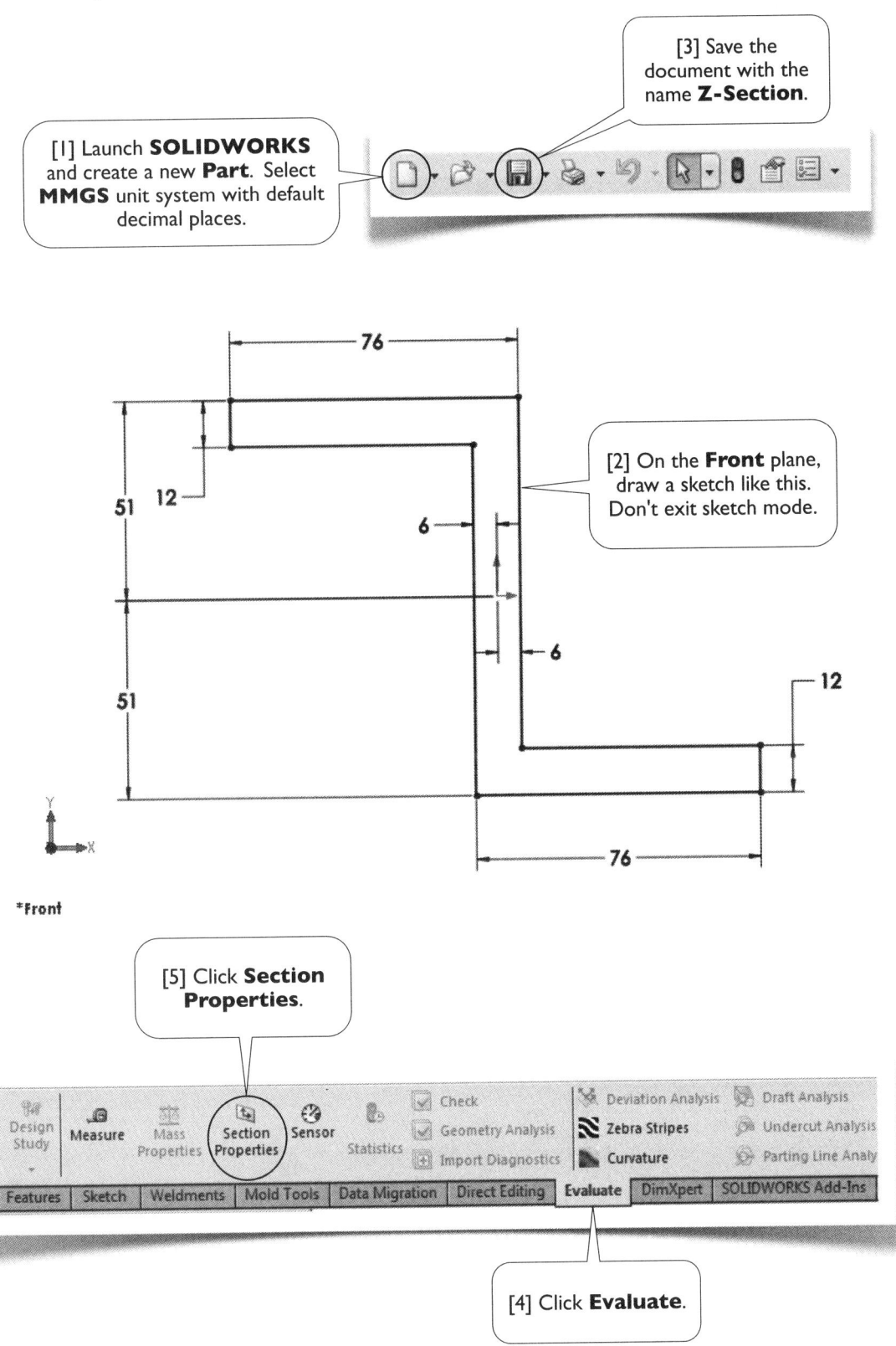

[3] Save the document with the name **Z-Section**.

[1] Launch **SOLIDWORKS** and create a new **Part**. Select **MMGS** unit system with default decimal places.

[2] On the **Front** plane, draw a sketch like this. Don't exit sketch mode.

*Front

[5] Click **Section Properties**.

[4] Click **Evaluate**.

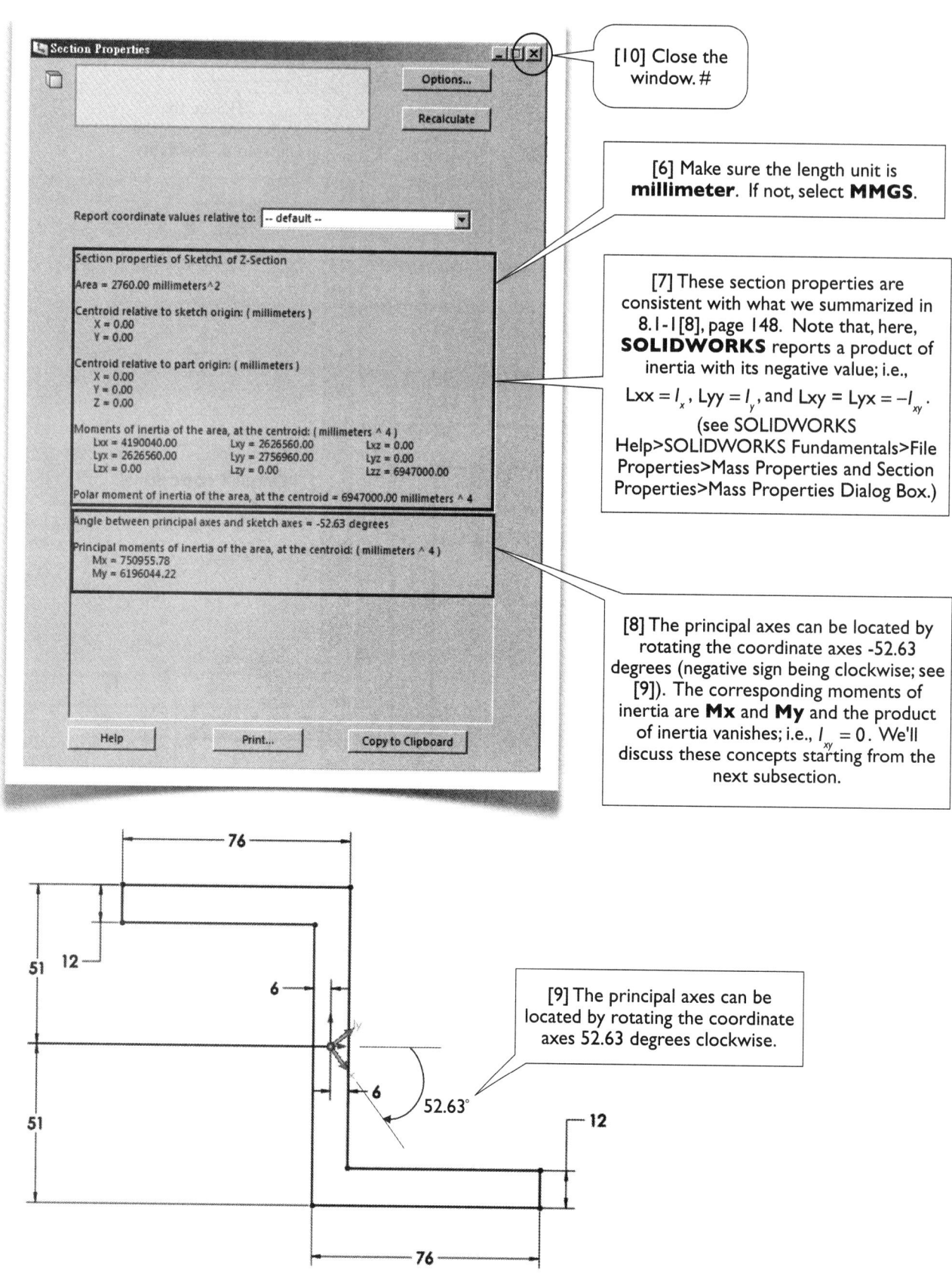

[10] Close the window. #

[6] Make sure the length unit is **millimeter**. If not, select **MMGS**.

[7] These section properties are consistent with what we summarized in 8.1-1[8], page 148. Note that, here, **SOLIDWORKS** reports a product of inertia with its negative value; i.e.,

$Lxx = I_x$, $Lyy = I_y$, and $Lxy = Lyx = -I_{xy}$.

(see SOLIDWORKS Help>SOLIDWORKS Fundamentals>File Properties>Mass Properties and Section Properties>Mass Properties Dialog Box.)

[8] The principal axes can be located by rotating the coordinate axes -52.63 degrees (negative sign being clockwise; see [9]). The corresponding moments of inertia are **Mx** and **My** and the product of inertia vanishes; i.e., $I_{xy} = 0$. We'll discuss these concepts starting from the next subsection.

[9] The principal axes can be located by rotating the coordinate axes 52.63 degrees clockwise.

8.1-3 Mohr's Circle

Constructing a Mohr's Circle

[1] By rotating the coordinate axes, a Mohr's circle can be constructed by plotting all possible (I_x, I_{xy}) pairs in an I_x - I_{xy} space. Let's denote the rotating angle θ, positive for counter-clockwise. With the data in 8.1-1[8] (page 148), we may plot two points in the I_x - I_{xy} space [2-5]. Note that I_{xy} reverses its sign after rotating x-axis 90° [4].

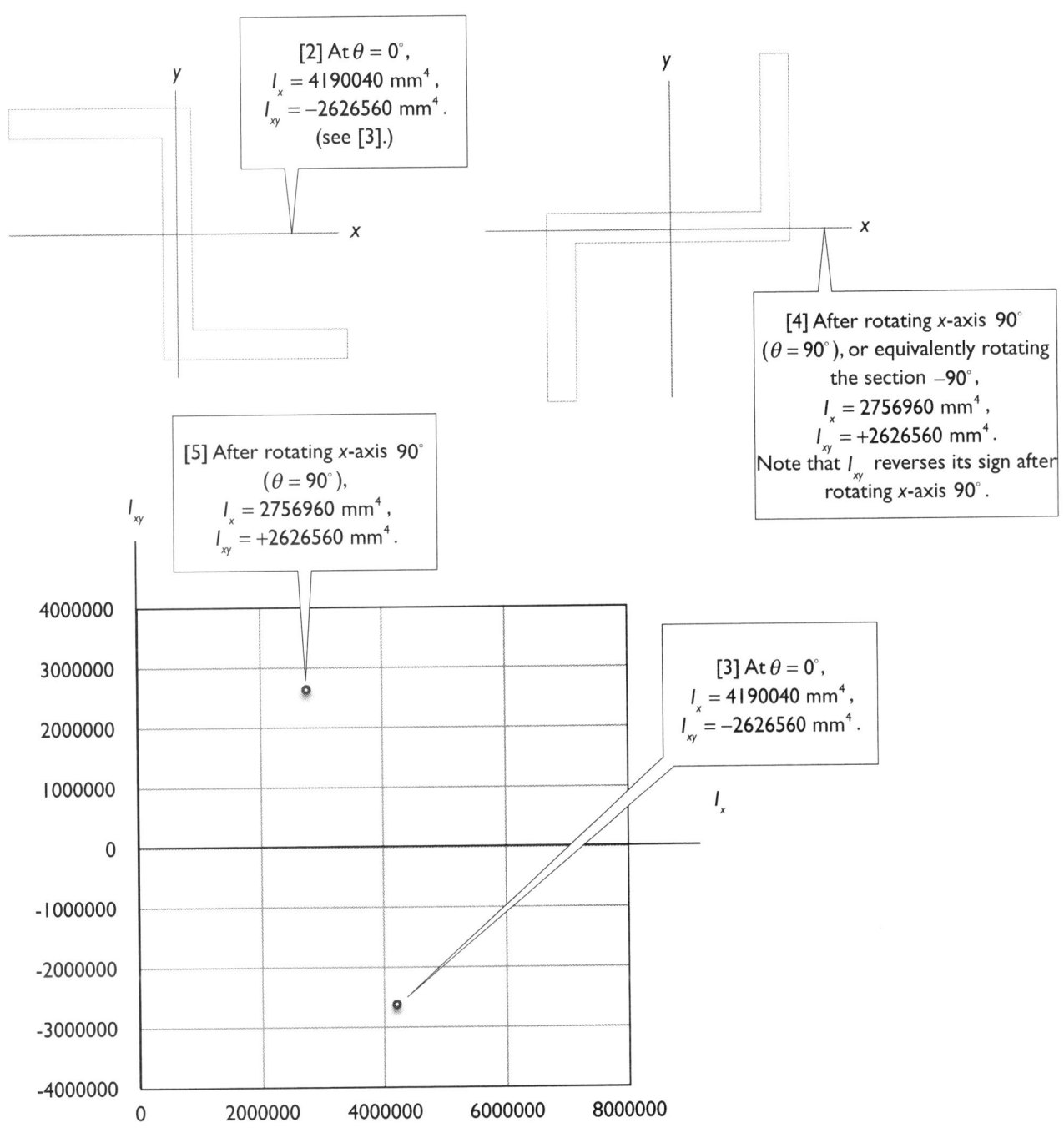

[2] At $\theta = 0°$,
$I_x = 4190040$ mm^4,
$I_{xy} = -2626560$ mm^4.
(see [3].)

[4] After rotating x-axis 90°
($\theta = 90°$), or equivalently rotating the section −90°,
$I_x = 2756960$ mm^4,
$I_{xy} = +2626560$ mm^4.
Note that I_{xy} reverses its sign after rotating x-axis 90°.

[5] After rotating x-axis 90°
($\theta = 90°$),
$I_x = 2756960$ mm^4,
$I_{xy} = +2626560$ mm^4.

[3] At $\theta = 0°$,
$I_x = 4190040$ mm^4,
$I_{xy} = -2626560$ mm^4.

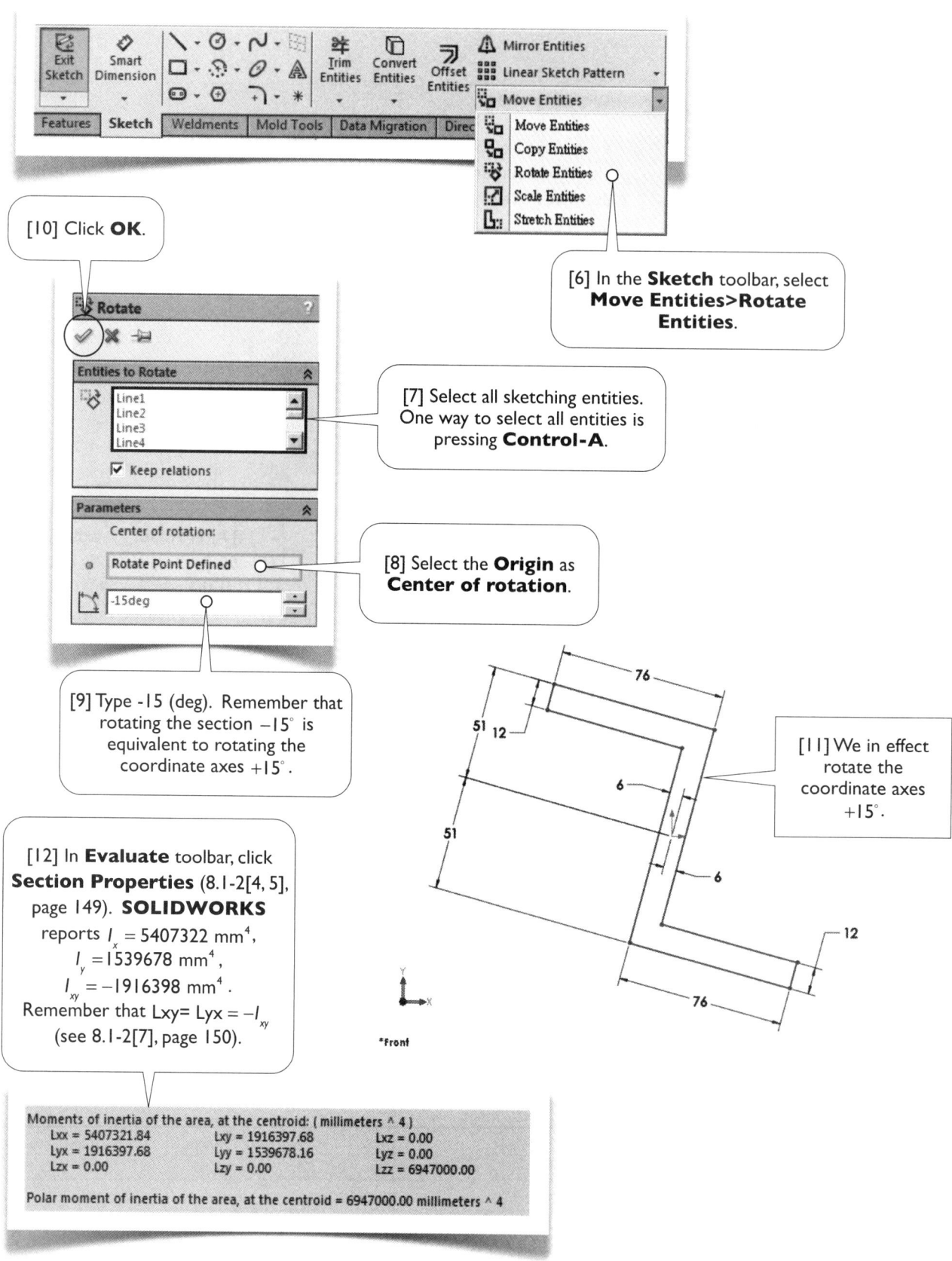

[10] Click **OK**.

[6] In the **Sketch** toolbar, select **Move Entities>Rotate Entities**.

[7] Select all sketching entities. One way to select all entities is pressing **Control-A**.

[8] Select the **Origin** as **Center of rotation**.

[9] Type -15 (deg). Remember that rotating the section −15° is equivalent to rotating the coordinate axes +15°.

[11] We in effect rotate the coordinate axes +15°.

[12] In **Evaluate** toolbar, click **Section Properties** (8.1-2[4, 5], page 149). **SOLIDWORKS** reports $I_x = 5407322$ mm^4, $I_y = 1539678$ mm^4, $I_{xy} = -1916398$ mm^4. Remember that Lxy= Lyx $= -I_{xy}$ (see 8.1-2[7], page 150).

```
Moments of inertia of the area, at the centroid: ( millimeters ^ 4 )
   Lxx = 5407321.84          Lxy = 1916397.68          Lxz = 0.00
   Lyx = 1916397.68          Lyy = 1539678.16          Lyz = 0.00
   Lzx = 0.00                Lzy = 0.00                Lzz = 6947000.00

Polar moment of inertia of the area, at the centroid = 6947000.00 millimeters ^ 4
```

θ (deg)	I_x (mm^4)	I_{xy} (mm^4)
0	4190040	-2626560
15	5407322	-1916398
30	6106438	-692738
45	6100060	716540
60	5389898	1933822
75	4166238	2632938
90	2756960	2626560
105	1539678	1916398
120	840562	692738
135	846940	-716540
150	1557102	-1933822
165	2780762	-2632938

[13] Repeat steps [6-12]. Each time rotate the section −15° (which is equivalent to rotating the coordinate axes +15°) and record I_x and I_{xy}.

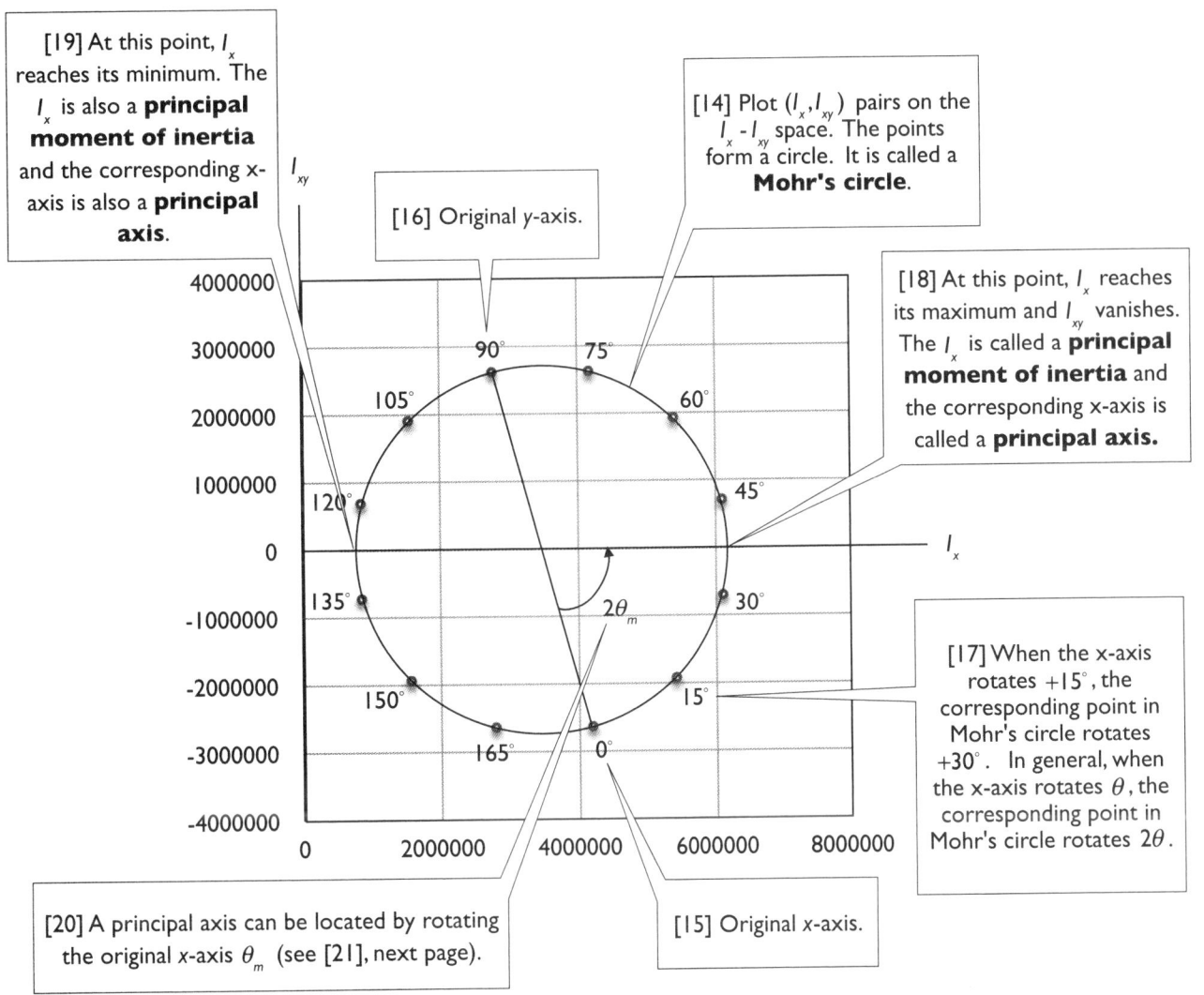

[19] At this point, I_x reaches its minimum. The I_x is also a **principal moment of inertia** and the corresponding x-axis is also a **principal axis**.

[14] Plot (I_x, I_{xy}) pairs on the I_x-I_{xy} space. The points form a circle. It is called a **Mohr's circle**.

[16] Original y-axis.

[18] At this point, I_x reaches its maximum and I_{xy} vanishes. The I_x is called a **principal moment of inertia** and the corresponding x-axis is called a **principal axis.**

[17] When the x-axis rotates +15°, the corresponding point in Mohr's circle rotates +30°. In general, when the x-axis rotates θ, the corresponding point in Mohr's circle rotates 2θ.

[20] A principal axis can be located by rotating the original x-axis θ_m (see [21], next page).

[15] Original x-axis.

Locating Principal Axes

[21] Knowing the original I_x, I_y, and I_{xy}, The angle θ_m can be calculated according to the geometry of the Mohr's circle:

$$\tan 2\theta_m = -\frac{2I_{xy}}{I_x - I_y} = -\frac{2(-2626560)}{4190040 - 2756960}$$

$$2\theta_m = 74.74° \text{ or } -105.26°$$

$$\theta_m = 37.37° \text{ or } -52.63°$$

We conclude that the **principal axes** can be located by rotating the original x-axis either $37.37°$ counter-clockwise or $52.63°$ clockwise, consistent with the angle reported by **SOLIDWORKS** (8.1-2[8], page 150).

Calculation of Principal Moments of Inertia

[22] Knowing the original I_x, I_y, and I_{xy}, The principal moments of inertia can be calculated according to the geometry of the Mohr's circle:

$$I_{max, min} = \frac{I_x + I_y}{2} \pm \sqrt{\left(\frac{I_x - I_y}{2}\right)^2 + I_{xy}^2}$$

$$= \frac{4190040 + 2756960}{2} \pm \sqrt{\left(\frac{4190040 - 2756960}{2}\right)^2 + 2626560^2}$$

$$= 3473500 \pm 2722544$$

$$I_{max} = 6196044 \text{ mm}^4, \quad I_{min} = 750956 \text{ mm}^4$$

The above results are consistent with the results reported by **SOLIDWORKS** (8.1-2[8], page 150).

In the next subsection, we'll locate the principal axes by rotating the x-axis $52.63°$ clockwise and show you that the **principal moments of inertia** reported by **SOLIDWORKS** are indeed these values. #

8.1-4 Principal Axes and Principal Moments of Inertia

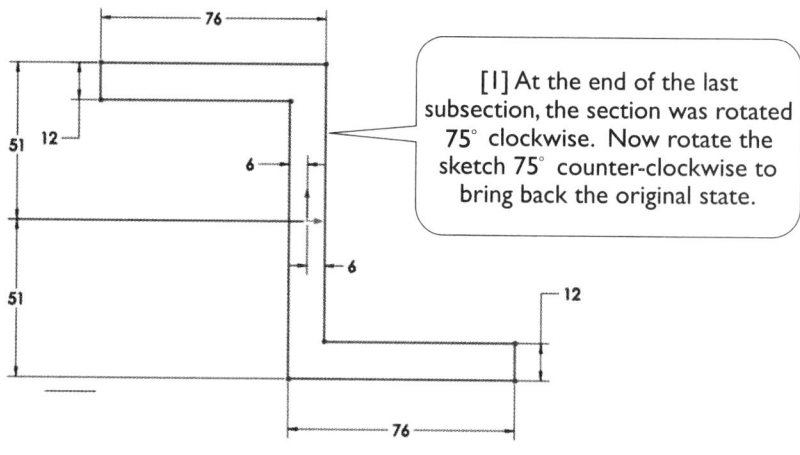

[1] At the end of the last subsection, the section was rotated 75° clockwise. Now rotate the sketch 75° counter-clockwise to bring back the original state.

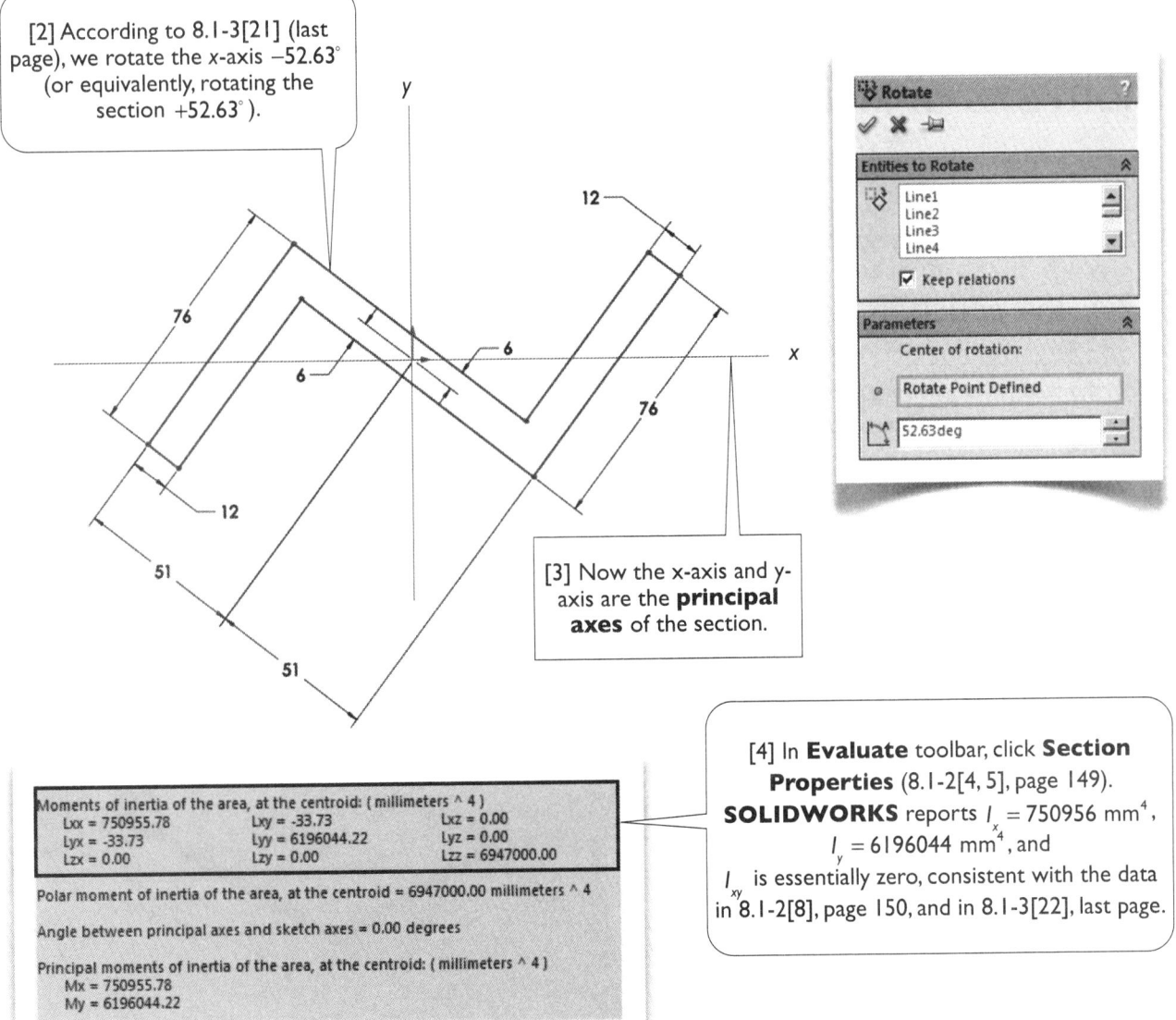

[2] According to 8.1-3[21] (last page), we rotate the x-axis −52.63° (or equivalently, rotating the section +52.63°).

[3] Now the x-axis and y-axis are the **principal axes** of the section.

Moments of inertia of the area, at the centroid: (millimeters ^ 4)
Lxx = 750955.78   Lxy = -33.73   Lxz = 0.00
Lyx = -33.73   Lyy = 6196044.22   Lyz = 0.00
Lzx = 0.00   Lzy = 0.00   Lzz = 6947000.00

Polar moment of inertia of the area, at the centroid = 6947000.00 millimeters ^ 4

Angle between principal axes and sketch axes = 0.00 degrees

Principal moments of inertia of the area, at the centroid: (millimeters ^ 4)
   Mx = 750955.78
   My = 6196044.22

[4] In **Evaluate** toolbar, click **Section Properties** (8.1-2[4, 5], page 149). **SOLIDWORKS** reports $I_x = 750956$ mm^4, $I_y = 6196044$ mm^4, and I_{xy} is essentially zero, consistent with the data in 8.1-2[8], page 150, and in 8.1-3[22], last page.

[5] Quit **SOLIDWORKS**. Click **Save all**. #

Section 8.2

Moments of Inertia of a Mass

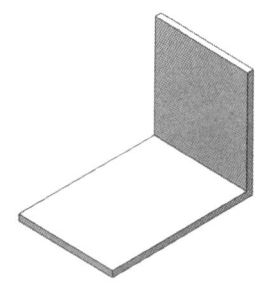

8.2-1   Introduction

[1] In the study of engineering mechanics, many formulas contain moments of inertia and products of inertia of a mass.

Given a mass and a coordinate system $Oxyz$, the moments of inertia of the mass I_x, I_y, I_y are defined as

$$I_x = \int (y^2 + z^2)\, dm$$

$$I_y = \int (z^2 + x^2)\, dm$$

$$I_z = \int (x^2 + y^2)\, dm$$

where dm is the differential mass element. Note that $(y^2 + z^2)$ is the distance from dm to x-axis, $(z^2 + x^2)$ is the distance from dm to y-axis, and $(x^2 + y^2)$ is the distance from dm to z-axis.

The products of inertia I_{xy}, I_{yz}, I_{zx} are defined as

$$I_{xy} = \int xy\, dm$$

$$I_{yz} = \int yz\, dm$$

$$I_{zx} = \int zx\, dm$$

The concepts and the related parallel-axis theorem and Mohr's circle are similar to those for an area, introduced in the last section, except the calculations are much more involved. In this section, using the L-plate introduced in Section 1.2 as an example [2], we'll show you how to obtain these mass properties with **SOLIDWORKS**.

In Section 8.1, the section properties are calculated for a sketch. In this section, we'll demonstrate that section properties can also be calculated for a face of a solid body.

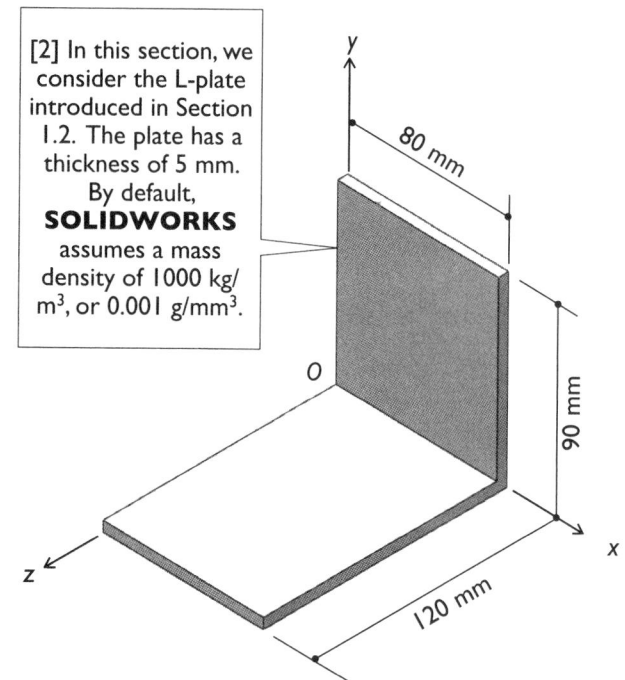

[2] In this section, we consider the L-plate introduced in Section 1.2. The plate has a thickness of 5 mm. By default, **SOLIDWORKS** assumes a mass density of 1000 kg/m^3, or 0.001 g/mm^3.

8.2-2 Start Up

[1] Launch **SOLIDWORKS** and click **Open** to open the file **Plate**, which was saved in 1.2-2[7], page 22.

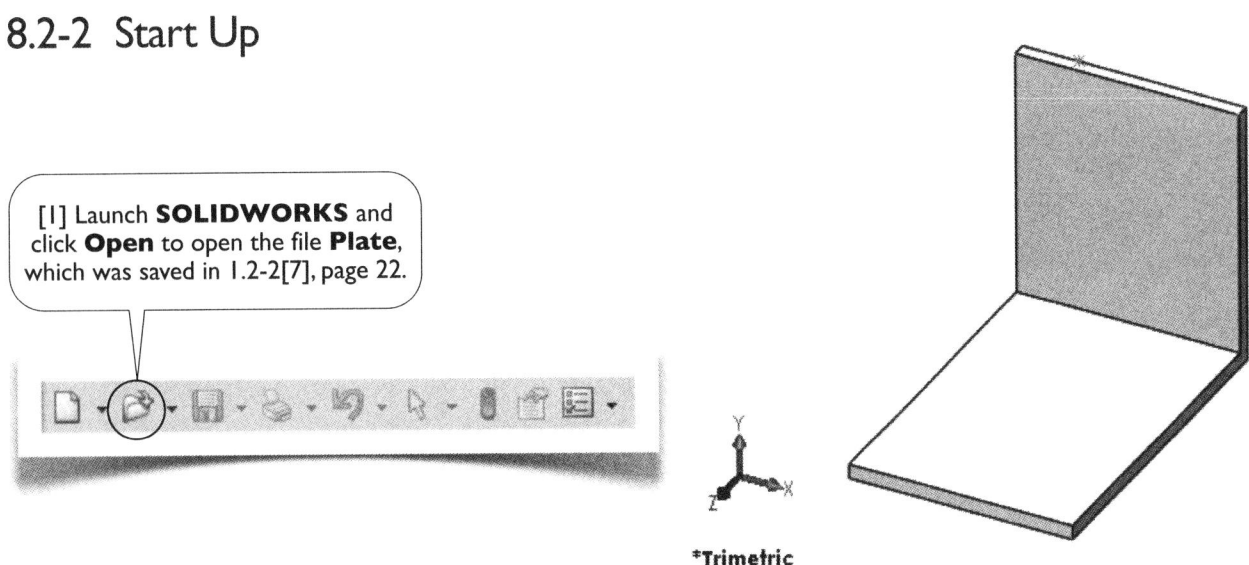

*Trimetric

8.2-3 Set Up Unit System

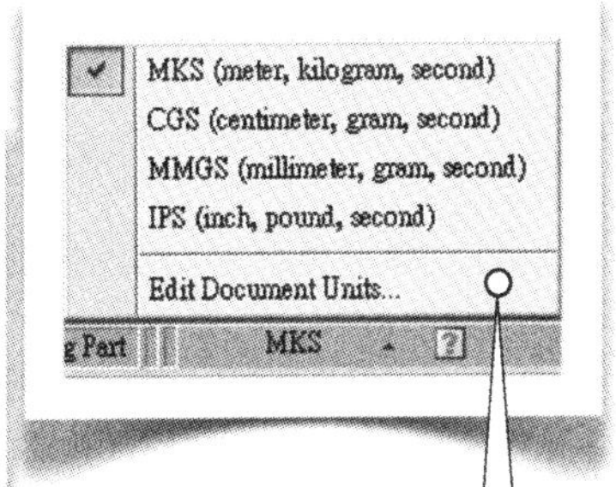

[2] Select **MMGS**.

✓ MKS (meter, kilogram, second)
CGS (centimeter, gram, second)
MMGS (millimeter, gram, second)
IPS (inch, pound, second)

Edit Document Units...

g Part | MKS

[1] Select **Edit Document Units...** (see 1.1-3[1], page 7)

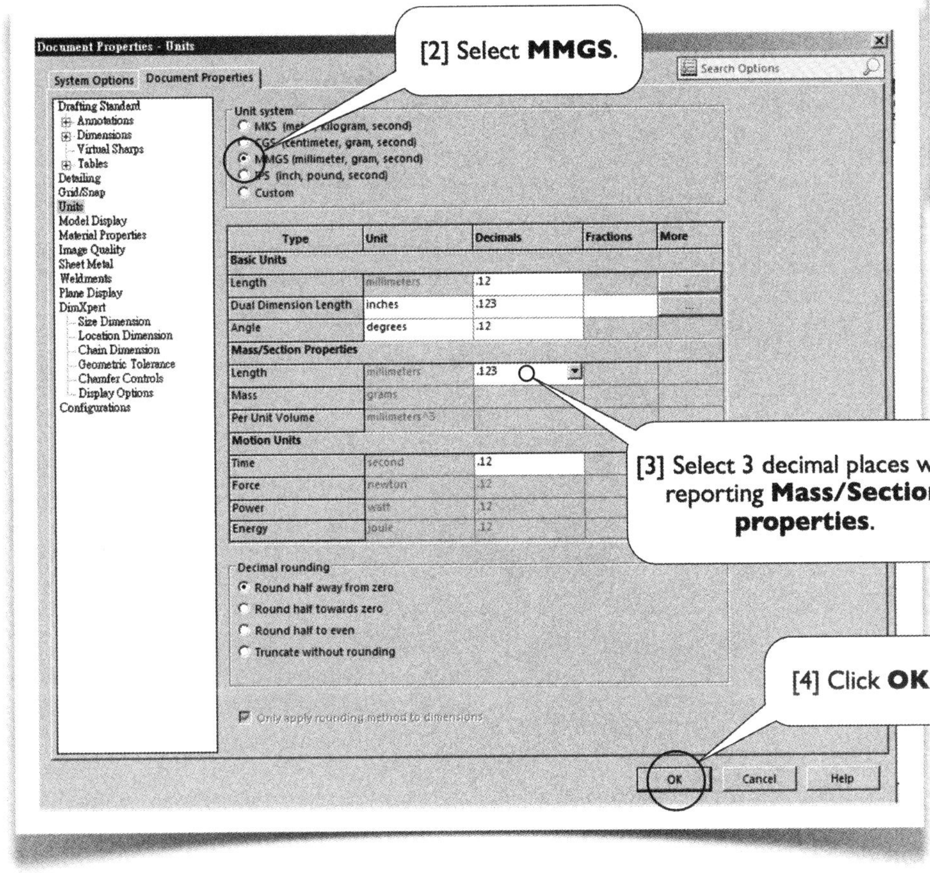

Document Properties - Units

System Options | Document Properties |

Drafting Standard
⊕ Annotations
⊕ Dimensions
  Virtual Sharps
⊕ Tables
Detailing
Grid/Snap
Units
Model Display
Material Properties
Image Quality
Sheet Metal
Weldments
Plane Display
DimXpert
  Size Dimension
  Location Dimension
  Chain Dimension
  Geometric Tolerance
  Chamfer Controls
  Display Options
Configurations

Unit system
○ MKS (meter, kilogram, second)
○ CGS (centimeter, gram, second)
◉ MMGS (millimeter, gram, second)
○ IPS (inch, pound, second)
○ Custom

Type	Unit	Decimals	Fractions	More
Basic Units				
Length	millimeters	.12		
Dual Dimension Length	inches	.123		
Angle	degrees	.12		
Mass/Section Properties				
Length	millimeters	.123		
Mass	grams			
Per Unit Volume	millimeters^3			
Motion Units				
Time	second	.12		
Force	newton	.12		
Power	watt	.12		
Energy	joule	.12		

[3] Select 3 decimal places when reporting **Mass/Section properties**.

Decimal rounding
◉ Round half away from zero
○ Round half towards zero
○ Round half to even
○ Truncate without rounding

☑ Only apply rounding method to dimensions

[4] Click **OK**. #

OK | Cancel | Help

8.2-4 Section Properties

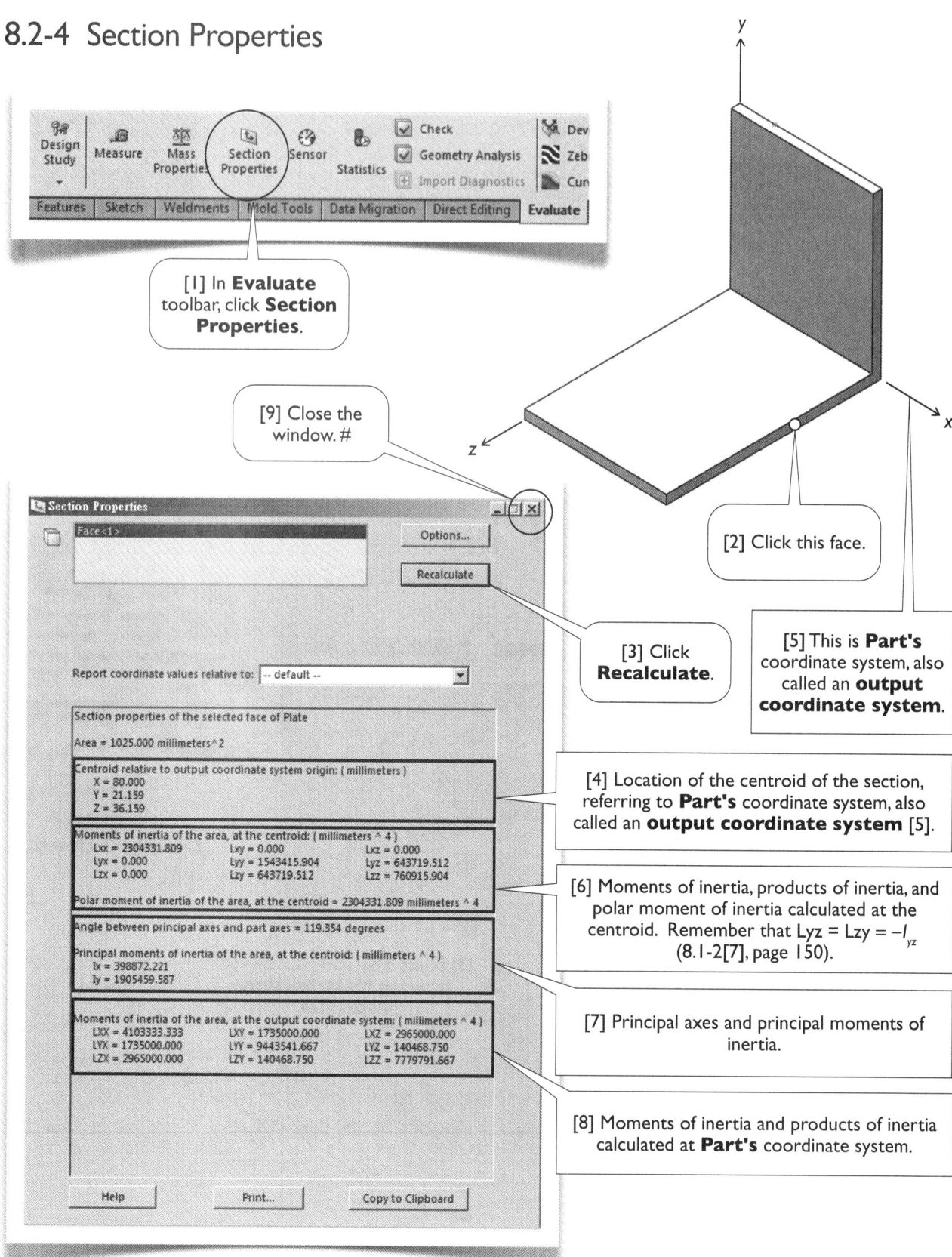

[1] In **Evaluate** toolbar, click **Section Properties**.

[9] Close the window. #

[2] Click this face.

[5] This is **Part's** coordinate system, also called an **output coordinate system**.

[3] Click **Recalculate**.

[4] Location of the centroid of the section, referring to **Part's** coordinate system, also called an **output coordinate system** [5].

[6] Moments of inertia, products of inertia, and polar moment of inertia calculated at the centroid. Remember that Lyz = Lzy = $-I_{yz}$ (8.1-2[7], page 150).

[7] Principal axes and principal moments of inertia.

[8] Moments of inertia and products of inertia calculated at **Part's** coordinate system.

Section Properties dialog:

Face<1>

Options...

Recalculate

Report coordinate values relative to: -- default --

Section properties of the selected face of Plate

Area = 1025.000 millimeters^2

Centroid relative to output coordinate system origin: (millimeters)
 X = 80.000
 Y = 21.159
 Z = 36.159

Moments of inertia of the area, at the centroid: (millimeters ^ 4)
 Lxx = 2304331.809   Lxy = 0.000   Lxz = 0.000
 Lyx = 0.000   Lyy = 1543415.904   Lyz = 643719.512
 Lzx = 0.000   Lzy = 643719.512   Lzz = 760915.904

Polar moment of inertia of the area, at the centroid = 2304331.809 millimeters ^ 4

Angle between principal axes and part axes = 119.354 degrees

Principal moments of inertia of the area, at the centroid: (millimeters ^ 4)
 Ix = 398872.221
 Iy = 1905459.587

Moments of inertia of the area, at the output coordinate system: (millimeters ^ 4)
 LXX = 4103333.333   LXY = 1735000.000   LXZ = 2965000.000
 LYX = 1735000.000   LYY = 9443541.667   LYZ = 140468.750
 LZX = 2965000.000   LZY = 140468.750   LZZ = 7779791.667

Help   Print...   Copy to Clipboard

8.2-5 Mass Properties

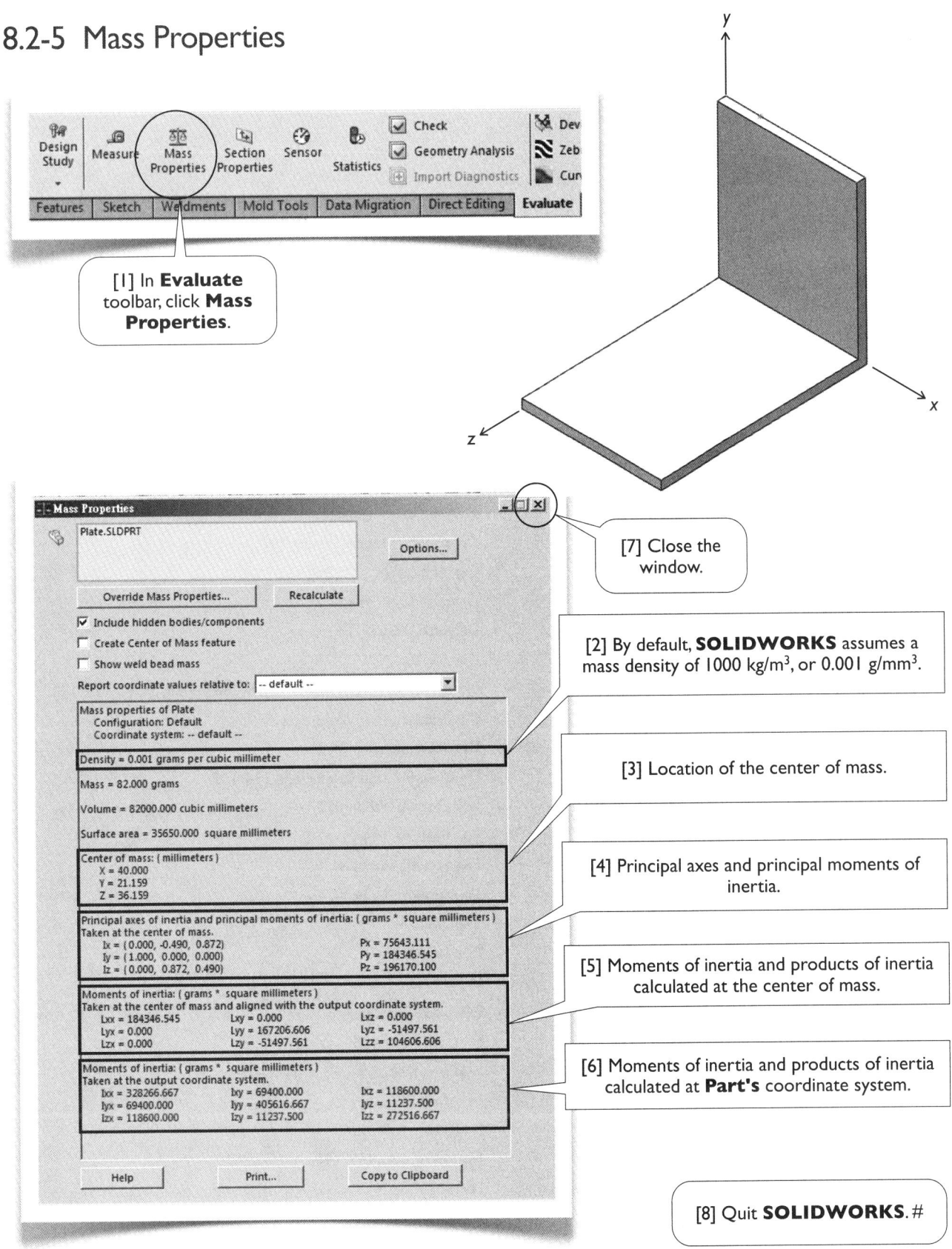

[1] In **Evaluate** toolbar, click **Mass Properties**.

[7] Close the window.

[2] By default, **SOLIDWORKS** assumes a mass density of 1000 kg/m^3, or 0.001 g/mm^3.

[3] Location of the center of mass.

[4] Principal axes and principal moments of inertia.

[5] Moments of inertia and products of inertia calculated at the center of mass.

[6] Moments of inertia and products of inertia calculated at **Part's** coordinate system.

[8] Quit **SOLIDWORKS**. #

Index